Textbook of Analytical Biochemistry

Edited by **Jessica Carol**

SYRAWOOD
PUBLISHING HOUSE

New York

Published by Syrawood Publishing House,
750 Third Avenue, 9th Floor,
New York, NY 10017, USA
www.syrawoodpublishinghouse.com

Textbook of Analytical Biochemistry
Edited by Jessica Carol

© 2016 Syrawood Publishing House

International Standard Book Number: 978-1-68286-129-5 (Hardback)

Printed in the United States of America.

Contents

Preface

Analytical biochemistry as a field of study incorporates principles, concepts and techniques of biological and biochemical sciences to understand and analyze chemical structures and processes. This book includes various researches and case studies by internationally acclaimed experts from around the globe that aim to provide a comprehensive overview of the discipline. It discusses current advancements in equipment and analytical procedures for determining and evaluating various materials, monitoring and analyzing various chemical and physical processes, etc. Students, researchers and academicians would find this book immensely helpful.

Various studies have approached the subject by analyzing it with a single perspective, but the present book provides diverse methodologies and techniques to address this field. This book contains theories and applications needed for understanding the subject from different perspectives. The aim is to keep the readers informed about the progress in the field; therefore, the contributions were carefully examined to compile novel researches by specialists from across the globe.

Indeed, the job of the editor is the most crucial and challenging in compiling all chapters into a single book. In the end, I would extend my sincere thanks to the chapter authors for their profound work. I am also thankful for the support provided by my family and colleagues during the compilation of this book.

Editor

Measurement of ice thickness on vitreous ice embedded cryo-EM grids: investigation of optimizing condition for visualizing macromolecules

Hye-Jin Cho[†], Jae-Kyung Hyun[†], Jin-Gyu Kim, Hyeong Seop Jeong, Hyo Nam Park, Dong-Ju You and Hyun Suk Jung[*]

Abstract

Background: Cryo-electron microscopy is an excellent method for the structural analysis of biological materials. Advantage of its use over conventional electron microscopy techniques is the preservation of the sample in a near-native, hydrated state. To achieve the analysis with greatly improved structural details, optimization of various parameters involved in sample vitrification is required. Most considerable parameter is the thickness of ice: thick and thin layers are ideally in favor for larger and smaller target objects.

Findings: We measured the thickness of vitreous ice from different types of widely used holey carbon grids using cryo-EM and electron energy loss spectroscopy. It showed that Quantifoil grids are suitable for the structural analysis of large biological macromolecules (>100 nm in size), whereas the use of lacey and C-flat grids are ideal for smaller particles.

Conclusions: This report provides informative details that may help increasing chances of obtaining optimal vitreous ice for various biological objects with different sizes, hence facilitate the successful application of cryo-electron microscopy.

Keywords: Transmission electron microscopy, Cryo-electron microscopy, Vitreous Ice thickness

Findings

Transmission electron microscopy (TEM) is a powerful and versatile technique that enables direct visualization of biological samples of sizes ranging from whole cell to near-atomic resolution details of a single protein molecule (Frank 2006). For structural analysis of protein macromolecular assemblies (e.g. virus particles) and proteins that harbor multiple conformations (e.g. ribosome), TEM has evolved into a popular tool despite lower achievable resolving power than X-ray crystallography and nuclear magnetic resonance (NMR) spectroscopy. Major advantages of TEM over other structure determination techniques include minimization of artifacts derived from harsh crystallization conditions, direct visualization of proteins in near-physiological environment, and relatively larger molecular mass limit for the analysis (van Heel et al. 2000). In addition, relatively small amount of protein sample is required for the TEM analysis, and therefore overcomes limited availability of target macromolecules.

Among many specialized TEM techniques, cryo-electron microscopy (cryo-EM) allows for direct imaging and structural analysis of fully hydrated biological specimen in near-physiological environment (Frank 2006). The technique involves preservation of the sample in vitrified aqueous solution (Adrian et al. 1984; Cavalier et al. 2008) and therefore prevents artifacts that can be derived from chemical fixation, staining and dehydration (Frank 2006; van Heel et al. 2000). Moreover, in comparison to stain-embedded samples of which the resolution is

* Correspondence: hyunsukjung@kbsi.re.kr
[†]Equal contributors
Division of Electron Microscopic Research, Korea Basic Science Institute, 113 Gwahangno, Daejeon 305-333, Korea

limited to the size of heavy metal salts, cryo-EM can achieve structure determination of macromolecular assemblies at near-atomic resolution because the image formation directly results from electron scattering of biological specimen (Gonen et al. 2005). Inherently low contrast of cryo-EM images can be compensated by using image processing tools which greatly improve signal-to-noise ratio though image alignment and averaging (Frank et al. 1996; van Heel et al. 1996; Grigorieff 2007). Despite these advantages, cryo-EM is often technically demanding. One of the difficulties comes from preparation of vitrified specimen that is suitable for efficient data collection and high-resolution image processing. An ideal specimen (a) should be free of contaminants such as ethane and ice crystals, (b) should have maximum number of particles within a field of view, yet the particles are sufficiently spaced, and (c) has a vitrified solution that is thick enough to contain the protein particles but thin enough to prevent excessive beam interference that reduces image contrast (Orlova & Saibil 2011). Therefore, for a given macromolecular sample, both concentration and ice thickness need be optimized by an experienced researcher through trial-and-error.

Because specimen preparation for biological cryo-EM is critical for achieving high-resolution structure determination, understanding relationship between the practical use of instrumentation and the materials is important. In order to entrap biological specimen in vitreous ice that is free of supporting material, EM grid coated with a thin carbon film with perforated holes, namely "holey carbon grid", is used. Uniformity of sample adsorption and even ice thickness across the EM grid requires the surface of carbon film to be rendered hydrophilic, which is usually achieved by glow discharging in vacuum (Aebi & Pollard 1987). After loading aqueous sample onto a holey carbon grid, excess solution is blotted and the grid is plunged into cryogen that is cooled to liquid nitrogen temperature. This step is crucial for reproducible and consistent outcome of the experiment, which is aided by using automated vitrification devices (Iancu et al. 2006; Grassucci et al. 2008). In addition, such devices can reduce ice crystal contamination since manual handling of the sample is minimized. In combination with user-defined virification conditions, a principal step to find the optimal thickness of vitreous ice is the choice of a particular type of support film.

In this study, using cryo-EM and electron energy loss spectroscopy (EELS), we measure one of critical but basic parameter for successful application of cryo-EM, which is the thickness of vitreous ice from various types of holey carbon grids that are commercially available. The results suggest that larger and smaller biological objects are ideally suitable for the use of Quantifoil and C-flat grids (Lacey grids), respectively, and the thickness of ice in each type of grid is dependent on the thickness of carbon support film.

Materials and methods
Specimen preparation
Three types of widely used holey carbon grids were subjected to the experiments; (a) Quantifoil grid (Cu 300 mesh, R2/2, SPI Supplies, U.S), (b) Lacey carbon grid (Agar Scientific, U.K.) and (c) C-flat (Cu 200 mesh, CF-1.2/1.3, Protochips Inc., U.S.). The grids were rendered hydrophilic using plasma cleaner (PDC-32G-2, Harrick Plasma, U.S), followed by application of $5\mu l$ of distilled water. The sample grid was semi-automatically vitrified using Vitrobot Mark I (FEI, U.S), at 100% relative humidity and 4°C. Blot time of 3.5 seconds and blot offset of –0.5 mm were consistently applied to all viritirification processes. The entire procedure is described in (Grassucci et al. 2008).

Cryo-electron microscopy
Vitrified samples were maintained at liquid nitrogen temperature during sample transfer and image acquisition using a cryo transfer holder (cryo holder-626, Gatan, U.S). The sample temperature throughout the experiment was carefully monitored using SmartSet Controller (Gatan, U.S), where the temperature was typically kept at approximately –177°C within the electron microscope. Tecnai G^2 Sprit Twin equipped with lanthanum hexaboride (Lab$_6$) gun operating at 120kV was used to acquire images. Micrographs were recorded using Ultracan™ 4000 CCD detector (Gatan, U.S), under low dose imaging mode, with typical electron does of 10–20 e/Å^2.

Electron energy loss spectroscopy
Ice thickness was estimated using log-ratio method (Malis et al. 1998), which is based on following relationship:

$$t\big/\lambda_\rho = \ln\left(\frac{I_t}{I_0}\right)$$

where t is the specimen thickness, λ_ρ is the effective mean free path length for inelastic scattering, and I_0 and I_t are zero-loss peak in electron energy loss spectrum and the whole spectrum, respectively. Electron energy loss spectrum was obtained using Gatan Imaging Filter (GIF) system (T12, Gatan, U.S), and the objective aperture that corresponds to the collection angle of 1.639 mrad was used for data collection. Integrated ratio between I_t and I_0 was used for estimation of inelastic mean free path, from which the specimen thickness was calculated using Digital Micrograph 3.0 software (Gatan, U.S). Typically, EELS was performed at nominal magnifications of ×28,000 and ×68,000, suitable for obtaining the spectrum of the entire hole and the center of vitreous ice, respectively.

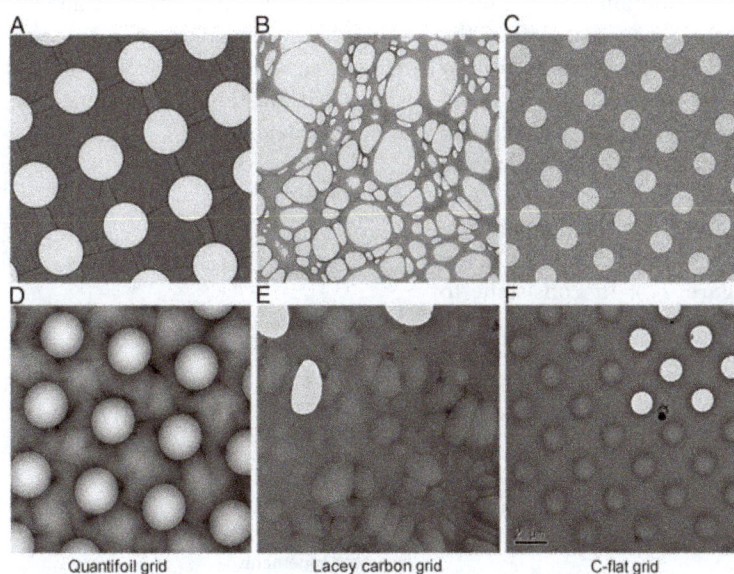

Figure 1 TEM images of holey carbon grids in the absence and the presence of vitreous ice. (A-C) Apparent views of Cryo-EM grids used in this study: Quantifoil grid (**A**), Lacey carbon grid (**B**), C-flat grid (**C**). (**D-F**) Appearances of each grid type with embedded vitreous ice. Note that C-flat grid is manufactured without plastics, thus it is known to be ultra-flat compared to other types of carbon grids.

Results and discussion

In order to correlate the thickness of supporting carbon film and that of the resulting vitreous ice, thickness of carbon film with no vitreous ice was measured using EELS. Average (represented by standard deviation) of 20 independent measurements from different areas of each grid type was used for evaluation (Figures 1 and 2). It was found that Quantifoil grid had the thickest carbon (49.11 ± 8.50 nm; mean ± S.D.) whereas C-flat grid had the thinnest (17.32 ± 0.82 nm). Reported carbon film thickness of C-flat grid (10–20 nm) agreed well with reported experimental data (Quispe et al. 2007). Lacey carbon grid had carbon film with an intermediate thickness (28.36 ± 2.95 nm). Variation of film thickness from different areas was minimal for C-flat grid (S.D. 0.82 nm), suggesting the most uniform carbon film amongst the three types of grids tested.

Distilled water with no solute was subjected to the analysis since adsorption of extra materials onto the grids may introduce further variability to the outcome. Thickness of vitreous ice was measured from each type of grid that was prepared under the same vitrification condition (Figure 2). Each measurement was obtained from the entire hole without surrounding carbon film (Figure 2B). Average of 20 independent measurements was used for evaluation (Figure 2D). Ice thickness ranged from the largest to the smallest in the order of Quantifoil grid (127.65 ± 12.42 nm), lacey carbon grid (99.05 ± 6.98 nm) and C-flat grid (92.35 ± 6.37 nm). In addition, ice thickness variation observed for Quantifoil grid (S.D. 12.42 nm) was almost two fold higher than that of lacey carbon

grid (S.D. 6.98 nm) and C-flat grid (S.D. 6.73 nm), suggesting the variation of carbon film thickness is directly reflected on the uniformity of resulting vitreous ice.

Variation of vitreous ice thickness across the hole was often apparent from electron micrographs, forming a smooth density gradient from the center toward the edge of the hole. Such thickness variation, so called 'lens effect', is well known for vitrified specimen, and may

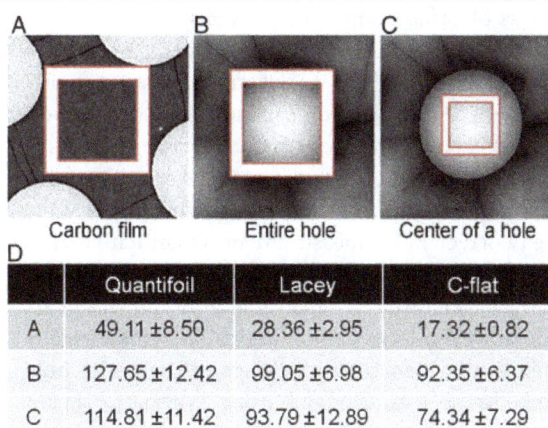

	Quantifoil	Lacey	C-flat
A	49.11 ±8.50	28.36 ±2.95	17.32 ±0.82
B	127.65 ±12.42	99.05 ±6.98	92.35 ±6.37
C	114.81 ±11.42	93.79 ±12.89	74.34 ±7.29

Figure 2 Thickness values measured from carbon film and vitreous ice embedded cryo-EM grids. (A-C) Chosen areas for thickness measurements are represented as inside rectangle: area of carbon film (**A**), entire area of vitreous ice within a hole (**B**), central region of vitreous ice (**C**). (**D**) Measured thickness values of each type grid according to different area as described in (**A-C**). For each condition, 20 independent areas was chosen and used to measure the thickness. Mean values are in nm and measured thickness is represented by standard deviation.

contribute to uneven particle distribution that can hamper efficient data collection. In order to characterize the lens effect for each grid type, micrographs were visually examined. All the grid types showed typical lens effect with varying extent although lacey carbon grid did not show distinctive, circular lens effect due to irregular size and shape of the holes (Figure 1).

For a more quantitative analysis, ice thickness of the center of a hole, which had the minimal electron density variation, was estimated (Figure 2C). In consistent to bare carbon thickness and ice thickness of the entire hole, the thickness of the center of a hole varied from thickest to the thinnest in the order of Quantifoil grid (114.81 ± 11.42 nm), lacey carbon grid (93.79 ± 12.89 nm) and C-flat grid (74.34 ± 7.29 nm). However, the extent of lens effect, as characterized by the difference between the ice thickness between the entire hole and the center, was most pronounced for C-flat grid (18.01 nm), followed by Quantifoil grid (12.84 nm) and lacey carbon grid (5.26 nm).

In this study we have used a simple method for estimating the thickness of vitreous ice using electron energy loss spectroscopy (EELS) and the log-ratio method. Thicknesses of supporting carbon film and embedded vitreous ice for three types of widely used holey carbon grids were efficiently measured. It was found that thickness of resulting vitreous ice was different in each type grid, possibly dependent of the thickness and of supporting carbon film. In addition, the extent of continuous variation of the ice thickness within the hole was characterized by estimating difference between the ice thickness of the entire hole and the central region.

Thickness of carbon support film and vitreous ice varied from thickest to thinnest in the order of Quantifoil grid, lacey carbon grid and C-flat grid. Substantial discrepancy between measured thickness of holy carbon grids from this study and that of manufacturer's description is possibly due to residual plastic layer underneath carbon film (Ermantraut et al. 1998), which may have hampered correct EELS measurement. Quantitative analysis showed that the lens effect was most pronounced for C-flat grid, followed by Quantifoil grid and lacey carbon grid. However, lacey carbon grid lacks symmetrically circular hole, and hence the thickness of the entire hole is unlikely to be estimated by using symmetric beam illumination.

Literatures suggest thinnest possible ice thickness that does not alter the integrity of the protein structure is optimal for cryo-EM (Orlova & Saibil 2011). With this respect, control of ice thickness for a given sample is extremely important. For extreme cases, ice thickness of 700–800 nm is required as exemplified by cryo-EM study of giant Mimivirus (Xiao et al. 2005; Xiao et al. 2009), whereas the thinnest achievable ice thickness is necessary for 7 nm DNA tetrahedron (Kato et al. 2009). Therefore characterization of vitreous ice thickness shown in this study would be beneficial, along with other vitrification parameters, for obtaining optimal cryo-EM data.

Abbreviations
TEM: Transmission electron microscopy; NMR: Nuclear magnetic resonance; Cryo-EM: Cryo-electron microscopy; EELS: Electron energy loss spectroscopy; Lab_6: Lanthanum hexaboride; GIF: Gatan imaging filter.

Competing interests
The authors declare that they have no competing interests.

Authors' contributions
H.S.J. designed research; H-J.C. and J-K.H. performed researches; J-G.K. contributed spectroscopy analysis. H-J.C. J-K.H. H.S.J. H.N.P. and D-J.Y. analzed EM data. J-K.H. and H.S.J. wrote the paper. All authors read and approved the final manuscript.

Acknowledgements
This work was supported by Korea Basic Science Institute grant (T33415) to JK Hyun.

References
Frank J (2006) Three-dimensional electron microscopy of macromolecular assemblies: visualization of biological molecules in their native state, 2nd edition. Oxford University Press, New York

van Heel M, Gowen B, Matadeen R, Orlova EV, Finn R, Pape T, Cohen D, Stark H, Schmidt R, Schatz M, Patwardhan A (2000) Single-particle electron cryo-microscopy: towards atomic resolution. Q Rev Biophys 33:307–369

Adrian M, Dubochet J, Lepault J, McDowall AW (1984) Cryo-electron microscopy of viruses. Nature 308:32–36

Cavalier A, Spehner D, Humbel BM (2008) Handbook of cryo-preparation methods for electron microscopy. CRC Press, Boca Raton

Gonen T, Cheng Y, Sliz P, Hiroaki Y, Fujiyoshi Y, Harrison SC, Walz T (2005) Lipid-protein interactions in double-layered two-dimensional AQP0 crystals. Nature 438:633–638

Frank J, Radermacher M, Penczek P, Zhu J, Li Y, Ladjadj M, Leith A (1996) SPIDER and WEB: processing and visualization of images in 3D electron microscopy and related fields. J Struct Biol 116:190–199

van Heel M, Harauz G, Orlova EV, Schmidt R, Schatz M (1996) A new generation of the IMAGIC image processing system. J Struct Biol 116:17–24

Grigorieff N (2007) FREALIGN: high-resolution refinement of single particle structures. J Struct Biol 157:117–125

Orlova EV, Saibil HR (2011) Structural analysis of macromolecular assemblies by electron microscopy. Chem Rev 111:7710–7748

Aebi U, Pollard TD (1987) A glow discharge unit to render electron microscope grids and other surfaces hydrophilic. J Electron Microsc Tech 7:29–33

Iancu CV, Tivol WF, Schooler JB, Dias DP, Henderson GP, Murphy GE, Wright ER, Li Z, Yu Z, Briegel A, Gan L, He Y, Jensen GJ (2006) Electron cryotomography sample preparation using the Vitrobot. Nat Protoc 1:2813–2819

Grassucci RA, Taylor D, Frank J (2008) Visualization of macromolecular complexes using cryo-electron microscopy with FEI Tecnai transmission electron microscopes. Nat Protoc 3:330–339

Malis T, Cheng SC, Egerton RF (1998) EELS log-ratio technique for specimen-thickness measurement in the TEM. J Electron Microsc Tech 8:193–200

Quispe J, Damiano J, Mick SE, Nackashi DP, Fellmann D, Ajero TG, Carragher B, Potter CS (2007) An improved holey carbon film for cryo-electron microscopy. Microsc Microanal 13:365–371

Ermantraut E, Wohlfart K, Tichelaar W (1998) Perforated support foils with pre-defined hole size, shape and arrangement. Ultramicroscopy 74:75–81

Xiao C, Chipman PR, Battisti AJ, Bowman VD, Renesto P, Raoult D, Rossmann MG (2005) Cryo-electron microscopy of the giant Mimivirus. J Mol Biol 353:493–496

Xiao C, Kuznetsov YG, Sun S, Hafenstein SL, Kostyuchenko VA, Chipman PR, Suzan-Monti M, Raoult D, McPherson A, Rossmann MG (2009) Structural studies of the giant mimivirus. PLoS Biol 7:e92

Kato T, Goodman RP, Erben CM, Turberfield AJ, Namba K (2009) High-resolution structural analysis of a DNA nanostructure by cryoEM. Nano Lett 9:2747–2750

2

Critical importance of the correction of contrast transfer function for transmission electron microscopy-mediated structural biology

Hyeong-Seop Jeong, Hyo-Nam Park, Jin-Gyu Kim and Jae-Kyung Hyun[*]

Abstracts

Background: Transmission electron microscopy (TEM) is an excellent tool for studying detailed biological structures. High-resolution structure determination is now routinely performed using advanced sample preparation techniques and image processing software. In particular, correction for contrast transfer function (CTF) is crucial for extracting high-resolution information from TEM image that is convoluted by imperfect imaging condition. Accurate determination of defocus, one of the major elements constituting the CTF, is mandatory for CTF correction.

Findings: To investigate the effect of correct estimation of image defocus and subsequent CTF correction, we tested arbitrary CTF imposition onto the images of two-dimensional crystals of Rous sarcoma virus capsid protein. The morphology of the crystal in calculated projection maps from incorrect CTF imposition was utterly distorted in comparison to an appropriately CTF-corrected image.

Conclusion: This result demonstrates critical importance of CTF correction for producing true representation of the specimen at high resolution.

Keywords: Contrast transfer function; Transmission electron microscopy; Electron crystallography; Structural biology

Introduction

Transmission electron microscopy (TEM) offers direct visualization of fine details of biological specimen. Recent advancements in sample preparation techniques and developments in algorithms for sophisticated image processing as well as availability of computation power pivoted rapid improvements in achievable resolution of the analysis and widened the range of biological systems that can be studied (Crowther 2010). In particular, structural analysis of protein macromolecules by TEM, either in the form of ordered arrays such as protein two-dimensional (2D) crystals or individual protein macromolecules, has improved greatly as evidenced by an increasing number of structure determination at near-atomic resolutions (Armache et al. 2010; Ge and Zhou 2011; Gonen et al. 2005; Yu et al. 2011). In addition, TEM analysis of biological system at moderate resolution can directly complement high-resolution structures obtained by X-ray crystallography and nuclear magnetic resonance (NMR) spectroscopy, providing pseudo-atomic resolution structure determination of large, multi-subunit complexes.

One of the key factors for successful structure determination of a biological specimen at high resolution is the correction for contrast transfer function (CTF) of a microscope. While images obtained from the electron microscope suffers from loss of faithful representation of the true object due to phase and amplitude modulation derived from imperfect imaging conditions, CTF models how an electron microscope transfers the actual specimen into a recorded image hence allowing for distortions present in the micrograph to be estimated (Frank 2006). Under weak-phase approximation, that is, electron scattering and the subsequent phase shift, is small as in the case for biological specimen, TEM image, and CTF can be described by the following relationships (Penczek et al. 1997):

$$I(k) = H(k)\Phi(k) \tag{1}$$

* Correspondence: hjk002@kbsi.re.kr
Division of Electron Microscopic Research, Korea Basic Science Institute, 169-148 Gwahangno, Daejeon 305-333, Republic of Korea

$$H(k) = \sin \gamma(k) - W \cos \gamma(k) \qquad (2)$$

where k is the spatial frequency vector, $I(k)$ is Fourier transform of micrograph, $\Phi(k)$ is Fourier transform of true object and $H(k)$ is CTF. W is amplitude contrast ratio, which denotes contribution of image contrast that result from inelastic scattering in the image formation that is dominated by elastic electron scattering, and $\gamma(k)$ is phase shift produced by spherical aberration and defocus that can be described by the Scherzer formula (Williams and Carter 1996),

$$\gamma(k) = \frac{\pi}{2}\left(C_s\lambda^3 k^4 - 2\Delta z\lambda k^2\right) \qquad (3)$$

where k is the scattering vector, λ is the wavelength of electron beam, C_s is the spherical aberration coefficient of a microscope, and Δz is the defocus value. While other parameters are constant for a given instrument, defocus is manually adjusted by an operator in order to produce image with optimal phase contrast. When considering elastic electron scattering alone, $\sin\gamma(k)$ is also referred to as phase contrast transfer function (PCTF). When plotted as a function of spatial frequency, PCTF oscillates sinusoidally, hence producing alternating negative phase contrast at higher spatial frequencies (Ruprecht and Nield 2001). If the negative contrast is left uncorrected, structural features of the object at high resolution is compromised, and therefore the image fails to represent true features of the object.

Additional complication with regard to precise CTF estimation comes from continuous attenuation of amplitude towards higher spatial frequency, termed envelope function, which is described by a simplified relationship below:

$$H(k) = E(k)H_{\text{ideal}}(k) \qquad (4)$$

where the experimental CTF, $H(k)$, results from ideal CTF, $H_{\text{ideal}}(k)$, multiplied by envelope function, $E(k)$. Major contributors of envelope function include beam energy envelope (E_{spread}), beam coherence envelope ($E_{\text{coherence}}$) and sample drift envelope (E_{drift}). Each envelope function is described by complex formula which takes account into parameters such as chromatic aberration of the microscope, semi-angle of aperture, energy spread of emitted beam, lens current stability and specimen drift (Frank 2006; Sorzano et al. 2007). In addition, the performance of image recorder, as defined by modulation transfer function, also contributes to the degradation of high-resolution information in the micrograph.

Due to its importance, there have been a significant number of studies dedicated to precisely determine CTF from micrographs. These works provided comprehensive description and algorithms for specific aspects such as defocus determination (Mindell and Grigorieff 2003), amplitude contrast (Toyoshima and

Unwin 1988; Toyoshima et al. 1993) and envelope function (Saad et al. 2001; Sander et al. 2003) as well as generation of reliable power spectrum density (Fernandez et al. 1997; Zhu et al. 1997) from which CTF of experimental data can be modeled from. As a result, CTF correction is now widely incorporated in various image processing software packages, and routinely performed for structure determination of vitrified biological specimen.

In the present work, the effect of precise estimation of image defocus, one of the most critical parameters required for CTF determination, in the preservation of structural integrity of specimen is demonstrated. Although the effects of alterations in critical parameters have been thoroughly investigated for CTF determination in the past (Sorzano et al. 2009), the main purpose of this short technical note is to illustrate visually the effect of appropriate CTF correction. Therefore, for simplicity, detailed theories of image formation in TEM and algorithms employed in CTF correction are omitted.

Availability and requirements
Specimen preparation and electron microscopy

2D crystals of Rous sarcoma virus (RSV) capsid protein with C-terminal truncation mutation (CA_{1-226}) was produced using mild acidification as described previously (Bailey et al. 2012; Hyun et al. 2010). In short, purified protein stored in low molarity buffer at neutral pH was jump-diluted with high molarity buffer at pH 4.9, followed by a 48-h incubation at 18°C.

Five microliters of the assembly product was applied onto a glow-discharged grid that was held by self-closing, anti-capillary tweezers (EMS, Hatfield, PA, USA). 60 to 90 s were allowed for the specimen to be adsorbed onto the carbon support film. Excess salt was washed off using three droplets of filtered, deionized water. Then, 5 µL of 2% (w/v) uranyl acetate solution was applied onto the grid. After staining for 60 s, excess stain solution was blotted away using a piece of filter paper.

A Tecnai 12 TEM (FEI, Hillsboro, OR, USA) equipped with lanthanum hexaboride (Lab_6) gun operating at 120 kV was used to acquire images of CA_{1-226} 2D crystals. The images were recorded on SO-163 photographic films (Kodak, Rochester, NY, USA) at nominal magnification of ×42,000, at approximately 1.5 µm under-focus. The films were developed in D19 developer (Kodak, USA) diluted 1:1 (v/v) with water for 10 min.

Image processing

Micrographs that displayed minimal astigmatism and drift were scanned using a Super Coolscan 9000 film scanner (Nikon, Tokyo, Japan). The micrographs were digitized at a step size of 10.0 µm (i.e., 2,549 dpi), which corresponded to 2.38 Å/pixel on the specimen. Within the micrograph, 2,048 × 2,048 pixel box that displayed best crystalline order was cropped for image processing.

2dx program suite was employed for image processing of the 2D crystals (Gipson et al. 2007a, b). The overall process of image processing is summarized in Figure 1. Fourier transform of the micrograph was used to determine lattice parameters by assigning each diffraction spot with Miller index. Also, astigmatism and defocus of the image was estimated using the CTFFIND3 software (Mindell and Grigorieff 2003) that is implemented in 2dx. For correct estimation of image defocus, spherical aberration of the microscope (2 mm), acceleration voltage (120 kV) and amplitude contrast for negatively stained specimen (20% contribution) were provided.

Inherent distortions and imperfection of the raw image was corrected using unbending routine, which employs cross-correlation between a small reference area and the rest of the image (Crowther et al. 1996). Background noise from Fourier transform was eliminated by masking diffraction spots, and inverse Fourier transformation was performed in order to produce noise-filtered image. Prior to the generation of a projection map from the noise-filtered and unbent data, CTF was corrected using Weiner filtering, based on the estimated defocus and astigmatism. For a given defocus value, CTF was simulated using the ctfExplorer software (Sidorov 2002). Instrumental parameters for appropriate CTF simulation were estimated by comparing computed intensity profile

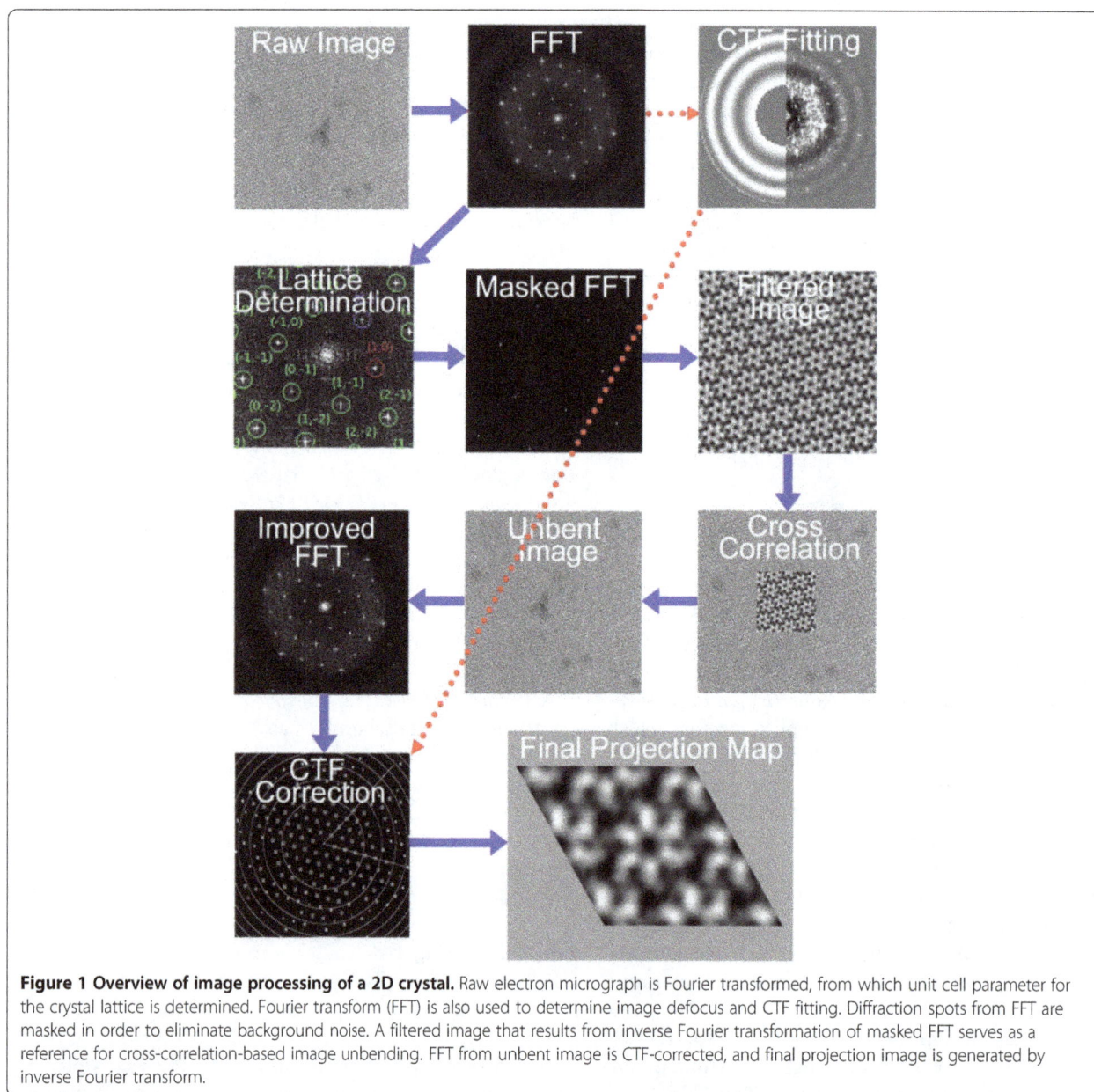

Figure 1 Overview of image processing of a 2D crystal. Raw electron micrograph is Fourier transformed, from which unit cell parameter for the crystal lattice is determined. Fourier transform (FFT) is also used to determine image defocus and CTF fitting. Diffraction spots from FFT are masked in order to eliminate background noise. A filtered image that results from inverse Fourier transformation of masked FFT serves as a reference for cross-correlation-based image unbending. FFT from unbent image is CTF-corrected, and final projection image is generated by inverse Fourier transform.

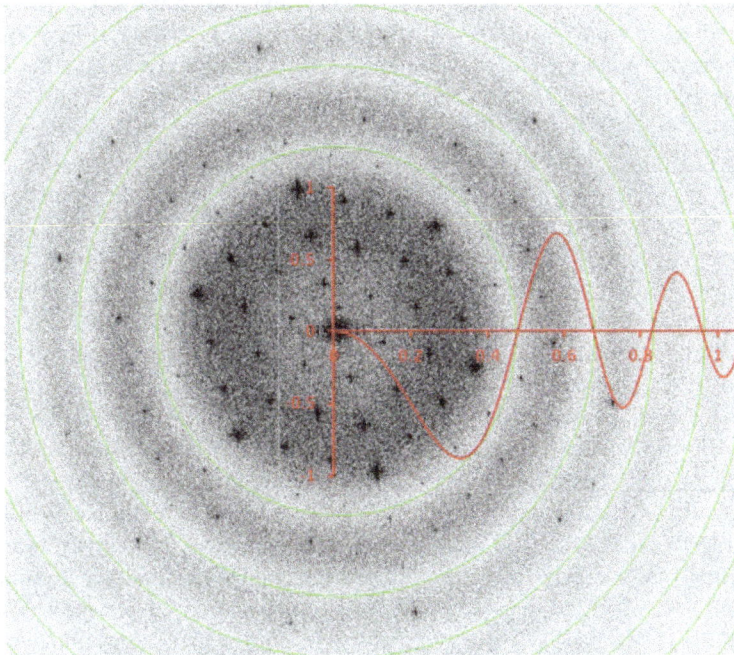

Figure 2 CTF fitted Fourier transform of RSV CA 2D crystal. FFT of RSV CA$_{1-226}$ 2D crystal clearly indicates Thorn rings that represent CTF plotted as a function of spatial frequency (red graph). Positions of zero CTF are indicated by green circles in the image. Diffraction spots that belong in every second Thorn rings in the FFT (i.e. positive CTF in the plot) must be phase-flipped in order to correctly extract information to a highest possible resolution.

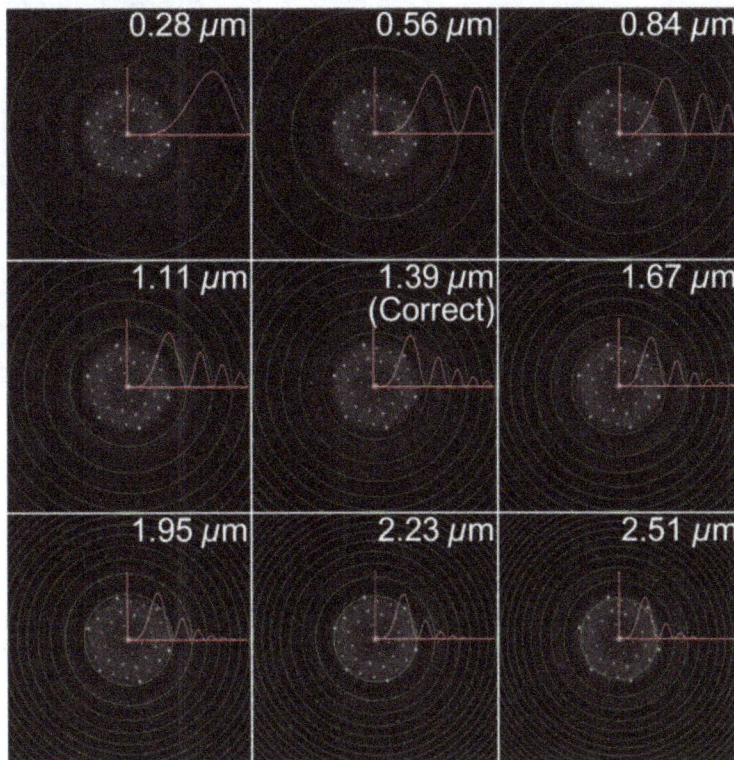

Figure 3 Fitting of arbitrary CTFs to the Fourier transform of RSV CA 2D crystal. CTFs derived from arbitrary defocus values are fitted onto FFT of RSV CA 2D crystal. Positions of calculated zero CTF are indicated by green circles in the image, and clear misfit is observed when the wrong CTF is fitted. Simulated CTF (red graph) was plotted based on the absolute values (i.e., intensity profile mode) for the simplicity.

obtained from a radial average of experimental Fourier transform with the calculated simulation. Satisfactory agreement between the experimental data and the simulation was achieved using 0.26 mrad convergence angle, 14.3 nm focal spread and, 2 eV energy spread. Subtle astigmatism in the image was not taken account into for the simulation.

Findings

Raw electron micrograph of RSV CA$_{1\text{-}226}$ 2D crystal exhibited hexagonal lattice symmetry with hexameric subunits clearly defined. Fourier transform of the image displayed amplitude modulation in the form of Thorn ring that represents CTF, and the diffraction spots extended far beyond the first node of CTF, up to approximately 14 Å. Estimation of defocus indicated that that the image was slightly astigmatic, where defocus of the image was estimated to be 13,917 and 13,447 Å in the longest and shortest dimensions that are perpendicular. Fitted CTF is shown in Figure 2, in which an independently simulated CTF function is overlaid. Correctly

estimated defocus in the image enabled precise positioning of minimal CTF in the Fourier transform, and the simulated CTF indicates oscillating positive and negative CTF as a function of resolution as well as resolution-dependent attenuation of phase amplitude. The plot indicates necessity for phase correction to diffraction spots (i.e., structure factors) that belong in every second node of Thorn rings.

Lattice parameters in reciprocal space were obtained by indexing diffraction spots, from which real space unit cell dimension was calculated (a = 96.203 Å, b = 96.385 Å, and γ = 119.964°). Initial phase and amplitude were extracted from unique reflections of h and k Miller indices in the Fourier transform. In order to distinguish diffraction from the 2D crystal from random background noise in the Fourier transform, the quality of the structural information was measured in terms of 'intelligence quotient (IQ)' number, where the signal-to-noise ratio of the reflection from the background is determined. IQ = 1 indicates signal-to-noise ratio of 8, whereas IQ = 9 is undistinguished spot from background. After initial structure

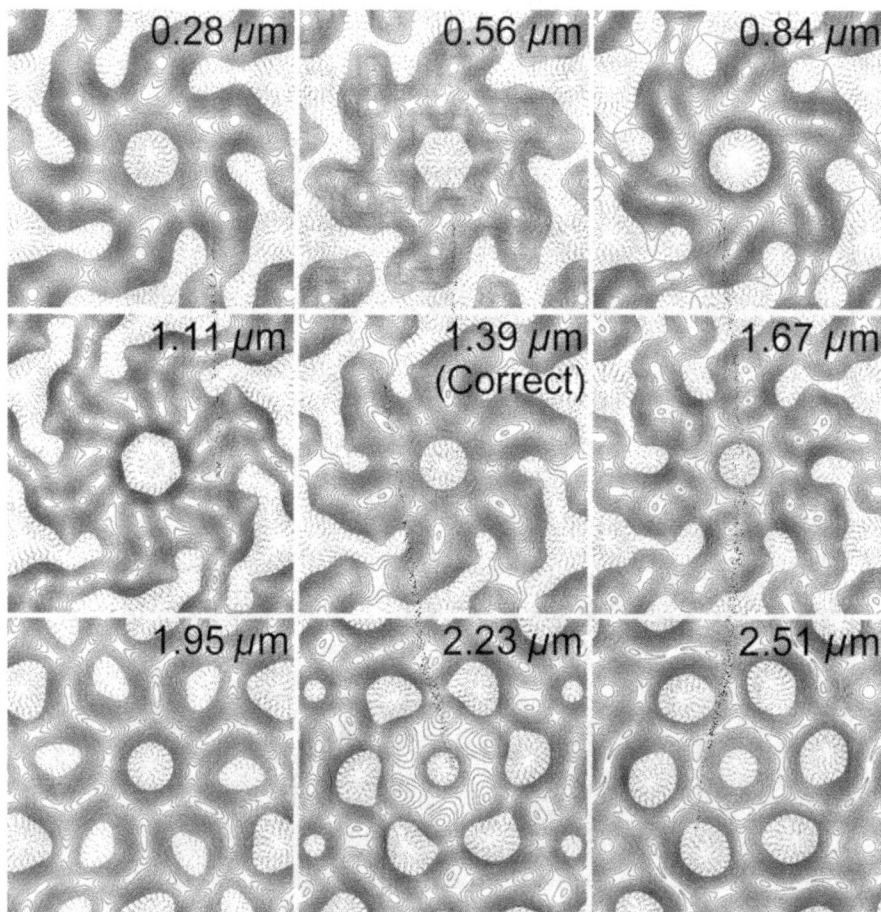

Figure 4 Projection maps resulting from wrong CTF fitting. Projection maps were produced from the data that had been fitted with arbitrary CTF (Figure 3). Structural alterations from the correct structure are apparent, although the degree of deformation varies between the maps.

factor extraction from diffraction spots with IQ score higher than 3 (good spots), filtered image was generated. Then, using a small area of the filtered image as a reference, lattice distortions in the whole image was corrected through cross-correlation between the reference and the whole image. After the unbending process, the number of 'good spots' increased from initial data by 194%.

In order to characterize the effect of CTF correction on the preservation of structural integrity, the data was either corrected for CTF from estimated defocus, or by applying arbitrary defocus values (Figure 3). For both CTF-corrected data and the data that was processed with incorrect CTF estimation, symmetry of the lattice was searched from one of 17 possible plain group symmetries. Phase residual in all cases was the lowest (28°) for p6, which agreed well with visual inspection of raw electron micrograph. Based on the symmetry, phase origin of the image was determined, and then projection maps with p6 symmetry imposition were generated (Figure 4).

Projection map of CTF-corrected image exhibited RSV CA hexamer with a central cavity, and weak density connection that links neighboring hexamers. In contrast, some projection maps produced from incorrect CTF imposition exhibited drastically altered structure. The degree of structural alteration was most dramatic when $\pi/2$ phase shift from the correct CTF was imposed, causing local phase contrast reversal in the structure (2.23 and 2.51 μm under-focus in Figure 3). In particular, such phase reversal at lower spatial frequency (resolution) compromised overall shape of the protein, leading to a complete misrepresentation of the actual specimen. Also, rapid attenuation of amplitude of those images, as shown by simulated CTF plots (red graphs in Figure 3), resulted in the loss of high-resolution structural features in the projection maps. Incorrect phase reversal at spatial frequency higher than the first node of CTF (0.56 and 0.84 μm under-focus in Figure 3) resulted changes in detailed features above approximately 20 Å resolution, of which close inspection of the structure or side-by-side comparison with known reference (i.e., atomic model) must be carried out in order to ensure structural integrity. When no phase correction was performed within the resolution limit of the data (0.28 μm in Figure 2), structural deformation of final map was subtle although the density distribution within the map varied significantly from the correct map.

As shown in the simulated CTF, it is to be noted that the attenuation of signal at higher spatial frequency changes significantly depending on the image defocus due to envelope function. Since envelope function is directly related to CTF, signal attenuates more rapidly at higher defocus (Zhou and Chiu 1993). Therefore, optimization of image defocus is necessary in order to capture high-resolution details up to a potential information limit of a given electron microscope, while producing enough image contrast for weak phase objects such as vitrified biological specimen. Such optimization, which may be based on simulated CTF or by trial-and-error, can be omitted when using highly coherent beam source such as field emission gun since the envelope function is greatly improved and amplitude attenuation is reduced at the resolution range that is adequate for structural analysis at molecular level (Zhou and Chiu 1993). In addition, development of phase plate may displace need for CTF correction for high-resolution imaging because the device allows for in-focus images with high image contrast (Nagayama 2011).

Conclusions

Structure determination of protein macromolecules using TEM is advancing rapidly, both in the range of applications and in the achievable resolution. Development of image processing algorithms made major contribution in such a rapid growth, and estimation of correct defocus and CTF correction played essential role for extending the resolution of the analysis. In this technical note, the importance of correct image defocus, a key parameter for CTF determination, was addressed through a simple image processing experiment. By illustratively demonstrating drastic misinterpretation of true structural features that result from wrong CTF correction, this study is expected to guide researchers, especially the beginners in the field, for careful monitoring of image processing steps in order to extract high resolution structural data.

Competing interests
The authors declare that they have no competing interests.

Authors' contributions
HJK designed and coordinated the study. HSJ and PHN carried out experiments and image processing. HJK, JHS and JGK refined the data, and drafted the manuscript. All authors read and approved the final manuscript.

Acknowledgements
This work was supported by the Korea Basic Science Institute grant (T33518) to J-K Hyun.

References
Armache JP, Jarasch A, Anger AM, Villa E, Becker T, Bhushan S, Jossinet F, Habeck M, Dindar G, Franckenberg S, Marquez V, Mielke T, Thomm M, Berninghausen O, Beatrix B, Söding J, Westhof E, Wilson DN, Beckmann R (2010) Cryo-EM structure and rRNA model of a translating eukaryotic 80S ribosome at 5.5-A resolution. Proc Natl Acad Sci USA 107(46):19748–19753
Bailey GD, Hyun JK, Mitra AK, Kingston RL (2012) A structural model for the generation of continuous curvature on the surface of a retroviral capsid. J Mol Biol 417(3):212–223
Crowther RA (2010) From envelopes to atoms: the remarkable progress of biological electron microscopy. Adv Protein Chem Struct Biol 81:1–32
Crowther RA, Henderson R, Smith JM (1996) MRC image processing programs. J Struct Biol 116(1):9–16
Fernandez JJ, Sanjurjo J, Carazo JM (1997) A spectral estimation approach to contrast transfer function detection in electron microscopy. Ultramicroscopy 68:267–295

Frank J (2006) Three-dimensional electron microscopy of macromolecular assemblies. Oxford University Press, New York

Ge P, Zhou ZH (2011) Hydrogen-bonding networks and RNA bases revealed by cryo electron microscopy suggest a triggering mechanism for calcium switches. Proc Natl Acad Sci USA 108(23):9637–9642

Gipson B, Zeng X, Stahlberg H (2007a) 2dx_merge: data management and merging for 2D crystal images. J Struct Biol 160(3):375–384

Gipson B, Zhang ZY, Stahlberg H (2007b) 2dx–user-friendly image processing for 2D crystals. J Struct Biol 157(1):64–72

Gonen T, Cheng Y, Sliz P, Hiroaki Y, Fujiyoshi Y, Harrison SC, Walz T (2005) Lipid-protein interactions in double-layered two-dimensional AQP0 crystals. Nature 438(7068):633–638

Hyun JK, Radjainia M, Kingston RL, Mitra AK (2010) Proton-driven assembly of the Rous Sarcoma virus capsid protein results in the formation of icosahedral particles. J Biol Chem 285(20):15056–15064

Mindell JA, Grigorieff N (2003) Accurate determination of local defocus and specimen tilt in electron microscopy. J Struct Biol 142(3):334–347

Nagayama K (2011) Another 60 years in electron microscopy: development of phase-plate electron microscopy and biological applications. J Electron Microsc (Tokyo) 60(Suppl 1):S43–S62

Penczek PA, Zhu J, Schröder R, Frank J (1997) Three dimensional reconstruction with contrast transfer compensation from defocus series. Scanning Microsc 11:147–154

Ruprecht J, Nield J (2001) Determining the structure of biological macromolecules by transmission electron microscopy, single particle analysis and 3D reconstruction. Prog Biophys Mol Biol 75:121–164

Saad A, Ludtke SJ, Jakana J, Rixon FJ, Tsuruta H, Chiu W (2001) Fourier amplitude decay of electron cryomicroscopic images of single particles and effects on structure determination. J Struct Biol 133(1):32–42

Sander B, Golas MM, Stark H (2003) Automatic CTF correction for single particles based upon multivariate statistical analysis of individual power spectra. J Struct Biol 142(3):392–401

Sidorov MV (2002) CtfExplorer: interactive software for 1d and 2d calculation and visualization of TEM phase contrast transfer function. Microsc Microanal 8:1572–1573

Sorzano CO, Jonic S, Núñez-Ramírez R, Boisset N, Carazo JM (2007) Fast, robust, and accurate determination of transmission electron microscopy contrast transfer function. J Struct Biol 160(2):249–62

Sorzano CO, Otero A, Olmos EM, Carazo JM (2009) Error analysis in the determination of the electron microscopical contrast transfer function parameters from experimental power Spectra. BMC Struct Biol 9:18

Toyoshima C, Unwin N (1988) Contrast transfer for frozen-hydrated specimens: determination from pairs of defocused images. Ultramicroscopy 25(4):279–291

Toyoshima C, Yonekura K, Sasabe H (1993) Contrast transfer for frozen-hydrated specimens II: amplitude contrast at very low frequencies. Ultramicroscopy 48:165–176

Williams DB, Carter CB (1996) Transmission electron microscopy: a textbook for material science. Plenum Press, New York

Yu X, Ge P, Jiang J, Atanasov I, Zhou ZH (2011) Atomic model of CPV reveals the mechanism used by this single-shelled virus to economically carry out functions conserved in multishelled reoviruses. Structure 19(5):652–61

Zhou ZH, Chiu W (1993) Prospects for using an IVEM with a FEG for imaging macromolecules towards atomic resolution. Ultramicroscopy 49(1–4):407–416

Zhu J, Penczek PA, Schröder R, Frank J (1997) Three-dimensional reconstruction with contrast transfer function correction from energy-filtered cryoelectron micrographs: procedure and application to the 70S Escherichia coli ribosome. J Struct Biol 118(3):197–219

3

A rapid assessment method for determination of iodate in table salt samples

Preeti S Kulkarni[1*], Satish D Dhar[2] and Sunil D Kulkarni[3]

Abstract

Background: In the present work, a simple and rapid method for determination of iodate is described.

Methods: Iodometric reaction between iodate, excess iodide, and acid has been used, and the iodine liberated is allowed to react with variamine blue (VB) dye in the presence of sodium acetate to yield a violet-colored species.

Results: A calibration curve was obtained in the concentration range of 2 to 30 μg of iodate in a final equilibration volume of 10 mL. The effect of different interfering anions on determination of iodate was also studied.

Conclusions: The developed method was applied to iodate determination in various iodized salt samples obtained from local markets in and around Pune city, India. The amount of iodate in various table salt samples was in the range of 10 to 25 ppm.

Keywords: Iodate; Table salt; Variamine blue dye; Iodometric reaction; Spectrophotometry

Background

Iodine is an essential trace element for human nutrition. The safe dietary intake of iodine as recommended by the World Health Organization (WHO) is 100 μg day^{-1} for infants and 150 μg day^{-1} for adults (Hetzel 1983). Iodine is required by the thyroid gland for the synthesis of T_3 and T_4 hormones (Visser 2006). The storehouse of iodine in the human body is the thyroid gland. Inadequate intake of iodine leads to iodine deficiency symptoms and disorders like goiter, extreme fatigue, mental retardation, and depression which are collectively called as iodine deficiency disorders (IDDs). In India, about 71 million people suffer from iodine deficiency disorders. Statistics furnished by the Ministry of Health and Family Welfare in its report revealed that Uttar Pradesh, Bihar, Madhya Pradesh, Maharashtra, and Gujarat states contributing to almost 70% population have maximum IDD cases.

The natural dietary sources of iodine include milk, vegetables, fruits, cereals, eggs, meat, spinach, and sea foods (Zimmermann 2009). However, natural sources of iodine may not satisfy its requirement by the body as iodine from these sources may not be in bioavailable form and also the concentration of iodine may be less.

Adequate intake of iodine can be achieved by consumption of iodized salt. Iodization of salt is done by addition of iodate to salt samples due to its good stability and bioavailability (Bürgi et al. 2001). Thus, determination of iodate in salt samples is of considerable importance as the amount of iodate in the salt samples may vary with environmental conditions, the nature of transport, packing conditions, and cooking methods (Bruchertseifer et al. 2003).

There are various analytical methods for determination of iodate in seawater and iodized salt samples. Some of the recent methods include kinetic spectrophotometric methods (Ni and Wang 2007), flow injection analysis (Shabani et al. 2011), microspectrophotometry after liquid-phase microextraction (Pereira et al. 2010), using cadmium sulfide quantum dots as fluorescence probes (Tang et al. 2010), liquid-liquid microextraction by high-performance liquid chromatography-diode array detection (Gupta et al. 2011), ion chromatography with integrated amperometric detection (Babulal et al. 2010), transient isotachophoresis-capillary zone electrophoresis (Wang et al. 2009), gas chromatography–mass spectrometry (Das et al. 2004), using polymer membrane selective for molecular iodine (Bhagat et al. 2008), and neutron activation analysis method (Bhagat et al. 2009). A non-suppressed ion chromatography with inductively coupled mass spectrometry (ICP-MS) has been

* Correspondence: bhagatpreeti@gmail.com
[1]Department of Chemistry, Postgraduate and Research Centre, MES Abasaheb Garware College, Pune-411005, India
Full list of author information is available at the end of the article

developed for the simultaneous determination of iodate and iodide in seawater (Zul et al. 2007). Most of the techniques are complex and involve sophisticated instruments and complex procedures. It is also observed that application of these analytical methods for iodate determination in table salt is complicated due to the presence of huge excess of chloride, for example, in the case of anion exchange chromatography with conductometric detection which requires the removal of large excess of chloride from the sample matrix (Kumar et al. 2001). Hence, development of a method that is selective for iodate and sensitive and requires simple and inexpensive experimental setup is of considerable scientific interest. Also, accurate determination of the contribution of iodine from table salt to total dietary intake requires novel methods. With this objective in the present work, a simple and rapid method for determination of iodate is described. Iodometric reaction between iodate, excess iodide, and acid has been used, and the liberated iodine is allowed to react with variamine blue (VB) dye to yield a violet-colored species with absorbance maxima at 550 nm. The developed method was applied to determine the iodate concentration in table salt samples obtained from local markets in and around Pune city in India. The kinetics of the method is very fast, and a large number of table salt samples can be screened for their iodate content in a short time. The iodate content thus determined by the developed method was compared with the iodate content determined by conventional iodometric titration. The method developed in the present work has advantages over conventional methods, for example, it is free from losses of iodine and it is interference free.

Methods

Apparatus

A computer-based spectrophotometer (Systronics, Ahmedabad, India) was used for all the absorbance measurements. A pH meter (Labtronics, Panchkula, India) was used to monitor the pH of the equilibrating solutions. The pH meter was standardized using pH 4, 7, and 10 buffer solutions. A digital balance (Contech, Mumbai, India) was used for weighing all the reagents. Double-distilled water was used throughout all the work which was prepared using Equitron's instrument (Mumbai, India).

Reagents and solutions

All reagents used were of analytical reagent grade (A.R. grade) and used without further purification. Variamine blue (Merck, Mumbai, India), potassium iodate (S.M Chemicals, Mumbai, India), potassium iodide (Loba Chemie, Mumbai, India), sodium chloride (Qualigens, Mumbai, India), potassium bromate (Qualigens), ammonium oxalate (Qualigens), potassium chloride (Qualigens), sodium bicarbonate (Qualigens), potassium nitrate (Qualigens), zinc sulfate (Qualigens), methyl alcohol (Qualigens), and

magnesium carbonate (Qualigens) were used. A variamine blue dye solution was prepared by dissolving 20 mg of the dye in methyl alcohol and diluting the solution to 50 mL using distilled water. A potassium iodate solution was prepared by dissolving 0.0122 g of KIO_3 in distilled water and diluting it to 100 mL [1 mL = 100 μg IO_3^-]. Sulfuric acid (1 M) was prepared by diluting 6.95 mL of stock H_2SO_4 to the mark in a 250-mL volumetric flask with distilled water. A solution of potassium iodide was prepared by dissolving 25 mg potassium iodide in water and diluting it up to 100 mL [1 mL = 250 μg]. A solution of sodium acetate (2 M) was prepared by dissolving 13.608 g of A.R. grade sodium acetate in distilled water and diluting the solution to 100 mL in a volumetric flask The different interfering ion solutions such as potassium chloride (KCl), sodium bicarbonate ($NaHCO_3$), potassium nitrate (KNO_3), zinc sulfate ($ZnSO_4$), potassium bromate ($KBrO_3$), etc. were prepared by dissolving and diluting suitable amounts of the respective salts in distilled water to make a concentration of 1 mL = 100 μg.

Samples for iodate determination

A total of 12 different brands of iodized salt samples were analyzed in the present work. The samples were purchased from local markets in and around Pune city. The samples were stored in cool and dry conditions. The contents of the packets were transferred immediately upon opening into an air tight container.

Optimization of parameters for the iodometric reaction

Various parameters associated with the iodometric reaction were optimized. The amount of potassium iodide and the concentration of acid were optimized in a similar manner as reported in our previous work (Bhagat et al. 2008) The concentration of iodate was fixed as 10 μg during the optimization experiments. Experiments were performed to optimize the dye concentration and pH of the reaction mixture. pH adjustments were done using either 2 M NaOH or 2 M HCl. The time required for the completion of the reaction was measured by studying the changes in absorbance as a function of time. The absorbance values were recorded in the intervals of 30 s till 30 min.

Measurement of iodate in the aqueous solution

An aliquot of iodate solution containing 2 to 30 μg of iodate was taken in 10-mL volumetric flasks. Excess of KI (250 μg) was added to each flask followed by 1 mL of H_2SO_4 (1 M). The solution turned yellow due to liberation of iodine. At this stage, 1 mL of dye solution was added followed by addition of 2 mL sodium acetate (2 M). The solutions were diluted to 10 mL with distilled water and kept for 5 min to allow the reaction to complete. After 5 min, the absorbances of all the solutions were recorded at 550 nm against water as a reagent blank. A calibration plot of absorbance values of VB dye was plotted against the amount of iodate in solution.

Interference studies

The effect of common interfering anions like Cl^-, SO_4^{2-}, NO_3^-, Br^-, PO_4^{3-} HCO_3^-, $C_2O_4^{2-}$, and BrO_3^- on determination of iodate by the VB method was studied by the following procedure. The concentration of iodate in the reaction mixture was kept fixed as 5.72×10^{-8} M, and the concentration of interfering anions in the equilibrating solution was varied.

Application to table salt samples

Before application of the method to table salt samples, it was applied to A.R. grade laboratory reagent NaCl to study the effect of sodium chloride on the absorbance values. In the case of iodized table salt samples, a homogenous portion of 2 g of sample was weighed accurately on a balance and dissolved in distilled water. The final volume was made up to 25 mL, and the solution was used for further analysis. The concentration of iodate in the samples was calculated using a calibration curve. Each sample was analyzed five times, and the standard deviation was calculated. The iodate content in these salt samples was also analyzed by conventional iodometric titration using $Na_2S_2O_3$ with starch as an indicator.

Results and discussion

Iodometric reaction

When an oxidizing agent (analyte) is added to excess iodide in an acidic medium to produce iodine which is determined by titration, the method is called iodometry. Iodometry provides a simple and rapid method of analysis. It also provides chemical amplification of signals (Zhang et al. 1998). However, conventional iodometric titrations suffer from several limitations like losses of iodine, titration error, lack of suitable indicators, and poor detection limits. These limitations can be overcome by converting the liberated iodine into an appropriate signal to prevent losses of iodine. Iodate is a good oxidizing reagent, hence oxidizes iodide to iodine in the presence of mineral acid (Pierce and Haenisch 1945). The reaction offers good signal enhancement as it can be seen from reaction 1 that the iodine content in the product side of the reaction is increased three times on a molar basis. This chemical amplification was conveniently used to determine the concentration of iodate in the present work.

$$IO_3^- + 5I^- + 6H^+ \rightarrow 3H_2O + 3I_2. \qquad (1)$$

This modified iodometric reaction is selective towards iodate, and huge excess of chloride in the table salt samples does not interfere the determination of iodate as chloride cannot oxidize iodide to iodine. The common

difficulty with the iodometric reaction for analytical purposes is the trapping of liberated iodine. In this work, the liberated iodine was allowed to react with the VB dye in the presence of sodium acetate (reaction 2). Variamine blue is known to be a suitable chromogenic reagent for iodine (Revansiddapa and Kumar 2001; Narayana and Cherian 2005; Coo and Martinez 2004). The liberated I_2 oxidizes the dye to a violet color whose absorbance maxima is at 550 nm. The visible spectrum of the oxidized variamine blue dye is shown in Figure 1. The use of this dye offered an advantage of quick reaction kinetics so that complete utilization of liberated iodine is ensured.

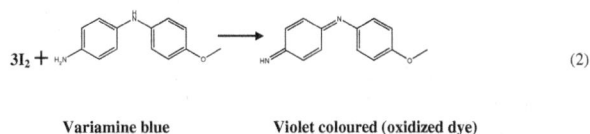

$$3I_2 + \qquad \longrightarrow \qquad \qquad (2)$$

Variamine blue   Violet coloured (oxidized dye)

Optimization of parameters for iodometric reaction

The optimization of different parameters related to the reaction was done. The concentration of the dye was varied in the range of 0.61×10^{-6} to 12.2×10^{-6} mol L^{-1}. Figure 2 shows the effect of dye concentration on its absorbance. It was observed that the maximum absorbance value was obtained for the dye concentration of 2.44×10^{-6} mol L^{-1}. However, the absorbance values were lower at dye concentrations below and above this value. This concentration of the dye was used for all further experiments. The amount of iodide was varied between 250 and 1,000 µg. It was found that 250 µg of I^- was enough to convert IO_3^- to I_2 quantitatively (Bhagat et al. 2008). The pH of the reaction mixture was varied between 2.5 and 11. It was found that the maximum absorbance was recorded at pH 5 as the oxidized form of

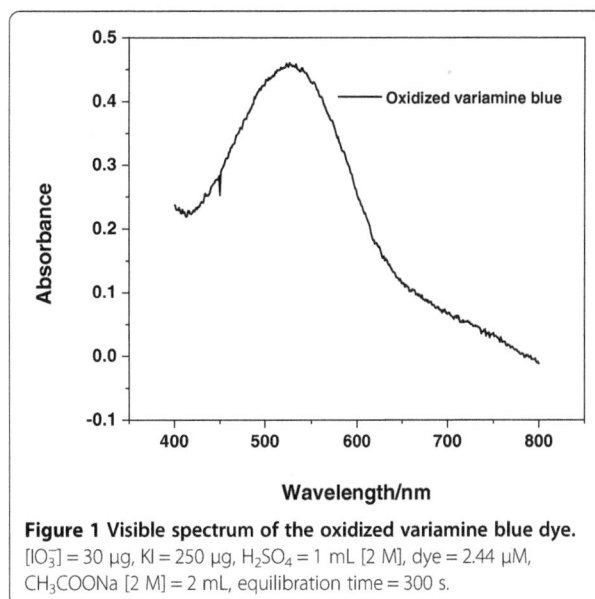

Figure 1 Visible spectrum of the oxidized variamine blue dye. $[IO_3^-] = 30$ µg, KI = 250 µg, $H_2SO_4 = 1$ mL [2 M], dye = 2.44 µM, CH_3COONa [2 M] = 2 mL, equilibration time = 300 s.

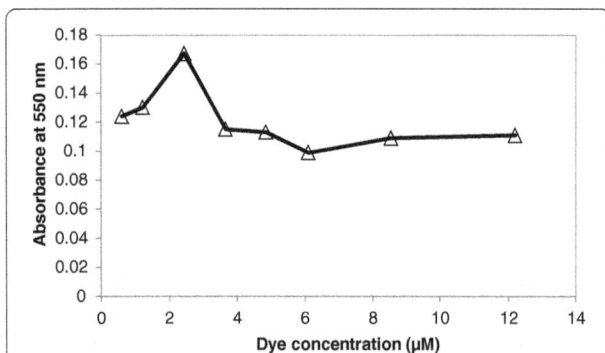

Figure 2 Effect of the concentration of variamine blue dye on absorbance at 550 nm. $[IO_3^-]$ = 10 µg, KI = 250 µg, H_2SO_4 = 1 mL [2 M], CH_3COONa [2 M] = 2 mL, equilibration time = 300 s.

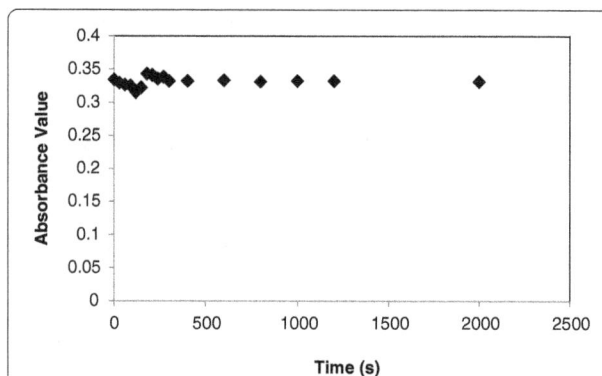

Figure 4 Absorbance of variamine blue dye as a function of the time of reaction. $[IO_3^-]$ = 10 µg, KI = 250 µg, H_2SO_4 = 1 mL [2 M], dye = 2.44 µM, CH_3COONa [2 M] = 2 mL.

variamine blue is stable at this pH (Figure 3). The effect of acid concentration on the uptake of iodate was studied by varying the amount of H_2SO_4 (1 M). It was found that 1 mL of H_2SO_4 gave maximum absorbance. To study the effect of time on the absorbance of VB dye, all other conditions were kept fixed and the absorbance was recorded at intervals of 30 s for 30 min. It was found that the time required for completion of the reaction was 300 s, after which the absorbance values remained constant (Figure 4). The optimized parameters in the method are given in Table 1.

Calibration curve and detection limits for iodate determination

The quantitative determination of iodate in aqueous solutions was done by constructing a calibration plot. The reaction was carried out using an iodate amount in the range of 2 to 30 µg in a final volume of 10 mL. Then addition of excess iodide was done followed by addition of acid, VB, and sodium acetate in that sequence. The absorbance values of the solutions were plotted as a

function of the iodate amount in the solutions and used as a calibration plot (Figure 5) for further quantifications. The plot was found to be linear till 30 µg of iodate. The regression value obtained for the calibration plot is 0.987, and the equation of the calibration plot calculated using 15 standards in the range of 2 to 30 µg in a final volume of 10 mL is $y = 0.013x$. A good linear relationship between absorbance values and amount of iodate suggests the use of the present method for quantitative determination of iodate in aqueous samples. The detection limit of the method is 0.25 µg, calculated using the relation DL = $3 s/S$, where s is the standard deviation of the reagent blank and S is the slope of the calibration curve.

Effect of interfering anions on determination of iodate

The effect of interfering anions on determination of iodate in aqueous samples like Cl^-, SO_4^{2-}, NO_3^-, Br^-, PO_4^{3-}, HCO_3^-, $C_2O_4^{2-}$, and BrO_3^- was studied. The amount of iodate in aqueous samples was kept as 5.72×10^{-8} M,

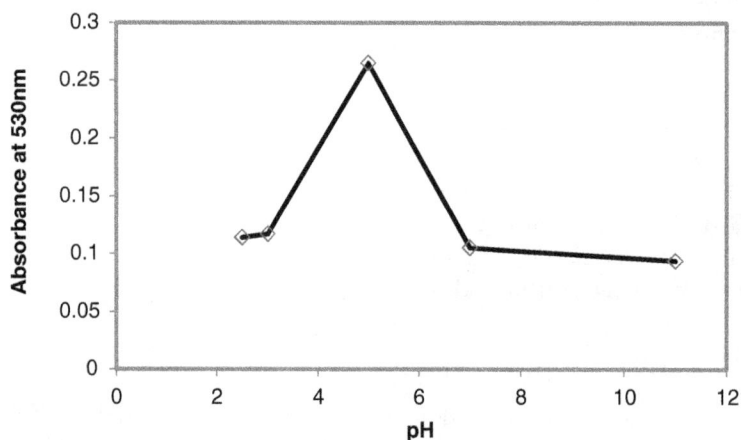

Figure 3 Effect of the pH of reaction mixture on the absorbance value. $[IO_3^-]$ = 10 µg, KI = 250 µg, H_2SO_4 = 1 mL [2 M], dye = 2.44 µM, CH_3COONa [2 M] = 2 mL, equilibration time = 300 s.

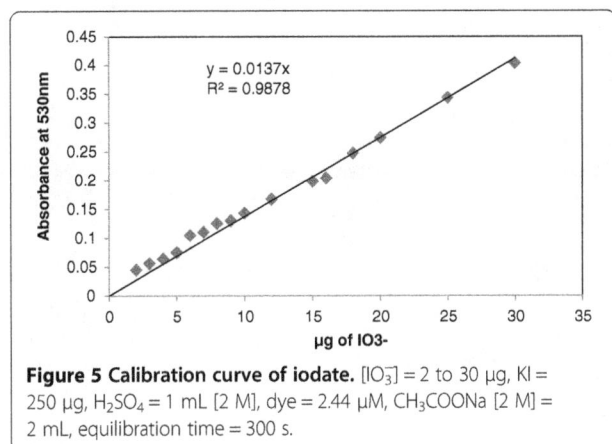

Figure 5 Calibration curve of iodate. $[IO_3^-]$ = 2 to 30 μg, KI = 250 μg, H_2SO_4 = 1 mL [2 M], dye = 2.44 μM, CH_3COONa [2 M] = 2 mL, equilibration time = 300 s.

Table 2 Effect of some interfering anions on iodate determination

Serial number	Interfering anion $[X^-]$	Tolerated ratio $[X^-]/[IO_3^-]$
1	Cl^-	500
2	NO_3^-	100
3	SO_4^{2-}	50
4	Br^-	50
5	PO_4^{3-}	50
6	HCO_3^-	5
7	$C_2O_4^{2-}$	25
8	BrO_3^-	5

and the amount of interfering anions in the equilibrating solution was varied. The uptake of IO_3^- in the presence of excess of anions likely to be encountered in the food samples and other aqueous samples was calculated by measuring the absorbance of VB in the solution as a function of the ratio $[X^-]/[IO_3^-]$, where $[X^-]$ is the concentration of the interfering anion. The tolerance limits of various anions are given in Table 2.

Determination of iodate in iodized salt samples

Fortification of table salt is done with KIO_3 in order to meet the requirements of iodine in human beings. Iodate is added in the range of 15 to 50 ppm to table salt. Various brands of iodized salt samples were subjected to the method described above for determination of iodate content. Before analyzing the iodized salt samples, the protocol was used for iodate determination in A.R. grade NaCl (lab reagent), and it was found that there was no effect of NaCl on the absorbance of VB. Consequently, the analysis method was applied to various iodized salt brands. The iodate concentrations obtained in these samples are given in Table 3. It is found that the iodate content was found to be in the range of 10 to 22 ppm. A set of five measurements were carried out for each sample, and the mean values are reported along with standard deviations. The standard deviation is in the range of 0.40 to 3.8. Brand number 3 and 5 showed less values of

iodate as compared to the quoted value when analyzed by this method. The iodine content as quoted by the manufacturer is also given in Table 3. The amount of iodine obtained by conventional iodometric titration is also given in the table.

According to WHO, the daily dietary intake of iodine is 150 μg for adults. Considering the spicy food habits in India, the average intake of salt through food may be in the range of 2 to 3 g per day. Using this approximation, the contribution of iodine from table salt can be calculated to be around 30 to 40 μg per day. The bioavailability of iodine may be considered to be 100% due to its solubility in digestive fluids. The contribution of iodine from table salt is estimated to be 20% to 40% of the total iodine requirement. This contribution may not be enough in regions where the soil is deficient in iodine content. Consequently, iodine deficiency disorders will be prevalent in these regions.

Table 1 Optimized parameters used for analysis

Serial number	Optimized parameters	Value
1	IO_3^-	2 to 30 μg
2	KI	250 μg
3	H_2SO_4	1 mL [2 M]
4	pH	5.0
5	Dye	2.44 μM
6	Sodium acetate	2 mL of 2 M
7	Time	300 s

Table 3 Iodate values obtained in local brands of table salt

Serial number	Brand name	Iodate (ppm) ± SD (n = 3) by variamine blue method	Iodate (ppm) ± SD (n = 3) by iodometry
1	Brand 1	15.31 ± 2.645	14.3 ± 0.2
2	Brand 2	14.67 ± 2.336	13.6 ± 0.1
3	Brand 3	11.83 ± 1.98	10.8 ± 0.2
4	Brand 4	15.76 ± 3.80	16.3 ± 0.1
5	Brand 5	10.78 ± 0.46	9.8 ± 0.1
6	Brand 6	15.22 ± 1.90	15.5 ± 0.2
7	Brand 7	16.01 ± 1.46	14.3 ± 0.1
8	Brand 8	21.09 ± 2.50	23.2 ± 0.1
9	Brand 9	16.02 ± 0.48	15.6 ± 0.2
10	Brand 10	16.20 ± 0.56	15.9 ± 0.1
11	Brand 11	18.20 ± 0.66	19.3 ± 0.2
12	Brand 12	17.60 ± 2.34	18.5 ± 0.2

Conclusions

A simple and rapid method has been developed for determination of iodate in aqueous samples. The method is applicable to iodate determination in the concentration range of 2 to 30 µg in a final equilibration volume of 10 mL. Optimized parameters for the method are IO_3^- (2 to 30 µg), KI (250 µg), 2 M H_2SO_4 (1 mL), pH (5.0), time of equilibration (20 min), 2.44 µM dye (20 µg, 1 mL), and 2 mL sodium acetate (2 M). The concentration of IO_3^- obtained in the salt samples was in the range of 10 to 22 ppm. The results obtained by the present method are in good agreement with those obtained by conventional iodometry, thus validating the method.

Competing interests

Authors declare that there are no competing interests.

Authors' contributions

PSK: Original idea, design of work, SDD: Execution of experiments, SDK: Data interpretation and manuscript writing. All authors read and approved the final manuscript.

Author details

[1]Department of Chemistry, Postgraduate and Research Centre, MES Abasaheb Garware College, Pune-411005, India. [2]Modern College, Shivajinagar, Pune 411005, India. [3]Department of Chemistry, Sir Parashurambhau, Pune-411030, India.

References

Babulal R, Parimal P, Ghosh PK (2010) Determination of iodide and iodate in edible salt by ion chromatography with integrated amperometric detection. Food Chem 123:529–534

Bhagat PR, Pandey AK, Acharya R, Natarajan V, Rajurkar NS, Reddy AVR (2008) Molecular iodine selective membrane for iodate determination in salt samples: chemical amplification and preconcentration. Anal Bioanal Chem 391:1081–1089

Bhagat PR, Acharya R, Nair AGC, Pandey AK, Rajurkar NS, Reddy AVR (2009) Estimation of iodine in food, food products and salt using ENAA. Food Chem 115:706–710

Bruchertseifer H, Cripps R, Guentay S, Jaeckel B (2003) Analysis of iodine species in aqueous solutions. Anal Bioanal Chem 375:1107–1110

Bürgi H, Schaffner T, Seiler JP (2001) The toxicology of iodate: a review of the literature. Thyroid 11:449–456

Coo LD, Martinez IS (2004) Nafion-based optical sensor for the determination of selenium in water samples. Talanta 64:1317–1322

Das P, Gupta M, Jain A, Verma KK (2004) Single drop microextraction or solid phase microextraction-gas chromatography–mass spectrometry for the determination of iodine in pharmaceuticals, iodized salt, milk powder and vegetables involving conversion into 4-iodo-N,N-dimethylaniline. J Chromatogr A 1023:33–39

Gupta M, Pillai AKKV, Singh A, Jain A, Verma KK (2011) Salt assisted liquid-liquid microextraction for the determination of iodine in table salt by high-performance liquid chromatography-diode array detection. Food Chem 124:1741–1746

Hetzel BS (1983) Iodine deficiency disorders (IDD) and their eradication. Lancet 2:1126–1127

Kumar SD, Maiti B, Mathur PK (2001) Determination of iodate and sulphate in iodized table salt by ion chromatography with conductivity detection. Talanta 53:701–705

Narayana B, Cherian T (2005) Rapid spectrophotometric determination of trace amounts of chromium using variamine blue as a chromogenic reagent. J Brazil Chem Soc 16:197–201

Ni Y, Wang Y (2007) Application of chemometric methods to the simultaneous kinetic spectrophotometric determination of iodate and periodate based on consecutive reactions. Microchem J 86:216–226

Pereira FP, Ferreiro SS, Lavilla I, Bendicho C (2010) Determination of iodate in waters by cuvetteless UV–vis micro-spectrophotometry after liquid-phase microextraction. Talanta 81:625–629

Pierce WC, Haenisch EL (1945) Quantitative analysis, 2nd edition. Wiley, New York, pp 199–216

Revansiddapa H, Kumar TLK (2001) A facile spectrophotometric method for the determination of selenium. Anal Sci 17:1309–1312

Shabani AMH, Ellis PS, McKelvie ID (2011) Spectrophotometric determination of iodate in iodised salt by flow injection analysis. Food Chem 129:704–707

Tang CR, Su Z, Lin B, Huang H, Zeng Y, Li S, Huang H, Wang Y, Li C, Shen G, Yu R (2010) A novel method for iodate determination using cadmium sulfide quantum dots as fluorescence probes. Anal Chim Acta 678:203–207

Visser TJ (2006) The elemental importance of sufficient iodine intake: a trace is not enough. Endocrinol 147:2095–2097

Wang T, Zhao S, Shen C, Tang J, Wang D (2009) Determination of iodate in table salt by transient isotachophoresis-capillary zone electrophoresis. Food Chem 112:215–220

Zhang M, Zhan G, Chen Z (1998) Iodometric amplification methods for the determinations of microgram amounts of manganese (II), manganese (VII), chromium (III) and chromium (VI) in aqueous solution. Anal Sci 14:1077–1083

Zimmermann MB (2009) Iodine deficiency. Endocrine Rev 30:376–408

Zul C, Megharaj M, Naidu R (2007) Speciation of iodate and iodide in seawater by non-suppressed ion chromatography with inductively coupled plasma mass spectrometry. Talanta 72:1842–1846

4

Trichloroacetic acid assisted synthesis of gold nanoparticles and its application in detection and estimation of pesticide

Gadadhar Barman, Swarnali Maiti and Jayasree Konar Laha[*]

Abstract

Background: Many analytical methods are available for detection of methyl parathion in water but they are not handy for on-site analysis. An attempt has been made to utilize stable GNP for methyl parathion detection by sensing the peak at 400 nm generated due to the interaction between methyl parathion and GNP.

Methods: GNP was produced by reduction of chloroauric acid solution by trichloroacetic acid in alkaline medium in presence of CTAB. Sensor properties of GNP were studied by varying the concentration of methyl parathion in gold sol from 0 to 500 ppm.

Results and discussion: GNP stabilized by CTAB showed only one peak at 532 nm and one broad peak near 300 nm was observed for pure methyl parathion. But as soon as methyl parathion was added in the GNP solution, one new peak at 400 nm developed in addition to the other two peaks. More interestingly, a quantitative decrease of the absorbance at 532 nm of GNP and increase of the absorbance at 400 nm, the new peak, were observed when methyl parathion concentration increased from 10 to 500 ppm.

Conclusions: The UV-VIS measurement and TEM images confirmed that the surfactant capped GNP can act as a colorimetric sensor for detection and estimation of methyl parathion pesticide present in water in ppm level.

Keywords: Gold nanoparticles, Cetyl trimethyl ammonium bromide, Sensor, Detection of methyl parathion, Estimation of pesticide, Spectroscopy

Background

The exponential growth in research in gold nanoparticles has been mainly due to the following reasons (i) stability of gold nanoparticles, (ii) many relatively easy preparation methods, (iii) its role in nanoscience and nanotechnology. For preparation of gold nanoparticles, many different reducing agents and stabilizing compounds have been employed to control shape and size of the particles. When gold nanoparticles approach each other and aggregate, the colour changes from red to blue because of the shift of the surface plasmon band to longer wavelength. The distinctive colors of gold nanoparticles have inspired people for years to use them as sensors because of several advantages over conventional electrochemical sensors. The literature shows that the use of conventional analytical methods for detection of pesticide is getting replaced by sensors due to ease of application. The detection of pesticide residues is done using different biosensors (Airoldi et al. 2007). Several enzymes have been used in the development of electrochemical biosensor based on the inhibition mode of the enzyme for the determination of pesticides (Upadhyay et al. 2009). GNP based dipstick immunoassay was developed for the rapid detection of organochlorine pesticides such as DDT at nanogram level (Lisa et al. 2009). Along similar lines, in the present work we have developed a method for preparation of gold nanoparticles (GNP) using trichloroacetic acid as reducing agent in alkaline medium in presence of surfactant which acts as capping agent. This surfactant capped GNP has been used as colorimetric sensor for detection and estimation of pesticide (methyl parathion) present in a sample. Methyl parathion is chosen because it is a

* Correspondence: j.laha@yahoo.co.in
Department of Chemistry, , Midnapore College, Midnapore 721101, W.B, India

highly neurotoxic agricultural chemical that is used extensively worldwide to control a wide range of insect pests. Its residue in the soil causes pollution in the environment and poses a risk to human health. We have employed cetyl trimethyl ammonium bromide (CTAB) capped GNP as sensor and the change in UV-Visible spectra was monitored when pesticide (methyl parathion) was added at ppm level.

Many analytical methods are available for detection of methyl parathion in water. The common analytical techniques include HPLC (high performance liquid chromatography) with UV detection (Huang et al. 2002), GC-Mass spectrometry (Ferrer et al. 2005; Garcia-Reyes et al. 2007), GC/ECD (gas chromatography coupled with electron capture) (Bicchi et al. 2003; Brito et al. 2002), GC/FPD (gas chromatography coupled with flame photometry) (Berijani et al. 2006), GC/FID (flame ionization detection) (Pinheiro & Andrade 2009). All these methods are very sensitive but each one requires its unique method for sample preparation and above all, they are not handy for on-site analysis. These are mainly applied in laboratory settings and restrict rapid analysis under field conditions. At the same time they are expensive and time consuming. For this reason, an attempt has been made to utilize stable GNPs for methyl parathion detection. The nano colloidial GNPs could detect methyl parathion by forming bond using sulphur present in methyl parathion and thereby generating a new peak near 400 nm. The height of this peak is proportional to the concentration of methyl parathion and this makes for a ready means of estimating the pesticide concentration.

Methods
Preparation

Trichloroacetic acid, chloroauric acid, cetyl trimethyl ammonium bromide, all AR grade, were purchased from Sigma-Aldrich Chemical Ltd. Sodium hydroxide and methyl parathion were purchased from Merck. Double distilled de-ionized water was used in all experiments.

GNP was produced by reduction of chloroauric acid solution by trichloroacetic acid in alkaline medium in presence of CTAB. The pH was varied during reduction by adding different amounts of NaOH solution.10 ml of $3 \times 10\text{-}3(M)$ trichloroacetic acid solution was added to an equal volume of the same concentration of alkaline CTAB. The mixture was cooled in ice cold water. Then 7 ml of $3 \times 10\text{-}3(M)$ aqueous chloroauric acid was added drop wise with continuous stirring. The mixture was cooled for 10 minutes and then it was heated at 850C for 1 hour. The colour of the solution gradually changed from yellow to violet to reddish violet. The reddish violet colour indicated the formation of gold nanoparticles (GNPs).

Characterization

The absorbance spectra of the GNPs were analyzed by using a 'SHIMADZU' UV 1800 spectrophotometer and TEM images were taken using JEOL-JEM 2100 high resolution transmission electron microscope (HR-TEM). Samples for the TEM studies were prepared by placing a drop of the aqueous suspension of particles on carbon-coated copper grids followed by solvent evaporation under vacuum. The crystalline nature of the GNPs was examined using X' Pert-PRO model (Analytical Holland) X-ray diffractometer.

Sensor properties of GNP were studied by varying the concentration of methyl parathion in gold sol from 0 to 500 ppm. 250 µl of solution containing different concentrations of methyl parathion was added to 5 ml of "as prepared" GNP. The sol was heated for 5 minutes with stirring. The reddish violet color changed into yellow. The intensity of yellow colour gradually increased with increase of pesticide concentration.

Results and discussion
Optimization of GNP preparation

GNP displays optical properties due to the presence of surface plasma resonance (SPR) band. For spherical gold particles the SPR occurs nearly at 540 nm. Small shifts in the SPR band occur perhaps due to the changes in the dielectric properties of the medium of the GNP or due to the specific adsorption of materials on the surface of the gold particle. Mie theory is generally used to explain the shifts in the SPR band (Mie 1908; Creighton & Eadon 1991). According to this theory, shifts in SPR may also occur when particles geometry changes from spherical to some other shapes. In the present work, the GNP as prepared produces SPR band at 532 nm. We have observed that the pH and concentration of the surfactant have important roles in the formation of GNP. Hence the effects of pH and concentration of CTAB were investigated thoroughly. To optimize the pH of the solution it was varied from 7.0 to 11.0 during the reaction. When the pH is higher than 8.0 and lower than 10.5, the GNP solution was stable for more than six months. However, if the pH is lower than 8.0, the precipitation occurred within several hours. One interesting observation should be pointed out that though pH and CTAB concentration have effect on the stability of the GNP formation yet the SPR band was fixed at 532 nm in all cases. From the extinction spectra (Figure 1A) it is seen that maximum absorbance of GNP peak i.e. 532 nm peak is found at pH 9.5. Furthermore, it has been observed that $3 \times 10\text{-}3$ (M) concentration of surfactant produced most stable sol, though $5 \times 10\text{-}3(M)$ concentration produced higher extinction coefficient of 532 nm peak (see Additional file 1: Figure S1B). At higher concentration of the surfactant, precipitation of the sol occurred. That is why in our experiments we

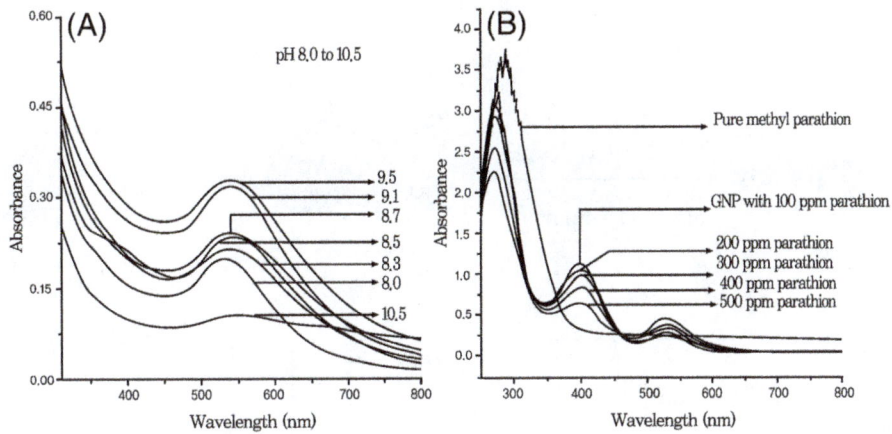

Figure 1 UV–vis spectra of (A) GNP at different pH and (B) Pure methyl parathion and GNP with different concentrations of methyl parathion.

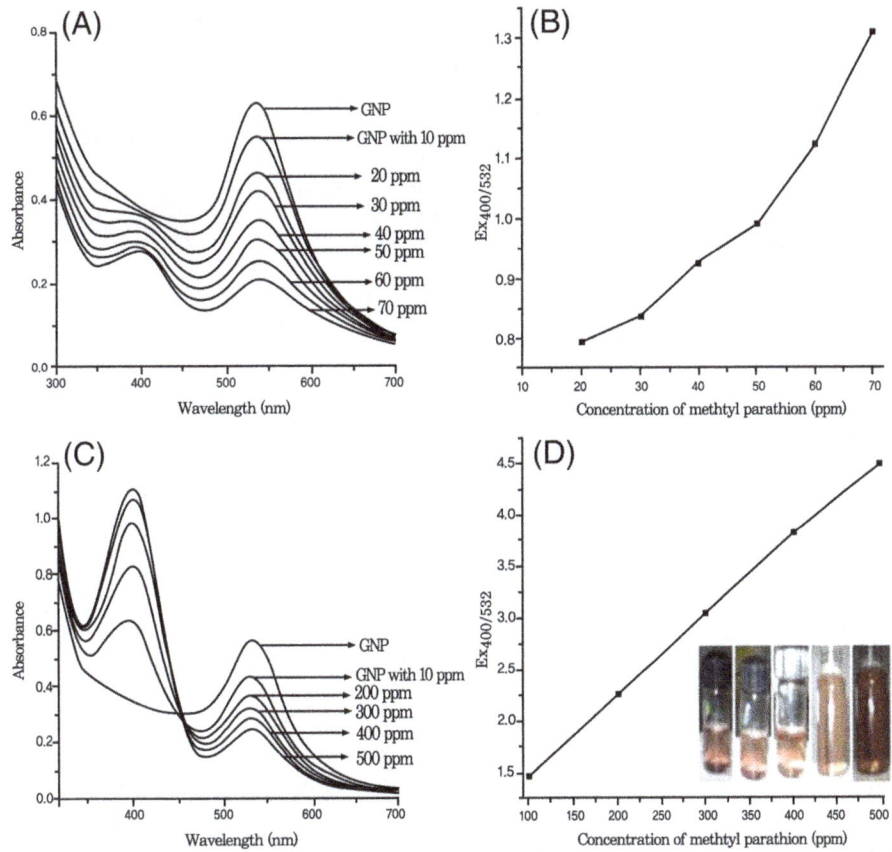

Figure 2 UV–vis spectra of GNP and GNP with various concentrations of methyl parathion (A) 10 to 70 ppm (C) 100 to 500 ppm and corresponding changes of absorption coefficients (Ex 400/532) with various concentrations of methyl parathion (B) 10 to 70 ppm (D) 100 to 500 ppm. 2(D) (inset): digital photographic images of color changes due to addition of methyl parathion.

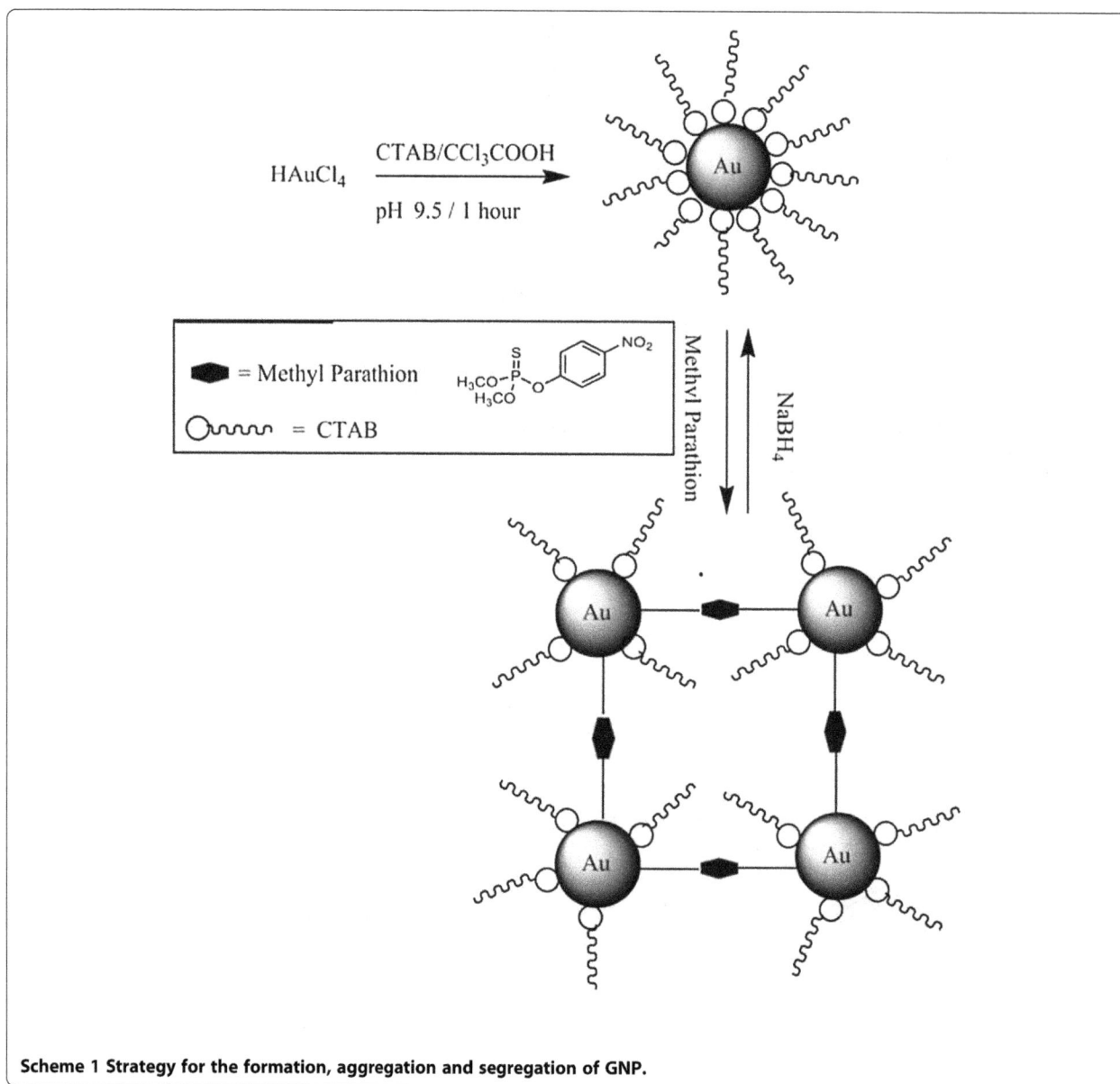

Scheme 1 Strategy for the formation, aggregation and segregation of GNP.

have used $3 \times 10\text{-}3(M)$ CTAB with equi-molar chloroaruric acid at pH 9.5 to get most stable GNP.

The change in the absorbance spectra due to the increasing methyl parathion concentration from 0 to c depicted in Figure 1(B), 2(A) & (C).

GNP stabilized by CTAB shows only one peak at 532 nm while pure methyl parathion shows one broad peak near 300 nm. But as soon as methyl parathion is added in the GNP solution, one new peak at 400 nm starts developing in addition to the other two peaks. More interestingly, a quantitative decrease of the absorbance peak at 532 nm of GNP and increase of the absorbance at 400 nm, the new peak, are observed when methyl parathion concentration increases from 10 to 500 - ppm. The appearance of the new peak near 400 nm might

be due to the bond formation between phosphorothioate group present in methyl parathion and the GNP. A control experiment shows that no such peak is obtained when methyl parathion is added in pure chloroauric acid (see Additional file 1: Figure S2B). It is well known that many compounds containing sulphur are specifically being employed for functionalization of GNP through ligand exchange reaction.

In recent years, mercaptopropionic acid (Huang & Chang 2007), mercaptoundecanoic acid (Templeton et al. 1998), glutathione (Chai et al. 2010), etc. were used for this purpose. From the chemical structure of the pesticide we can also predict that phosphorothioate group may definitely help the functionalization of GNP due to which the new peak appears near 400 nm (Figure 1(B), 2(A) & 2(C)).

The probable mechanism for bond formation and aggregation of sol particles are elaborated in Scheme 1.

The absorption coefficient ratio (the ratio of absorption coefficient of 400 and 532 nm peaks, Ex 400/532), which measures the changes in absorption peaks due to the addition of various concentrations of pesticide, has been investigated thoroughly. The ratio increases with the increase of concentration of the pesticide. This phenomenon suggests that more the phosphorothioate group more is the functionalization of GNP up to a certain concentration of pesticide. A calibration curve between the absorption coefficient ratios versus concentration of pesticide can help one to estimate quantitatively the presence of methyl parathion in a sample at ppm level (Figure 2(B) & 2(D)).

The TEM measurements of GNP produced at around pH 9.5 show almost spherical particles of different sizes (Figure 3(A)). The capping of the hydrophobic part of CTAB on GNP shows a layer surrounding the gold particle. In presence of methyl parathion, the size, shape altogether changes drastically and a completely different look with an exciting feature is observed (Figure 3(B), 3(C) & 3(D)).

It appears that restructuring of GNP occurs after addition of methyl parathion. Spherical particles become spheroidal and agglomeration of particles is observed. It is likely that the surface of the GNP forms an Au-S co-ordination bond (this can happen as the sol is being heated after addition of methyl parathion) and some methyl parathion molecules get adsorbed on the Au surface by replacing CTAB. As methyl parathion is anionic in alkaline medium, its adsorption on the GNP surface lowers the surface charge and thus they agglomerate and the interesting particle clustering is observed. Each cluster has an average size of 50 nm. These clusters segregate to give back GNP reversibly when reduction is performed with NaBH$_4$ (Scheme 1 and Figure 4).

Detailed studies are in progress to know the exact chemical nature of the bond formed between GNP and methyl parathion. A recent review discusses the recent progress in understanding of molecular structure of the gold-sulfur interface in thiolate-protected gold surfaces and interfaces from the viewpoint of theory and computations, with connections to relevant experiments (Hakkinen 2012). Though thiolate group is not present in this work, phosphorothioate group may be responsible for the Au-S interaction. We are also interested to know what sort of interactions is actually taking place on the GNP surface with methyl parathion so that new peak develops at 400 nm. To investigate this bonding XRD and FTIR have been employed. The XRD

Figure 3 TEM micrographs of GNP (A) before and (B), (C) & (D) after addition of methyl parathion.

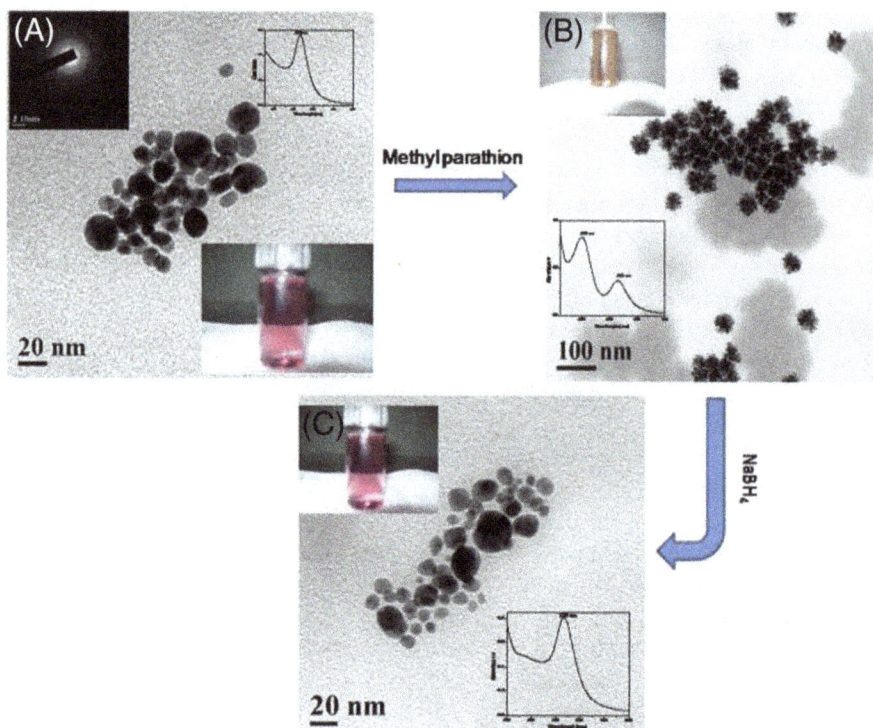

Figure 4 A strategy of aggregation of GNP (A) due to reaction with methyl parathion (B) and segregation of GNP by reduction with NaBH₄ (C).

pattern of the GNP shown in Figure 5(A) indicates the face centered cubic (fcc) structure of the bulk gold having peaks at 38.175o, 44.525o, 64.675o, 77.675o and 81.825o corresponding to (111), (200), (220), (311) and (222) planes, respectively. In literature we find Au2S shows peaks near 31o, 35.5o, 51.5o, 61.5o and 65o in XRD (Kuo & Huang 2008). The XRD spectrum of the GNP after reaction with methyl parathion are shown in Figure 5(B) and it is visible that the spectrum has all peaks due to Au as well as some

additional peaks at 50.5 o, 60.06o and 64.5 o which may be attributed to the Au-S co-ordination bond.

FTIR spectra (Additional file 1: Figure S5) of the GNP before and after addition of methyl parathion show almost no changes in the peak positions (slightly shifted). The peaks are mainly due to the CTAB present in the system (Liu et al. 2007). The comparison of the spectra shows the appearance of new peaks near 1480 cm-1 and 853 cm-1 which we are unable to assign. More work is needed to understand the

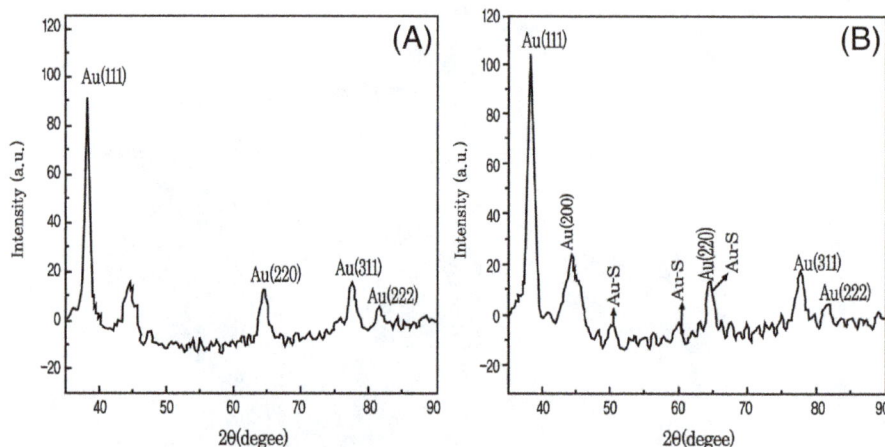

Figure 5 XRD of GNP (A) before and (B) after addition of methyl parathion.

nature of the bond which may be crucial for some applications of gold based nanoparticles for some other purposes.

Conclusion

We have employed CTAB capped GNP as sensor and the change in UV-Visible spectra was monitored when methyl parathion was added at ppm level. The UV–VIS measurements and TEM images confirm that the surfactant capped GNP can act as a colorimetric sensor for detection and estimation of methyl parathion pesticide present in ppm levels by utilizing the bonding between GNP and phosphorothioate group of the pesticide. Agglomeration of the GNPs occurs due to the adsorption of methyl parathion on the GNP surface, and hence the interesting particle clustering is observed. These clusters segregate to give back GNP reversibly when reduction with $NaBH_4$ is done.

Supporting information

UV–vis spectra of GNP formation with varying time and concentrations of chloroauric acid and CTAB result of control experiment with pure chloroauric acid, strategy and some TEM photographs for aggregation and segregation of GNP. These materials are available free of charge on the Web at http://www.jsac.or.jp/analsci/.

Additional file

> **Additional file 1: Figure S1.** UV-Vis spectra of GNP formation with varying concentrations of (A) chloroauric acid (B) CTAB. **Figure S2.** UV-Vis spectra of GNP formation varying time of reaction from 20 minutes to 2 hours keeping chloroauric acid concentration constant at 3×10^{-3} (M) and (B) Absorption spectra of pure chloroauric acid in presence of methyl parathion at various concentration. **Figure S3.** Different TEM micrographs of GNP (A) & (B) and GNP after reaction with methyl parathion (C) & (D). **Figure S4.** UV representation of aggregation of GNP (A) due to reaction with methyl parathion (B) and segregation of GNP by reduction with $NaBH_4$ (C). **Figure S5.** FTIR spectra of (A) GNP and (B) GNP with methyl parathion.

Competing interests
The authors declare that they have no competing interests.

Authors' contributions
GB carried out the experiment. GB and SM drafted the manuscript. JKL guided the research and modified the manuscript. All three authors read and approved the final manuscript.

Authors' information
JKL is Associate Professor and Head of Department of Chemistry, Midnapore College, West Bengal, India. GB and SM are research scholars of this Department.

Acknowledgements
We gratefully acknowledge the financial support received from UGC (Ref.No. F. PSW-096 / 10–11.) We are also thankful to Central Research Facility at IIT Kharagpur, India for HR-TEM and XRD measurements.

References

Airoldi FPS, Da Silva WTL, Crespilho FN, Rezende MOO (2007) Evaluation of the Electrochemical Behavior of Pentachlorophenol by Cyclic Voltammetry on Carbon Paste Electrode Modified by Humic Acids. Water Environ Res 79:63

Berijani S, Assadi Y, Anbia M, Hosseini MRM, Aghaee E (2006) Dispersive liquid–liquid microextraction combined with gas chromatography-flame photometric detection: Very simple, rapid and sensitive method for the determination of organophosphorus pesticides in water. J Chromatogr. A. 1123:1

Bicchi C, Cordero C, Iori C, Rubiolo P, Sandra P, Janete H, Yariwake JH, Zuin VG (2003) SBSE-GC-ECD/FPD in the Analysis of Pesticide Residues in Passiflora alata Dryander Herbal Teas. J Agric Food Chem 51:27

Brito NM, Navickiene S, Polese L, Jardim EFG, Abakerli RB, Ribeiro ML (2002) Determination of pesticide residues in coconut water by liquid–liquid extraction and gas chromatography with electron-capture plus thermionic specific detection and solid-phase extraction and high-performance liquid chromatography with ultraviolet detection. J Chromatogr. A. 957:201

Chai F, Wang C, Wang T, Li L, Su Z (2010) Colorimetric detection of Pb^{2+} using glutathione functionalized gold nanoparticles. ACS Appl Mater Interfaces 25:1466

Creighton JA, Eadon DG (1991) Ultraviolet-visible absorption spectra of the colloidal metallic elements J. Chem. Soc. Faraday Trans. 87:3881

Ferrer I, Thurman EM, Fernandez-Alba AR (2005) Quantitation and accurate mass analysis of pesticides in vegetables by LC/TOF-MS. Anal Chem 77:2818

Garcia-Reyes JF, Hernando MD, Ferrer C, Molina-Diaz A, Fernandez-Alba AR (2007) Large Scale Pesticide Multiresidue Methods in Food Combining Liquid Chromatography, Time-of-Flight Mass Spectrometry, and Tandem Mass Spectrometry. Anal Chem 79:7308

Hakkinen H (2012) The Gold-Sulfur Interface at the Nanoscale. Nature Chem. 4:443

Huang CC, Chang HT (2007) Parameters for selective colorimetric sensing of mercury (II) in aqueous solutions using mercaptopropionic acid-modified gold nanoparticles. Chem Commun 12:1215

Huang G, Ouyang J, Baeyens WRG, Yang Y, Tao C (2002) High-performance liquid chromatographic assay of dichlorvos, isocarbophos and methylparathion from plant leaves using chemiluminescence detection. Anal Chim Acta 474:21

Kuo CL, Huang MH (2008) Hydrothermal Synthesis of Free-Floating Au2S Nanoparticle Superstructures. J Phys. Chem. C. 112:11661

Lisa M, Chouhan RS, Vinayaka AC, Manonmani HK, Thakur MS (2009) Gold nanoparticles based dipstick immunoassay for the rapid detection of dichlorodiphenyltrichloroethane: An organochlorine pesticide. Biosens Bioelectron 25:224

Liu XH, Luo XH, Lu SX, Zhang JC, Cao WL (2007) A novel cetyltrimethyl ammonium silver bromide complex and silver bromide nanoparticles obtained by the surfactant counterion. J Colloid Interface Sci. 307:94

Mie G (1908) Contributions to the Optics of Turbid Media, Especially Colloidal Metal Solutions. Ann. Physik. 25:377

Pinheiro AS, Andrade JB (2009) Development, validation and application of a SDME/GC-FID methodology for the multiresidue determination of organophosphate and pyrethroid pesticides in water. Talanta 79:1354

Templeton AC, Hostetler MJ, Warmth EK, Chen SW, Hartsshorn CM, Krishnamurty VM, Forbes MDE, Murray RW (1998) Gateway Reactions to Diverse, Polyfunctional Monolayer-Protected Gold Clusters, J Am. Chem. Soc. 120:4845

Upadhyay S, Rao GR, Sharma MK, Bhattacharya BK, Rao VK, Vijayaraghavan R (2009) Immobilization of acetylcholineesterase–choline oxidase on a gold–platinum bimetallic nanoparticles modified glassy carbon electrode for the sensitive detection of organophosphate pesticides, carbamates and nerve agents. Biosens Bioelectron 25:832

5

Canvassing of thrombolytic, cytotoxic, and erythrocyte membrane-stabilizing attributes in *in vitro* screening of *Gynocardia odorata*

Faisal Asif[1][*], Arshida Zaman Boby[1], Nur Alam[1], Muhammad Taraquzzaman[1], Sharmin Reza Chowdhury[1] and Mohammad Abdur Rashid[2]

Abstract

Background: In the current study, different plant extracts of *Gynocardia odorata* such as methanol extract (ME), aqueous soluble fraction (AQSF), chloroform soluble fraction (CSF), carbon tetrachloride soluble fraction (CTCSF), and petroleum ether soluble fraction (PESF) were examined for the analysis of thrombolytic, cytotoxic, and erythrocyte membrane-stabilizing activities.

Methods: A well-explicated method was accomplished for plant extractives investigation. The plant extractives were involved in thrombolytic, cytotoxic, and erythrocyte membrane-stabilizing activity evaluation, on the basis of their ability of clot lysis, cytotoxic potentials, and stabilizing erythrocyte membrane under hypotonic solution and heat-induced conditions. Both thrombolytic and erythrocyte membrane-stabilizing activities were performed by using Swiss albino laboratory mice. In addition, plant cytotoxic activity was performed by using the nauplii of brine shrimp as *in vitro* model.

Results: The study of *G. odorata* extracts enumerated basic thrombolytic activity ($19.94 \pm 0.53\%$ to $10.64 \pm 0.46\%$; $p < 0.05$), and basic cytotoxic LC_{50} value (23.09 ± 2.01 µg mL^{-1} to 1.18 ± 0.14 µg mL^{-1}; $p < 0.05$) including statistical analysis confidence limit ranges, chi-square value, and regression equation was entailed. The erythrocyte membrane-stabilizing activity under hypotonic solution-induced hemolysis ($47.41 \pm 0.46\%$ to $18.445 \pm 0.095\%$; $p < 0.05$) and heat-induced hemolysis ($27.95 \pm 0.55\%$ to $17.84 \pm 0.59\%$; $p < 0.05$) was determined. All the statistical calculation was optimized by one-way ANOVA followed by Turkey's *post hoc* test including Dunnett t tests.

Conclusion: This study indicates that *G. odorata* could be a natural medication alternative of thrombolytic agents as well as source of potent bioactive compounds.

Keywords: *Gynocardia odorata*; Brine shrimp nauplii; Cytotoxic; Erythrocyte membrane-stabilizing activity

Background

According to the World Health Organization (WHO) estimate, more than 80% of the population of the developing countries rely on conventional plants for initial health care (Mulat et al. 2013). Only in Asia, medicinal plant has a big impact on economy and primary health care. There are approximately 6,500 species used for curative purpose in Asia. Bangladesh has a fertile with reputable inheritance of herbal medicines among the countries in South Asia. In Bangladesh, almost 500 medicinal species are assessed, and 250 species are used for preparation of traditional medicine. In addition, a large amount of pharmaceutical raw materials including medicinal plants and semi-processed plant are used. The majority of these plants have not yet been studied for their pharmacological and toxicological bioactivities (Rajaei et al. 2012; Karki et al. 1999; Uddin et al. 2011). Presently, thrombolytic drugs like urokinase (UK), alteplase, tissue plasminogen activator (t-PA), and streptokinase (SK) are widely used as clinical thrombolytic agents. Nowadays, streptokinase and urokinase are used in developing countries due to their reasonable price, but these drugs are coupled with side

* Correspondence: faisalasif1@gmail.com
[1]Department of Pharmacy, State University of Bangladesh, Dhaka 1205, Bangladesh
Full list of author information is available at the end of the article

effects which lead to anaphylactic reaction, hemorrhage, and systemic fibrinolysis. As a result of immunogenicity, multiple treatment of streptokinase is restricted in a given patient (Sherwani et al. 2013; Dewan et al. 2013). Traditional plants are relatively safe for use to treat different types of diseases, including various traditional plants that have thrombolytic, antiplatelet, anticoagulant, and antithrombotic activities which are successfully applied for therapeutic purposes (Das et al. 2013). Present phytochemical research scientists have isolated hundreds of bioactive chemical compounds, for example aconitine, acronycine, compounds from Amaryllidaceae plants, bisindole, camptothecine, cephalotaxus, colchine, ellipticine, emtine, phenanthroquinolizidine, and pyrrolizidine, which are ascertained cytotoxic against tumor cells (Geoffrey et al. 1993).

G. odorata, a medium-sized tree is commonly found in lower hill forest of South Asia, commonly in India, Bangladesh, Nepal, and China. This evergreen tree belongs to the Achariaceae family. People normally used fruit-extracted oils for cooking and lighting purposes. The fruit itself is poisonous before processing (Rai and Rai 1994; Shu 2014). This research aims to study the versatile extracts of G. odorata herbal plants of Bangladesh especially on its thrombolytic activity and in stabilizing erythrocyte membrane by using Swiss albino laboratory mice and its cytotoxic activity by using the nauplii of brine shrimp as in vitro model.

Methods
Plant collection and identification
The leaves and fruits of G. odorata were collected from different areas of Dhaka, Bangladesh. A voucher sample number (DACB-39206) has been issued from the Bangladesh National Herbarium, Mirpur, Dhaka, Bangladesh. The leaves were sun dried for several days and then oven dried for 24 h at considerably low temperature (not more than 40°C) for better grinding. The dried leaves were then crushed to a powder form using a grinding machine in the Phytochemical Research Laboratory, Faculty of Pharmacy, University of Dhaka.

Chemicals and reagents
All chemicals and solvents, i.e., methanol, carbon tetrachloride, n-hexane, chloroform, dimethyl sulfoxide (DMSO), and other reagents, used in these experiments were analytical grade and purchased from Merck (Darmstadt, Germany), and vincristine sulfate (VS) was purchased from Sigma-Aldrich (Steinheim, Germany). Commercially available lyophilized streptokinase (SK) vial (15,000,000 IU) was gifted from Beacon Pharmaceuticals Ltd. (Dhaka, Bangladesh). This suspension was used for in vitro thrombolysis as a stock from 100 μL (30,000 IU).

Extraction of the plant material
The desirable sample plants at first were sun dried for few days and then oven dried at 40°C for nearly 24 h for easy grinding. About 500 g of powdered sample plant materials were attenuated in 2.0 L of 95% methanol for 7 days and then filtered through a cotton plug accompanied by Whatman filter paper number 1. Applying a temperature of 40°C to 45°C and reducing the pressure, the extract was concentrated with the assistance of a rotary evaporator. The concentrated methanol extract was partitioned, and the sequent partitioning, i.e., n-hexane (1.2 g), chloroform (800 mg), carbon tetrachloride (1.0 g), and aqueous soluble (1.6 g) fractions, was utilized for the experiment (Anosike et al. 2012).

Animals
In this experiment, adult Swiss albino mice (25 to 30 g) of either sex were used which were grown in the State University of Bangladesh laboratory. They were placed under standard laboratory conditions of 12:12 h light and dark cycle, temperature of 23°C ± 2°C, and maintained relative humidity of 55 ± 5%. During the study, they were fed with standard animal feed and ad libitum water. Institutional animal ethical committee (IAEC) approved experimental protocol requirements was supervised by Department of Pharmacy, State University of Bangladesh.

Making of erythrocyte suspension
Whole blood was collected from the mice through the retro-orbital plexus. Anticoagulant ethylenediaminetetraacetic acid (EDTA) was added to preclude clotting. The blood was rinsed three times with 0.9% saline. The bulk of saline was assessed and reconstituted as a 40% (v/v) suspension with an isotonic buffer solution (pH 7.4) which comprised in 1 L of distilled water with NaH_2PO_4. $2H_2O$ (0.26 g), Na_2HPO_4 (1.15 g), NaCl (9 g), and 10 mM sodium phosphate buffer was centrifuged for 10 min at 3,000 rpm (Kuddus et al. 2012).

Thrombolytic activity
Commercially available lyophilized streptokinase vial (15,000,000 IU) was fused properly with 5 mL phosphate-buffered saline (PBS). This suspension was used as a stock from which appropriate dilutions were made to observe the thrombolytic activity. Five milliliters of blood was withdrawn from each mouse ($n = 10$) through the retro-orbital plexus. Fresh blood was collected in pre-weighed sterile micro-centrifuge tube (1 mL per tube) and incubated at 37°C for 45 min. The observations were taken in triplicate. After clot formation, the serum was completely removed without disturbing the clot, and each tube having the clot was again weighed to determine the clot weight (Clot weight = Weight of clot containing tube – Weight of tube alone). To each micro-centrifuge tube containing the pre-weighed clot, 100-μL aqueous solution of different

plant extracts such as ME, PESF, CTCSF, CSF, and AQSF of the methanolic extract of *G. odorata* were added individually. About 100 µL of SK considered as the positive control and about 100 µL of distilled water considered as the negative control were individually added to the control tubes. At 37°C for 90 min, all the tubes were incubated and observed for clot lysis. After the incubation process, the excess fluid was removed, and the tubes were again weighed to mark the deviation in weight after clot disruption. Deviation obtained in weight taken before and after clot lysis was conveyed as percentage of clot lysis using the following equation: % of clot lysis = (weight of released clot/clot weight) × 100 (Prasad et al. 2007; Rahman et al. 2013; Dhande et al. 2014).

Cytotoxicity activity

The brine shrimp eggs (*Artemia salina*) were amassed from the Fisheries Department, University of Dhaka, Bangladesh. The simulated seawater (3.8% NaCl) was allowed for 48 h for the brine shrimp eggs to hatch and develop into nauplii. All the plant sample extracts were filled in vials by dissolving them in 100 µL pure DMSO to acquire stock solutions. We consider a 50-µL stock solution with simulated sea water (3.8% NaCl) to reach a series of concentrations approximately 400 µg mL^{-1} each. Standard VS was used as the positive control, and 50-µL pure DMSO diluted to 5 mL was used as the negative control. The matured nauplii were applied to each of all experimental vials and control vial. After 24 h, the vials were scrutinized using a ×3 magnifying glass, and the number of living nauplii in each vial was counted (Meyer et al. 1982; Ali et al. 2013).

Erythrocyte membrane-stabilizing activity
Hypotonic solution-induced hemolysis

The experiment sample comprised of stock of mice erythrocyte (RBC) suspension (0.50 mL) with 5 mL of hypotonic solution (50 mM NaCl) in 10 mM sodium phosphate-buffered saline (pH 7.4) containing either the different methanolic extracts (2.0 mg mL^{-1}) or acetyl salicylic acid (0.10 mg mL^{-1}). The acetylsalicylic acid was used as a reference standard. The mixtures were incubated for about 10 min at 25°C and then centrifuged for 10 min at 3,000×g, and the absorbance (O.D.) of the supernatant was measured at 540 nm using a Shimadzu UV spectrophotometer (Tokyo, Japan). The percentage inhibition of hemolysis was calculated using the following equation: % inhibition of hemolysis = 100 × (OD$_1$ – OD$_2$)/OD$_1$, where OD$_1$ is the optical density of the hypotonic buffered saline solution alone (control) and OD$_2$ is the optical density of the test sample in hypotonic solution.

Heat-accelerated hemolysis

Approximately 5 mL of the isotonic buffer containing 1.0 mg mL^{-1} of the different extracts of plants was put into two twin sets of centrifuge tubes. The vehicle with the same quantity was added to another tube as the control. The erythrocyte suspension of mice (30 µL) was added to each tube and mixed gently by upending. One pair of the tubes was incubated at 54°C for 20 min in a water bath. The other pair was placed in an ice bath at 5°C. The whole reaction mixture was centrifuged for 3 min at 1,300 rpm, and the yielding supernatant was measured at 540 nm. The percentage inhibition in the tests was taken and was calculated according to the

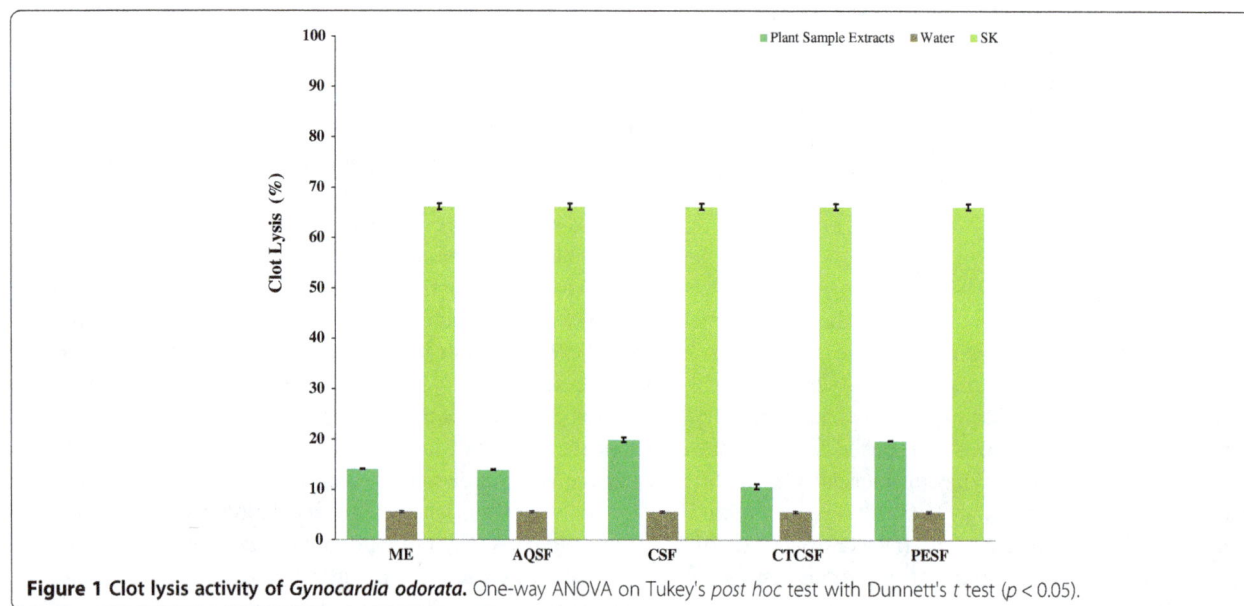

Figure 1 Clot lysis activity of *Gynocardia odorata*. One-way ANOVA on Tukey's *post hoc* test with Dunnett's *t* test ($p < 0.05$).

Canvassing of thrombolytic, cytotoxic, and erythrocyte membrane-stabilizing attributes in in vitro screening of...

29

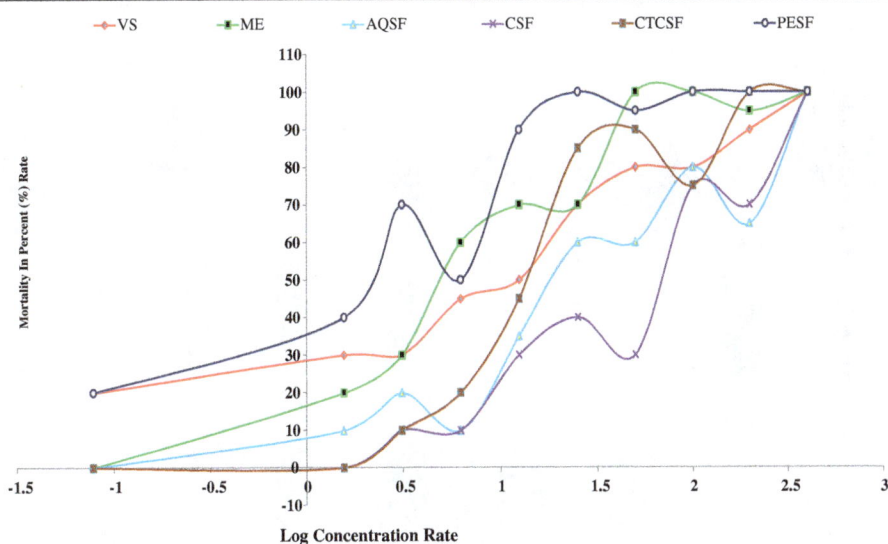

Figure 2 Cytotoxic activity of *Gynocardia odorata*. One-way ANOVA on Tukey's *post hoc* test with Dunnett's *t* test ($p < 0.05$).

following equation: % inhibition of hemolysis = 100 × [1 − (OD_2 − OD_1/OD_3 − OD_1)], where OD_1 is the unheated test sample, OD_2 is the heated test sample, and OD_3 is the heated control sample (Rashid et al. 2011; Sharma et al. 2013).

Statistical analysis

The implication between % clot lysis through SK and plant sample extracts, LC_{50} values through VS, extracts, hypotonic solution-accelerated % inhibition of hemolysis, and heat-accelerated % inhibition of hemolysis was tested using Dunnett's *t* test analysis using the software SPSS (SPSS for Windows, Version 19.0, IBM Corp., Armonk, New York, USA). The data were expressed as mean ± standard deviation. The mean difference between the positive and negative controls was considerably substantial at $p < 0.05$. The LC_{50} value of nauplii was calculated from a linear regression applying 'Biostat-2009' software (AnalystSoft Inc., Vancouver, Canada). All the experiment was performed *in vitro* and expressed as mean ± standard error of the mean.

Results and discussion

Platelets act as a crucial function in atherothrombosis through adhering to the disrupted region of the endothelial surface which further initiates plaque formation and growth. Plasmin, a natural fibrinolytic agent, helps break down fibrinogen and fibrin and initiates the lyses of clot. Typically, SK forms a stoichiometric complex with plasminogen to change plasmin (Chowdhury et al. 2011).

As part of the investigation on the thrombolytic activities of natural sources, the extractives of *G. odorata* were evaluated; the results are depicted in Figure 1. An addition of 100 μl SK, a positive control (30,000 IU) to the clots and subsequent incubation for 90 min at 37°C, showed 66.17 ± 0.59% lysis of clot. Consequently, distilled water was treated as the negative control which exhibited negligible percentages of lyses of clot (5.57 ± 0.23%). The mean difference in clot lysis percentage between the positive and negative controls was found very significant. In this study, *G. odorata* sample extracts CSF and PESF demonstrated the highest thrombolytic activity

Table 1 Cytotoxic activity potential values of *Gynocardia odorata*

Sample	LC_{50} (μg mL^{-1})	Confidence limit ranges (μg mL^{-1})	Chi-square value	Regression equation
VS	0.432 ± 0.02	0.47 to 0.39	25.63	$Y = 0.1008 + 0.0251 \times X$
ME	4.89 ± 0.09	5.06 to 4.70	61.61	$Y = 0.1648 + 0.0178 \times X$
AQSF	23.09 ± 2.01	25.93 to 20.25	87.04	$Y = 0.5682 + 0.0235 \times X$
CSF	43.55 ± 2.50	48.45 to 38.66	152.70	$Y = 0.6943 + 0.0255 \times X$
CTCSF	11.88 ± 0.42	12.70 to 11.06	79.72	$Y = 0.5803 + 0.0181 \times X$
PESF	1.18 ± 0.14	1.45 to 0.91	112.95	$Y = 0.1648 + 0.0178 \times X$

LC_{50} values are the mean of the triplicates (mean ± SEM, $p < 0.05$). VS, vincristine sulfate; ME, methanol extract; AQSF, aqueous soluble fraction of the methanol extract; CSF, chloroform soluble fraction; CTCSF, carbon tetrachloride soluble fraction; PESF, petroleum ether soluble fraction.

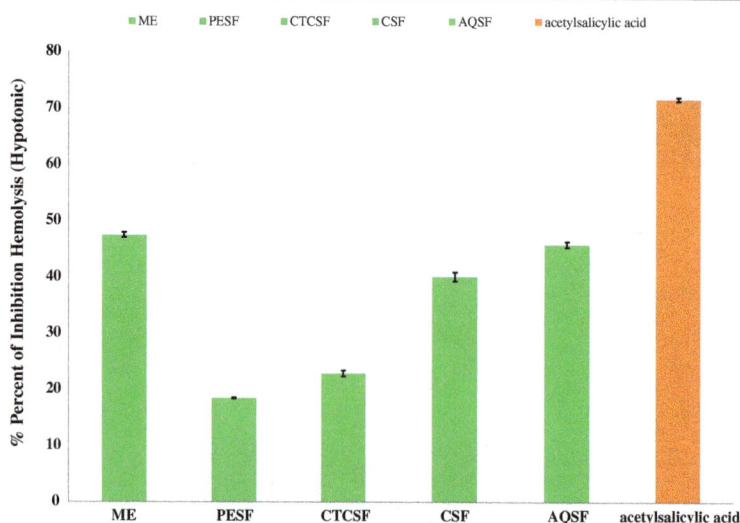

Figure 3 Hypotonic hemolysis of *Gynocardia odorata*. One-way ANOVA on Tukey's *post hoc* test with Dunnett's *t* test ($p < 0.05$).

$(19.94 \pm 0.53\%$ and $19.75 \pm 0.08\%$, respectively). However, significant thrombolytic activity was also evidenced by ME $(14.06 \pm 0.06\%)$, AQSF $(13.90 \pm 0.04\%)$, and CTCSF $(10.64 \pm 0.46\%)$ which showed moderate thrombolytic activity. The mean difference in clot lyses percentage between the positive and negative controls was very significantly $(p < 0.01)$.

Actually, investigation on the toxicity of certain plants allows safety issue for use in further medicinal purpose and focuses on anti-microbial, anti-fungal, and anti-tumor activities (Das et al. 2010). The brine shrimp lethality bioassay is widely used and is quite inexpensive for assaying cytotoxicity of medicinal plants. Bioactive compounds are almost always toxic at higher dose. Thus, *in vivo* lethality in a simple zoological organism can be used as a reliable information for screening and fractionation in the discovery of new bioactive natural products. In the present bioactivity study, all the crude extracts, n-hexane, carbon tetrachloride, chloroform, crude extract, and aqueous soluble fractions of methanolic extract showed positive results, indicating that the test samples are biologically active (Ping et al. 2013). The mortality rates against the concentration are depicted in Figure 2.

Here, the linear regression analysis of LC_{50} and chi-square of all the plant sample extracts with standard VS, as presented in Table 1, is significant $(p < 0.005)$. VS was used as the positive control, and LC_{50} was found to be 0.432 ± 0.02 μg mL^{-1} (Table 1) for VS compared with the negative control. VS (positive control) gave significant mortality, and the LC_{50} values of the different extractives were compared to this positive control. The LC_{50} values of the plant sample extracts were significant in PESF $(1.18 \pm 0.14$ μg mL$^{-1})$ and ME $(4.89 \pm 0.091$μg mL$^{-1})$. Also, CTCSF $(11.88 \pm 0.42$ μg mL$^{-1})$, AQSF $(23.09 \pm$

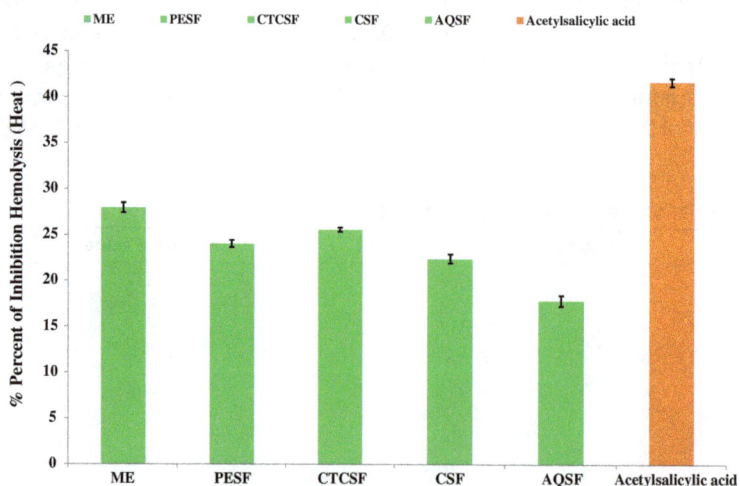

Figure 4 Heat-induced hemolysis of *Gynocardia odorata*. One-way ANOVA on Tukey's *post hoc* test with Dunnett's *t* test ($p < 0.05$).

2.01 µg mL^{-1}), and CSF (43.55 ± 2.50 µg mL^{-1}) showed moderate cytotoxic activity (Table 1).

Membrane-stabilizing attributes were acknowledged for their power to interpose with release of phospholipases that activate the establishment of inflammatory intercessors (Aitadafouri et al. 1996). The erythrocyte membrane matches to the lysosomal membrane, and the consequence of medication on the stabilization of erythrocyte would be inferred to the stabilization of the lysosomal membrane. Therefore, the membrane stabilizes which interferes in the release and the action of intercessors such as serotonin, histamine, leukotrienes, and prostaglandins (Latif et al. 2013). The main purpose of the anti-inflammatory agents is to reduce cyclooxygenase enzymes which convert arachidonic acid to prostaglandins. Different types of plants possess antiinflammatory activities. This study involved hypotonic solution- and heat-induced hemolysis in mice models on the basis of their membrane-stabilizing properties (Saleem et al. 2011; Umukoro et al. 2006).

The different plant sample extracts of *G. odorata* at a concentration of 1.0 mg mL^{-1} significantly saved the lysis of mice erythrocyte membrane induced by a hypotonic solution and at a temperature-induced condition (Figures 3 and 4), which is similar to that using the standard acetyl salicylic acid (0.10 mg mL^{-1}). At a concentration of 1.0 mg mL^{-1}, the reference standard acetylsalicylic acid has a 71.58 ± 0.33% inhibition. The hypotonic solution-induced percent inhibition of hemolysis is highest in ME (47.41 ± 0.46%), AQSF (45.77 ± 0.5%), and chloroform soluble fraction (40.055 ± 0.81). Also, the percent inhibition is relatively lesser in CTCSF (22.84 ± 0.52%) and PESF (18.445 ± 0.095%).

At the heat-induced condition, the lysis of mice erythrocyte membrane is the same as that using the standard acetyl salicylic acid (0.10 mg mL^{-1}). At a concentration of 1.0 mg mL^{-1}, the percent inhibition of the reference standard acetylsalicylic was 41.68 ± 0.46%. The hemolysis percent inhibition of the heat-induced solution is highest in ME (27.95 ± 0.55%), CSF (24.01 ± 0.41%), and AQSF (17.84 ± 0.59%). Also, the percent inhibition is relatively lesser in CTCSF (25.58 ± 0.22%) and PESF (24.01 ± 0.41%).

Conclusions

The different sample plant extracts of *G. odorata* were subjected to different biological investigations, i.e., thrombolytic activity, cytotoxic activity, and erythrocyte membrane-stabilizing activity. This plant is widely used in Bangladesh as a traditional folk medication. This plant has many significant cytotoxic and erythrocyte membrane-stabilizing activities. Also, some different plant sample extracts have abundant amount of thrombolytic activity. Therefore, considering the potential bioactivity, the plant materials can further be studied extensively to find out their undiscovered efficaciousness and to justify their use as traditional folk medication.

Competing interests
The authors declare that they have no competing interests.

Authors' contributions
AZB, MNA, and MT designed the experiment and carried out the experiment. FA contributed in composing the article. SRC and MAR supervised the work. All authors read and approved the final manuscript.

Acknowledgements
This work was affirmed by the Department of Pharmacy, State University of Bangladesh, for the technical and laboratory work.

Author details
[1]Department of Pharmacy, State University of Bangladesh, Dhaka 1205, Bangladesh. [2]Phytochemical Research Laboratory, Department of Pharmaceutical Chemistry, Faculty of Pharmacy, University of Dhaka, Dhaka 1000, Bangladesh.

References
Ali N, Aleem U, Shah SWA, Shah I, Junaid M, Ahmed G, Ali W, Ghias M (2013) Acute toxicity, brine shrimp cytotoxicity, anthelmintic and relaxant potentials of fruits of *Rubus fruticosus agg.*. BMC Complement Altern Med 13:138

Aitadafouri M, Mounnnieri C, Heyman SF, Binistic C, Bon C, Godhold J (1996) 4-Alkoxybenzamides as new potent phospholipase A2 inhibitors. Biochem Pharm 51:737–742

Anosike CA, Obidoa O, Ezeanyika LUS (2012) Membrane stabilization as a mechanism of the anti-inflammatory activity of methanol extract of garden egg (*Solanum aethiopicum*). DARU J Pharm Sci 20:76

Chowdhury NS, Alam MB, Haque ASMT, Zahan R, Mazumder MEH, Haque EH (2011) In vitro free radical scavenging and thrombolytic activities of Bangladeshi aquatic plant *Aponogeton undulatus* Roxb. Global J Pharmacol 5:27–32

Das A, Dewan R, Masudur S, Ali R, Debnath PC, Billah M (2013) Investigation of in vitro thrombolytic potential of ethanolic extract of *Momordica charantia* fruits: an anti-diabetic medicinal plant. Der Pharm Sin 4:104–108

Das R, Kaushik A (2010) Synergistic activity of *Fagonia arabica* and *Heteropneustes fossilis* extracts against myocardial, cerebral infarction, and embolism disorder in mice. J Pharm Bioallied Sci 2:100–104

Dewan SMR, Das A (2013) Investigation of in vitro thrombolytic potential and phytochemical nature of *Crinum latifolium* L. leaves growing in coastal region of Bangladesh. IJBPR 4:1–7

Dhande SR, Dongare PP, Kaikini AA, Patil KA, Kadam VJ (2014) In vitro antioxidant and thrombolytic activities of medicinal plants. IJUPBS 3:78–87

Geoffrey A, Cordell A, Kinghorn D, Pezzuto JM (1993) Separation, structure elucidation and bioassay of cytotoxic natural products. In: Colegate SM, Russell J (ed) Bioactive natural products: detection, isolation, and structural determination. Molyneux, Florida, pp 195–197

Karki M, Williams JT (1999) Priority species of medicinal plants in South Asia. Medicinal & Aromatic Plants Program in Asia (MAPPA), Jor Bagh, pp 34–36

Kuddus MR, Alam MS, Chowdhury SR, Rumi F, Sikder MAA, Rashid MA (2012) Evaluation of membrane stabilizing activity, total phenolic content, brine shrimp lethality bioassay, thrombolytic and antimicrobial activities of *Tagetes patula* L. J Pharmacog Phyto 1:57–62

Latif F, Islam F, Kuddus MR, Hossain MK (2013) Antioxidant, thrombolytic and membrane stabilizing activities of *Mussaenda roxburghii* Hook. f. iP-Planet 1:13–19

Meyer BN, Ferrigni NR, Putnam JE, Jacobsen JE, Nichols DE, McLaughlin JL (1982) Brine shrimp: a convenient general bioassay for active plants constituents. J Med Plant Res 45:31–34

Mulat Y (2013) The role of cultural practices in trees management practices among the Qemant community Chilga Woreda (district), Ethiopia. IJRD 7:177–180

Prasad S, Kashyap RS, Deopujari JY, Purohit HJ, Taori GM, Daginawala HF (2007) Effect of *Fagonia arabica* (Dhamasa) on in vitro thrombolysis. BMC Complement Altern Med 7:36

Ping KY, Darah I, Chen Y, Sasidharan S (2013) Cytotoxicity and genotoxicity assessment of *Euphorbia hirta* in MCF-7 cell line model using comet assay. Asian Pac J Trop Biomed 3:692–696

Rahman M, Khatun A, Islam MMI, Akter MA, Chowdhury SA, Khan MMA, Shahid IZ, Rahman AA (2013) Evaluation of antimicrobial, cytotoxic, thrombolytic, diuretic properties and total phenolic content of *Cinnamomum tamala*. Int J Green Pharm 7:236–243

Rai T, Rai L (1994) Trees of the Sikkim Himalaya. Indus Publishing Company, New Delhi, pp 64–65

Rajaei P, Mohamadi N (2012) Ethnobotanical study of medicinal plants of Hezar mountain allocated in south east of Iran. IJPR 11:1153–1167

Rashid MA, Sikder MMA, Kaisar MA, Miah MK, Parvez MM, Hossian AKMN (2011) Membrane stabilizing activity - a possible mechanism of action for the anti-inflammatory activity of two Bangladeshi medicinal plants: *Mesua nagassarium* (Burm. F.) and *Kigelia pinnata* (Jack) Dc. IJPRD 3:1–5

Saleem TKM, Azeem AK, Dilip C, Sankar C, Prasanth NV, Duraisami R (2011) Anti-inflammatory activity of the leaf extracts of *Gendarussa vulgaris* Nees. Asian Pac J Trop Biomed 1:147–149

Sherwani SK, Bashir A, Haider SS, Shah HA, Kazmi SU (2013) Thrombolytic potential of aqueous and methanolic crude extracts of *Camellia sinensis* (green tea): in vitro study. J Pharmacog Phyto 2:125–129

Sharma V, Singh M (2013) In vitro antiarthritic and hemolysis preventive: membrane stabilizing efficacy of ethanolic root extract of *Operculina turpethum*. WJPPS 2:302–312

Shu MDG (2014) Flora of China 13: 116. 2007, *Gynocardia* r. Brown in Roxburgh. Pl Coromandel 3(95):1820

Uddin SJ, Grice ID, Tiralongo E (2011) Cytotoxic effects of Bangladeshi medicinal plant extracts. Evid Based Complement Alternat Med 2011:7–8

Umukoro S, Ashorobi RB (2006) Stabilizing property of aqueous leaf extract of *Momordica charantia* in rats. Afr J Biomed 9:119–124

6

Systems biology from virus to humans

Youri Lee[1], Yu-Jin Kim[1], Yu-Jin Jung[1], Ki-Hye Kim[1], Young-Man Kwon[1], Seung Il Kim[2] and Sang-Moo Kang[1*]

Abstract

Natural infection and then recovery are considered to be the most effective means for hosts to build protective immunity. Thus, mimicking natural infection of pathogens, many live attenuated vaccines such as influenza virus, and yellow fever vaccine 17D were developed and have been successfully used to induce protective immunity. However, humans fail to generate long-term protective immunity to some pathogens after natural infection such as influenza virus, respiratory syncytial virus (RSV), and human immunodeficiency virus (HIV) even if they survive initial infections. Many vaccines are suboptimal since much mortality is still occurring, which is exampled by influenza and tuberculosis. It is critically important to increase our understanding on protein components of pathogens and vaccines as well as cellular and host responses to infections and vaccinations. Here, we highlight recent advances in gene transcripts and protein analysis results in the systems biology to enhance our understanding of viral pathogens, vaccines, and host cell responses.

Review

Introduction

Viruses contain all the essential factors necessary for initiating infection and replication in a new target cell. Thus, information on the protein composition of a virus particle often serves as an initial guide in determining functional roles for viral proteins as well as antiviral and/or vaccine antigen target molecules. With advances in proteomics techniques and the availability of annotated genomic sequences for several mammalian species, the view that a virion is a minimal package of its genome and essential viral proteins for the first round of genome replication is being changed. Proteomic analysis of virions identified host proteins that are packaged into the virus particles along with the viral components (Table 1). In particular, enveloped viruses have the capability to incorporate numerous host proteins, both into the interior of the virus particles as well as into the lipid envelope (Cantin et al. 2005; Bortz et al. 2003; Johannsen et al. 2004; Kattenhorn et al. 2004; Zhu et al. 2005). Similarly, host proteins have been detected in vaccinia virions, human immunodeficiency virus (HIV) type 1, and Moloney murine leukemia virus (MoMLV) vector particles (Chung et al. 2006; Chertova et al. 2006; Saphire et al. 2006).

It is expected that cellular proteins found within the virus particles would provide clues as to the virus assembly pathway and events that govern virus infectivity as well as vaccine development.

Vaccination is considered the most effective measure to prevent global infectious diseases. Smallpox and polio diseases are good successful cases of global threats that almost disappeared by effective global vaccination. However, there are still many infectious diseases that claim over 15 million deaths annually. Live attenuated influenza virus (LAIV) vaccine was approved in 2003 and is currently being used for human vaccination. LAIV is safe and effective in young children and adults (Rhorer et al. 2009). Recently, the use of noninfectious virus-like particles (VLPs) that self-assemble by spontaneous interactions of viral structural proteins has been suggested and developed as alternative approaches for developing advanced vaccines for a wide range of viruses that cause disease in humans (Roy and Noad 2009; Kang et al. 2009a; Kang et al. 2009b). It is worth noting that a VLP-based human papillomavirus (HPV) vaccine against HPV responsible for cervical cancer was produced in the yeast system and approved for the market in 2006 (Garland et al. 2007).

Influenza VLPs expressed by recombinant baculovirus (rBV) systems that present multi-component antigens, including HA and matrix 1 (M1), with or without NA, and that are capable of inducing cognate responses against homologous strains of influenza virus have been widely

---

* Correspondence: skang24@gsu.edu
[1]Center for Inflammation, Immunity & Infection, Institute for Biomedical Sciences, Georgia State University, Atlanta, GA 30303, USA
Full list of author information is available at the end of the article

Table 1 Proteomic analysis of representative host cell proteins incorporated into virus or virus-like particles

Virus	Host proteins in virus or VLP*
Influenza virus[a]	ADAR3, H2A, HSP90, nucleolin, ITGAV, hnRNPA1, glypican 4 (kglypican), ANXA4, CD9, CD81, cofilin 1, cyclophilin A, profilin, HSP27, integrin beta, ANXA11, CD59
Respiratory syncytial virus[b]	
Cytoskeleton	B-actin, keratin, moesin, filamin-A, TUBA1B, TUBB, HSP70, HSP90-alpha,beta
Energy pathways	ALPI, PK, GAPDH
Immune response	Putative annexin A2-like protein, ANXA1, CD55
H5N1 VLP[c]	ADP-ribosylation factor, alpha tubulin, Arp2/3 complex subunit, beta-tubulin 10, actin, 40S ribosomal protein S24, 60S acidic ribosomal protein P1,2, ribosomal protein L40, HSP90, heterotrimeric G-protein gamma subunit-like protein, putative Rho1, glutathione S-transferase sigma, ATPase subunit C, casein kinase II subunit alpha, fatty acid-binding protein, 14-3-3 zeta, Rbp1-like RNA-binding protein PB
H5N1/H3N1 VLP[d]	b-tubulin, Myosin, ECM proteins integrin alpha, HSP90, HSP70, HSP27, annexin, tetraspanin, CD9, glycolytic enzymes enolase 1, pyruvate kinase, glyceraldehyde-3-phosphate dehydrogenase, unclassified proteins, aldo-keto reductase family 1, WD repeat domain 1, gamma-glutamyltransferase, peroxiredoxin 2

[a]Influenza virus (H1N1,A/Cal/07/2009) was purified after growing in A549 cells (Shaw et al. 2008; Dove et al. 2012).
[b]Respiratory syncytial virus was purified after growing in A549 cells (Radhakrishnan et al. 2010).
[c]H5N1 VLPs (H5N1, A/Indonesia/102 5/2005) were purified after expressing in SF9 insect cells (Song et al. 2011).
[d]H3N2/H5N1 VLPs (H3N2, A/Taiwan/083/2006 and H5N1, A/Hanoi/30408/2005H5N1) were purified after expressing in Vero cells (Wu et al. 2010).
*Adenosine deaminase, RNA-specific, 3 (ADAR3), histone 2A (H2A), heterogeneous nuclear ribonucleoprotein A1 (hnRNP1), tubulin alpha-1B (TUBA1B), tubulin B (TUBB), intestinal type alkaline phosphatase (ALPI), heat shock protein 27 kDa (HSP27) peroxiredoxin (PRX), annexin A 4 (ANXA4), annexin A11 (ANXA11), annexin A11 (ANXA1), integrin alpha V (ITGAV), dynein 1, heat shock protein 90 kDa (HSP90), heat shock protein 70 kDa (HSP70), pyruvate kinase (PK), beta-actin (B-actin), alkaline phosphatase (ALPI), complement decay acceleration factor (CD55), glyceraldehyde-3-phosphate dehydrogenase (GAPDH).

described (Roy and Noad 2009; Kang et al. 2009a; Kang et al. 2009b). In particular, 2009 H1N1 new pandemic, H5N1, and H7N9 avian influenza VLP vaccines were produced by the insect cell rBV expression system, and tested in clinical trials, demonstrating their safety and efficacy (Khurana et al. 2011; Lopez-Macias 2012; Lopez-Macias et al. 2011; Klausberger et al. 2014; Smith et al. 2013; Fries et al. 2013). Also, influenza VLPs were engineered to express highly conserved influenza virus M2 ectodomains and found to induce cross immunity to heterologous influenza virus strains (Kim et al. 2013a; Kim et al. 2014; Kim et al. 2013b).

Any cellular proteins that may be incorporated into viral particles are also likely to be present at very low levels. Mass spectrometry of tryptic peptides combined with database searching for identification is now becoming the preferred and advanced method for such proteomic studies revealing details on protein components even at a very low level (Table 1).

Vaccinology and immunology were born from the pioneering work of scientists such as Jenner and Pasteur over 200 years ago. Despite the common origins of vaccination and immunology concepts, the two disciplines have evolved in parallel directions independently. Immunologists remain largely ignorant about the mechanisms of how vaccines work. Meanwhile, vaccines were made empirically, and until recently, vaccinologists have shown little interest in the immunological mechanisms by which the vaccine confers protective immunity. Better understanding of immunological mechanisms of vaccination is expected to provide informative insights into the rationale design of future vaccines against difficult pathogens.

The innate immune system including dendritic cells (DCs), macrophages, and other immune cells senses viruses, bacteria, parasites, and fungi through pathogen-recognizing receptors (PRRs) (reviewed in (Coffman et al. 2010; Kawai and Akira 2007; Iwasaki and Medzhitov 2010)). The nature of the DC subtypes and the particular PRR triggered plays an essential role in modulating the strength, quality, and memory of adaptive immune responses (Pulendran and Ahmed 2006; Steinman 2008). More than 26,000 genes are estimated to be present in human genomes. Exposure to a pathogen or vaccination introduces changes in the expression of a substantial fraction of these genes. Systems biological tools offer an informative insight into the complex network of the immune system in our body. That is, high-throughput data on the genes, mRNAs, microRNAs, and proteins that constitute the biological networks are providing new information in our overall understanding of complex immune systems. Systems biology capitalizes on several 'omic' technologies that might define and monitor all the components of the systems. Thus, recent advances in the innate immune system and the use of systems biological approaches are providing powerful tools to reveal the basic mechanisms by which the innate immune system modulates protective immune responses to vaccination (Pulendran and Ahmed 2006; Steinman 2008).

This review focuses on proteomic components of some viruses and vaccines (Table 1), cellular responses of host target cells and immune cells upon exposure to virus and vaccines (Table 2), as well as *in vivo* gene expression profiles in animals and humans in response to respiratory viral infection or certain successful vaccinations.

Table 2 Representative gene expression profiles in host cells as a result of virus infection or VLP stimulation

Host cells	Virus/VLPs	Gene expression profiles in host cells*
MDCK[a]	Influenza	TUBA2, CK-8, B-actin, keratin 10, ANXA1, KCIP-1
VERO[b]	Influenza	Keratin1, 8, 10, tubulin, TUBA, RBBP4, ITGA3, HSP27, Ndr1, ANXA4, PK, GAPDH, HSP105, ITGAV, dynein 1, HSP90, ITGA3, HSP 70
A549[c]	RSV	HRNR, HIST2H2BE, HIST1H2BC, H2AFX, DTX3L, IFI35, SYNE1, SUMO2, MYADM, PARP14, VIM, C21ORF70, TFAM, POLDIP3, DEK, EFHD2, KIAA1967, ENO1, ENY2
BEAS-2B[d]HBEC	RSV	ATP6V0D2, OTOA, SCNN1G, SMA5, CYP4F8, EPB41L4B, ELP2, PRKCE, HBP1, DUSP2, TUBB1, TRNT1, ELOVL5, PTPRG, MAP2K7, TSGA10, KRT12, SYN1, SIRPB1
MDDC[e]	HIV-VLP	Expression of genes involved in the morphological and functional changes characterizing the MDDCs activation and maturation: MEF2A, NFE2L2, DC-UbP, PGM3, DGKH, PTGS2, IL8, ARRB1, BTG3, SOSTM1, HLA-DOA - MHC class II, TADA3L, MTHFD2, PRKX, BAG3 KCNK10, member 10, PBEF1
PBMC[f]	HIV-VLP	Expression of genes involved in the morphological and functional changes characterizing the PBMCs activation and maturation: CTSL, IL3RA, SMOX, BCL2, G0S2, IER3, SERPINB2, LIMK2, IL6, IL8, PBFE1, PBX3, IL1-A,B, CCL3 (MIP 1α), CCL7, CCL18, CCL20, CXCL1, CXCL2, CXCL3, CXCL6, CXCL13, INHBA, ACTN1, AQP9, EMR1, SLC25A37, SLCO4A1, MAD, CCL4 (MIP1β)
RAW264.7[g]	HBc-VLP	HSP70, Eno1 protein chain A, peroxiredoxin 1, phosphoglycerate kinase, glutathione S-transferase A2, ferritin light chain 1, hypothetical protein, prohibitin, glyceraldehyde-3-phosphate dehydrogenase

[a]Madin-Darby canine kidney (MDCK) cells that were infected with influenza virus (H1N1, A/PR/8/34) (Vester et al. 2010).
[b]Vero (kidney epithelial cells) cells that were infected with influenza virus (H1N1, A/PR/8/34) (Vester et al. 2010).
[c]A549 (adenocarcinomic human alveolar basal epithelial cells) cells were infected with respiratory syncytial virus (Munday et al. 2010).
[d]BEAS-2B4 (human bronchial epithelial cell line, subclone S6), human bronchial epithelial cells (HBEC) were infected with respiratory syncytial virus (Huang et al. 2008).
[e]Human monocyte-derived dendritic cells (MDDCs) were stimulated with HIV-1 (clade A) Pr55gag virus-like particles (HIV-VLPs) (Arico et al. 2005).
[f]CD14+ human peripheral blood mononuclear cells (PBMCs) were stimulated with HIV-1 Pr55gag virus-like particles (HIV-VLPs) (Buonaguro et al. 2008).
[g]RAW264.7 (mouse macrophage cell line) cells were stimulated with hepatitis B virus core protein virus-like particles (HBc-VLPs) (Yang et al. 2008).
*Influenza: H1N1 (A/PR/8/34), HIV-VLp (HIV-1 Pr55gag virus), HBc-VLP (hepatitis B virus). Tubulin alpha (TUBA), protein kinase C inhibitor protein-1 (KCIP-1), nameretinoblastoma binding protein 4 (RBB4), serine/threonine kinase 38 (Ndr), MADS box transcription enhancer factor 2 (MEF2A), nuclear factor-like 2 (NFE2L2), dendritic cell-derived ubiquitin-like protein (DC-UbP), phosphoglucomutase 3 (PGM3), diacylglycerol kinase (DGKH), prostaglandin-endoperoxidase synthase 2 (PTGS2), interleukin 8 (IL-8), arrestin beta 1 (ARRB1), BTG familymember 3 (BTG3), sequestosome 1 (SOSTM1), MHC class II (HLA-DOA), butyrate-induced transcript (HSPC121), transcriptional adaptor 3 (TADA3L), Mst3 and SOK1-related kinase (MST4), UDP-Gal:betaGlcNAc bea 1.4-galactosyltransferase (B4GALT5), NAD-dependent methylenetetrahydrofolate dehydrogenase (MTHFD2), small glutamine-rich tetratricopeptide repeat (SGTB), protein kinase (PRKX), Nedd4 binding protein 1(X-linked N4BP1), pellino homologue 1(PELI1), sterol-C4-methyl oxidase-like (SC4MOL), BCL2-associated athanogene 3 (BAG3), potassium channel, subfamily K (KCNK10), pre-B-cell colony enhancing factor 1 (PBEF1), hornerin (HRNR), H2B type 2-E (HIST2H2BE), H2B type 1-C/E/F/G/ (HIST1H2BC), H2A (H2AFX), protein deltex-3-like (DTX3L), interferon-induced 35 kDa protein (IFI35), Nesprin-1 (SYNE1), small ubiquitin-related modifier 2 (SUMO2), myeloid-associated differentiation marker (MYADM), poly [ADP-ribose] polymerase 14 (PARP14), vimentin (VIM), uncharacterized protein C21orf70 (C21ORF70), transcription factor A, mitochondrial (TFAM), polymerase d-interacting protein 3 (POLDIP3), protein DEK (DEK), EF-hand domain-containing protein D2 (EFHD2), protein KIAA1967 (KIAA1967), a-enolase (ENO1), enhancer of yellow 2 transcriptionfactor homologue (ENY2), ATPase, H+ transporting, lysosomal 38 kDa, V0 subunit d isoform 2 (ATP6V0D2), otoancorin (OTOA), sodium channel non-voltage-gated 1 gamma (SCNN1G), cytochrome P450 family 4 subfamily F polypeptide 8 (CYP4F8), erythrocyte membrane protein band 4.1 like 4B (EPB41L4B), signal transducer and activator of transcription 3 interacting protein 1(ELP2), protein kinase C epsilon (PRKCE), HMG-box transcription factor 1 downregulated genes (HBP1), dual specificity phosphatase 2 (DUSP2), tubulin beta 1 (TUBB1), tRNA nucleotidyl transferase CCA-adding 1 (TRNT1), ELOVL family member 5 elongation of long chain fatty acids (ELOVL5), protein tyrosine phosphatase receptor type G (PTPRG), mitogen-activated protein kinase 7 (MAP2K7), testis specific 10 (TSGA10), keratin 12 (KRT12), synapsin I (SYN1), signal-regulatory protein beta 1 (SIRPB1), glucuronidase, beta pseudogene (SMA5), EGF-like module containing, mucin-like, hormone receptor-like sequence 1(EMR1).

Cellular and viral proteins within virus and virus-like particles

Influenza virus

With advances in proteomic analysis and studies, a virion is not simply a package of its genome and viral proteins. Numerous host cellular proteins were found to be incorporated into virus particles in particular enveloped virions. HIV virions were found to incorporate Tsg101, cyclophilin A, and APOBEC3G in addition to their viral proteins (Chertova et al. 2006; Saphire et al. 2006; Demirov et al. 2002; Franke et al. 1994; Mariani et al. 2003). Tsg101 host protein was reported to play a crucial role in virus assembly of HIV whereas cyclophilin A modulates HIV-1 infectivity and APOBEC3G is an anti-viral factor that promotes hypermutation of the viral genome (Franke et al. 1994; Mangeat et al. 2003). Among many other cellular proteins, identification of

these three host proteins has increased our understanding of how HIV would interact with its host proteins. Nine viral proteins out of the 11 influenza A virus proteins are present in the influenza virion (PB1, PB2, PA, HA, NP, NA, M1, M2, NEP). The glycoproteins hemagglutinin (HA) and neuraminidase (NA) are embedded into the lipid envelope of the influenza virus particle and form the spikes visible under the electron microscope. The ion channel protein M2 is also found within the virion but at a lower level. The matrix M1 protein lies beneath the viral membrane, surrounding the ribonucleoproteins, which consist of eight viral RNA segments coated with the nucleoprotein (NP) and bound by the trimeric polymerase complex (PB1, PB2, PA). The nuclear export protein (NEP) is also found within influenza virus particles.

By utilizing mass spectrometry of tryptic peptides (multidimensional protein identification technology liquid chromatography-tandem mass spectrometry (LC-MS/MS) analysis) combined with database searching for identification, 36 host-encoded proteins were identified in the influenza A/WSN/33 (H1N1) virus particles that were grown in Vero (African green monkey kidney) cells (Shaw et al. 2008). The following host cell proteins are also found in other enveloped viruses. (i) Cytoskeletal proteins: Cytoskeletal proteins such as tubulin and actin were present in the interior of influenza virions which most likely reflects their active participation in moving the viral components to the assembly site and possible cytoskeletal reorganization that occurs during bud formation. (ii) Annexins: Several annexin family members (A1, A2, A4, A5, and A11) were found in influenza virus particles. Annexins are calcium-dependent phospholipid-binding proteins and are proposed to act as scaffolding proteins at certain membrane domains. (iii) Tetraspanins: Two members of the tetraspanin family, CD9 and CD81, were present within influenza virions and are most likely inserted into the viral envelope. Tetraspanins have four transmembrane domains and two extracellular loops and are involved in specialized membrane domains. Tetraspanins (CD9 and CD81) have also been implicated in both fusion and egress pathways for a number of other viruses (Ho et al. 2006; Jolly and Sattentau 2007). (iv) Cyclophilin A: Cyclophilin A (a peptidyl-prolyl isomerase) was shown to be in the core of the influenza virion and in other different viruses. There is a precedent for the involvement of cyclophilin proteins in replication of different viruses. (v) CD59: CD59 is a complement regulatory protein that acts by inhibiting formation of the membrane attack complex. CD59 (glycosylphosphatidylinositol membrane anchored protein) was found to be associated the influenza virus envelope. Enveloped viruses are susceptible to direct complement-mediated lysis and incorporating CD59 (with DAF and CD46) into lipid envelope is considered to play a protective role from membrane attack by the host complement system. This has important implications for virus host-range. The virus produced and transmitted within the same host species would be protected since complement control proteins are highly species specific. However, virus transmitted to another host species would become susceptible to lysis by the complement system from a different host. (vi) Metabolic enzymes: Proteomic analysis of influenza virus particles identified a number of proteins involved in the glycolytic pathway (pyruvate kinase, enolase 1, GAPDH glyceraldehyde-3-phosphate dehydrogenase, phosphoglycerate kinase). It is possible that some of these cellular proteins might have other additional functions such as viral RNA genome transcription.

Respiratory syncytial virus

Respiratory syncytial virus (RSV) is the most important respiratory virus causing lower respiratory tract infection in young children and neonates. There is no vaccine against RSV. In contrast to influenza virus, formalin inactivated RSV (FI-RSV) in alum adjuvant formulation is known to cause vaccine-enhanced respiratory disease (Kapikian et al. 1969; Kim et al. 1969). Reinfections are common throughout life, indicating that natural RSV infection fails to establish long-lasting immunity (Hall et al. 1991; Piedra 2003; Bont et al. 2002; Nokes et al. 2008; Scott et al. 2006; Glezen et al. 1986). Identification of protein components of RSV might provide a unique clue to therapeutic intervention or vaccine design targets.

The RSV A2 strain was produced in the human respiratory airway cell-line Hep2 and purified using sucrose gradient ultracentrifugation. The RSV protein components were analyzed by one-dimensional nano-LC-MS/MS, resulting in identification of 26 cellular proteins in addition to all the major virus structural proteins (Radhakrishnan et al. 2010). Representative host cell proteins associated with purified RSV particles include proteins associated with the cortical actin network, energy pathways, and heat shock proteins (HSP70, HSC70, and HSP90). In particular, the HSP90 protein was suggested to play an important role in the RSV assembly process. The presence of virus-associated actin network proteins as well as cofilin-1, caveolin-1, and filamin-1 in the mature virus may indicate an important role in RSV assembly in lipid raft microdomains and in maintaining the RSV architecture. Unlike other viruses, high levels of heat-shock proteins associated with RSV particles remain unclear for their significance.

Protein components of VLPs as vaccine candidates

Influenza VLP vaccines that were produced using the rBV-insect cell expression system were demonstrated to be safe and immunogenic in the clinical trials of healthy adults (Khurana et al. 2011; Lopez-Macias et al. 2011). Insect cell culture derived influenza VLP vaccines were shown to be more immunogenic compared to the conventional egg-substrate split vaccines (a Phase II human clinical trial of the trivalent seasonal influenza VLP vaccine candidate, Novavax, Inc.). These clinical studies demonstrated the safety and efficacy of VLP vaccines produced in insect cells using the rBV expression system. Thus, it is highly significant to have information on protein components of VLP vaccines. Using one-DE-LC-MS/MS technology, comprehensive proteomic analysis of the insect cell derived, rBV expressed influenza H5 VLPs identified viral proteins as vaccine target antigens as well as 37 additional host-derived proteins (Song et al. 2011). Many of host-cell-derived proteins in influenza VLPs are known to be present in other enveloped viruses and involved in different cellular structures and functions including those from the cytoskeleton, translation, chaperone, and metabolism. Influenza H5 VLPs produced in insect cells were found to be associated with host proteins involved in actin cytoskeleton, vesicular trafficking (ADP-ribosylation factor, vesicle-associated membrane proteins, vacuolar protein sorting 28, myosin II essential light chain), heat-shock protein 90, ribosomal proteins, putative ubiquitin/ ribosomal protein S27Ae fusion protein, and cell-signaling-related proteins (heterotrimeric guanine nucleotide binding protein gamma subunit-like protein, Rho1). As expected, many rBV vector-derived proteins were also found to be in H5 VLPs. These structural proteins originated from rBV include occlusion derived and polyhedron associated proteins (AcOrf-102, –114), capsid or capsid associated proteins, and baculovirus envelope proteins.

Mammalian influenza VLPs may more closely mimic authentic virions in their morphology, in functional HA, and in other molecular constituents. Stably transformed Vero cells expressing influenza M1, M2, HA, and NA were used to produce mammalian influenza H3N2 VLPs and H5N1 VLPs (Song et al. 2011). Proteomic analysis of mammalian VLPs using LC-MS/MS technologies identified 22 VLP-associated cellular proteins that are analogous to those cellular proteins commonly found in the influenza virions (Song et al. 2011). These cellular proteins incorporated into mammalian influenza VLPs include cytoskeleton proteins, extracellular matrix proteins, heat shock proteins, annexins, tetraspanins, clathrin heavy chain, and glycolytic enzymes. Thus, the cellular proteins identified in VLPs without viral genomes are important in the normal virus life cycle during virus assembly and budding from the host cells.

Host cell gene expression profiles upon viral infections or in response to VLPs

A proteomic approach applying the quantitative 2-D DIGE and nanoHPLC-nanoESI-MS/MS analysis was used to investigate the dynamic cellular host cell response induced by influenza virus infection in two different cell lines, Madin-Darby canine kidney (MDCK) and Vero cells. Upon influenza virus (A/PR/8/34) infection, changes in gene expression of MDCK infected cells were observed in the interferon (IFN)-induced signal transduction, cytoskeleton remodeling, vesicle transport, and proteolysis (Vester et al. 2010). In Vero cells infected with influenza virus, alterations of gene expression include heat shock and oxidative stress response-related proteins.

To gain an understanding of the RSV associated host cell gene expression, differentially expressed genes in human respiratory epithelial cells (A549) were determined by cDNA microarray analysis after RSV infection (Martinez et al. 2007). Among 85 genes that were up-regulated at early times post infection (0 to 6 h post infection (pi)), most highly expressed genes are involved in chemotaxis, inflammation, and some integrins. Genes related to IFN-

stimulation and NF-ƙB pathways were up-regulated between 6 and 12 h pi. At later times post infection, immune response-related genes were expressed at high levels. These findings suggest a temporal relationship between RSV infection and the host response to RSV replication. Supplementary validation experiments using conventional methods are required to confirm these findings.

HIV-1 gag virus-like particles (HIV-VLPs) produced by the recombinant baculovirus expression system was used to stimulate CD14$^+$ monocyte-derived dendritic cells (MDDCs) enriched from peripheral blood mononuclear cells (PBMCs) of normal healthy donors (Buonaguro et al. 2008). Genomic transcriptional profile of HIV-VLPs activated MDDCs revealed high expression of genes that are responsible for activation and maturation of MDDCs. Representative genes up-regulated include antigen processing and presentation (IL3RA, BCL2), cell shape and extracellular matrix, chemokine and cytokines (IL-6), cytokine network (IL1A, B), cytokine-receptor interactions, immune response membrane proteins, chemokine receptors (CCL3, 4), and Toll-like receptor (TLR) signaling pathway (IL8). Similar data of gene expression profile were also reported to be observed using PBMCs activated by HIV-VLPs (Buonaguro et al. 2008). The same group of study demonstrated that the effects of HIV-VLPs on MDDCs are not mediated through TLR2 and TLR4 signaling (Buonaguro et al. 2006). Also, influenza VLP-loaded MDDCs that were obtained from human healthy donors were demonstrated to be effective in generating functional CD8 T cells (Song et al. 2010), implicating that VLP vaccines can induce both humoral and cellular host immune responses.

Hepatitis B virus core antigen VLP (HBc-VLP)-pulsed and control macrophage cells (RAW264.7) were subjected to two-dimensional electrophoresis and tandem MS (Yang et al. 2008). Analysis of differentially expressed proteins revealed that heat-shock protein 70 and prohibitin in addition to proteins in the glycolytic pathway were highly up-regulated upon stimulation with HBc-VLPs. It is speculated that stress-response proteins (HSP70, prohibitin) may contribute to the uptake, processing, and presentation of VLP vaccine particles.

Animal models and systems biology to better understand pathogenesis and vaccination

The dynamics of virus pathogenesis are multifaceted and can be better comprehended by looking at the system as a whole. Human patients with highly pathogenic avian influenza H5N1 virus typically develop a viral primary pneumonia progressing rapidly to acute respiratory distress syndrome (Abdel-Ghafar et al. 2008). An aberrant immune response is thought to play a significant role in the severe respiratory disease that may ultimately lead to death (Peiris et al. 2009). The term 'cytokine storm' referring to an uncontrolled inflammatory response is often associated with H5N1 virus pathogenesis (Tisoncik et al. 2012). Human patients infected with H5N1 virus were shown to have high serum levels of pro- and anti-inflammatory cytokines (IL-6, IL-10, IFN-γ), macrophage and neutrophil chemoattractant chemokines (CxCL10, CXCL2, IL-8) (de Jong et al. 2006; Peiris et al. 2004; To et al. 2001). Host response to influenza H5N1 virus has been investigated in non-human primate, mouse, and ferret models. Global transcriptional profiling of infected lungs has revealed that virulence of influenza virus is associated with increased early and sustained inflammatory responses. Genes that showed correlates with disease severity during H5N1 virus infection include inflammasome components, viral sensing, neutrophil activation, NF-κB signaling, and chemokine signaling (Ibricevic et al. 2006; Cilloniz et al. 2010; Baskin et al. 2009; Cameron et al. 2008; Chang et al. 2011; Shinya et al. 2012).

It is suggested that lung homeostasis is lost when the innate immune system reached a high level of activation but was unable to contain the pathogen before viral cytopathicity (Boon et al. 2011; Sanders et al. 2011). The depletion of innate immune cell types lowered inflammatory cytokine levels in mouse lung homogenates but resulted in elevated lung viral titers, systemic virus spread, and reduced survival (Tumpey et al. 2005). Mice with decreased myeloid infiltrates and lack of NLRP3 inflammasome activation exhibited high susceptibility to influenza infection (Allen et al. 2009; Thomas et al. 2009). Selective neutrophil targeting in infected mice caused enhanced mortality (Tate et al. 2009). Therefore, innate inflammatory cells have host-beneficial functions rather than a primary causal role in pathology (Brincks et al. 2008; Tate et al. 2012).

An alternative view is that lung function is largely dysregulated through the damaging effects of leukocytes on epithelial and endothelial cells (Aldridge et al. 2009; Le Goffic et al. 2006; Lin et al. 2008). In support for this idea, monocyte-derived inflammatory macrophages and dendritic cells contributed to fatality (Lin et al. 2008). The relative pathogenic contributions of direct viral cytopathic damage versus dysregulated host inflammatory responses to lethal influenza infections remain as an important question to be answered. By using extensive microarray analysis of multigene transcriptional signatures from infected mouse lungs, a recent study suggested that differential activation of inflammatory signaling networks distinguished lethal from sublethal infections. From combined flow cytometry and gene expression analysis of isolated cell subpopulations from infected mouse lungs showed that neutrophil influx was largely responsible for the predictive transcriptional signatures. Together with these gene expression and flow data, automated imaging analysis identified chemokine-driven proinflammatory

neutrophils, which might be activated by lethal viral loads. In line with these data, attenuation, but not ablation, of the neutrophil-driven response was shown to improve survival without changing viral spread. These findings with a possible a roadmap for the systematic dissection of infection-associated tissues support evidence for the primary contribution of damaging innate inflammation to some forms of influenza-induced lethality and provide. To more clearly differentiate host protective from damaging immunity, comprehensive data sets at both the organ and the cell level are needed.

As described above, RSV is also an important human respiratory pathogen against which there is no vaccine. Systemic gene expression signatures have been examined in lungs of mice infected with RSV (Pennings et al. 2011; Janssen et al. 2007; Tripp et al. 2013). A robust transcriptional response of interferon-associated and innate immunity genes was observed at day 1 (pi) but was reduced by day 3 pi, and the peak lung transcriptional response preceded the peak of viral replication. Host genes that were expressed were diverse and involved in the IFN response, inflammation, chemoattraction, and antigen processing. In particular, cytokine genes such as IL-1β orchestrate the proinflammatory response while others including intracellular reactive oxygen species (ROS) were effectors of inflammation (Janssen et al. 2007; Tripp et al. 2013; Segovia et al. 2012).

Host innate immune responses in response to vaccination are relatively not well understood yet. The live attenuated yellow fever vaccine 17D (YF-17D) is considered one of most effective vaccine with a 65-year history. YF-17D was shown to activate human monocyte-derived DCs by up-regulating CD80, CD86 markers and inflammatory cytokines (IL-6, TNF-α), and chemokines (IP-10, MCP-1) (Querec et al. 2006). Using CD11c[+] DCs derived from mutant mice, YF-17D was found to stimulate DCs via multiple TLRs 2, 7, 8, and 9 to elicit the proinflammatory cytokines IL-12p40, IL-6, and IFN-α. Mice with a transfer of OT-I T cell receptor transgenic T cells and inoculated with YF-Ova8 induced a mixed T helper cell (Th)1/Th2 cytokine profile and CD8 T cells. Thus, effective vaccines may need to activate multiple innate immune components.

Systems biology in humans for better understanding of disease and vaccination

Systems biology in humans with respiratory viral infections

Systems biology is a newly advancing field that uses an interdisciplinary approach aimed at understanding and predicting the properties of a living system through systematic quantification of all its components and intensive mathematical and computational modeling. Each component of the system is measured using high-throughput 'omic' techniques and in theory examined from the cellular level to the whole organism. Systems technologies include transcriptomic (microarray gene expression), modern mass spectrometry (proteomics, lipidomics, metabolomics), genomics, and protein-DNA interaction (chromatin immunoprecipitation).

A recent study (Mejias et al. 2013) reported whole blood gene expression profiles of microarray data to assess disease severity in infants (ages <6 months, 2 to 24 months) with respiratory syncytial virus infection in comparison with influenza and human rhinovirus (HRV), attempting to identify biomarkers that can objectively predict RSV disease severity. Despite the fact that influenza, RSV, and HRV infect common respiratory tracts, this study demonstrated that the degree of activation/suppression of specific immune-related genes was markedly different. Influenza stimulated a stronger activation of interferon, inflammation, monocyte, and innate immune response genes compared with RSV and rhinovirus. Neutrophil-related genes were significantly overexpressed in patients with RSV, followed by patients with rhinovirus, and were at a lower level in patients with influenza. Interestingly, RSV was associated with marked suppression of genes involved in B cell, T cell, lymphoid lineage, and antimicrobial responses. In contrast, this suppression was significantly milder or absent in children with influenza and rhinovirus.

The overexpression of interferon and innate immunity genes was similar in children with moderate and severe RSV but greater than that in children with mild RSV disease. The overexpression of neutrophil, monocyte, and innate immunity genes induced during the RSV acute disease faded over time. However, T cell lymphoid lineage and antimicrobial response genes were suppressed during the acute phase and then recovered back to normal levels. Remarkably, the suppression of B cell genes was persistent when patients' samples were analyzed 1 month after the acute infection. This low level of B cell genes might explain partially the less protective antibody responses after acute RSV infection. This study also indicates that RSV suppressed both the adaptive and innate responses more severely in younger infants less than 6 months old. Children with severe RSV demonstrated significantly greater underexpression of genes associated with T cells, cytotoxic and NK cells, and plasma cells.

Influenza triggers a more robust immune response than RSV, with greater induction of respiratory and systemic cytokines (Garofalo et al. 2005; Gill et al. 2008; Welliver et al. 2007). Antiviral responses against influenza and RSV were shown to be correlated with the interferon signature gene expression from peripheral blood mononuclear cells isolated from patients with acute influenza or RSV bronchiolitis (Ioannidis et al. 2012). This study provides evidence of the profound systemic dysregulation of both the innate and adaptive

immune response induced by RSV infection in children and supports systems biology of gene expression profiling as a practical and powerful strategy to objectively stratify children with viral infection such as RSV. Nonetheless, these observations will require further analysis, as they may have implications for RSV vaccine development.

Systems vaccinology

One potential application of systems biology is to predict vaccine efficacy in humans. Molecular patterns or signatures of genes in the blood after vaccination might predict the later development of protective immune responses, representing a strategy to prospectively determine vaccine efficacy. Blood cells provide a snap-shot of many lineages and differentiation states within the immune system including the sites of vaccination. Microarray analyses using the Affymetrix Human Genome U133 Plus 2.0 array of total PBMCs revealed a molecular signature comprised of genes that are involved in innate sensing of viruses and antiviral immunity, in most of the vaccines.

YF-17D is a live-attenuated yellow fever virus, one of most effective successful human vaccines ever developed (Querec et al. 2009; Gaucher et al. 2008) and would provide an excellent model for systems vaccinology study in humans. YF-17D single vaccination induces antigen-specific $CD8^+$ T cells and neutralizing antibody responses in humans that persist for several decades (Pulendran 2009; Pulendran et al. 2013). Recent studies (Querec et al. 2009; Gaucher et al. 2008) reported transcriptomic analysis of PBMCs isolated 3 to 7 day-post-vaccination of healthy adults with YF-17D. In these studies, a pattern of gene expression profile was revealed, which consists of genes encoding proteins involved in antiviral sensing and viral immunity, including the type I IFN pathway. It seems to be that the YF-17D vaccine is mimicking an acute viral infection.

Using computational analysis, signatures of gene expression in human PBMCs after vaccination appeared to be correlated with the magnitude of the antigen-specific $CD8^+$ T cell and neutralizing antibody responses afterward (Querec et al. 2009). The functional relevance of one of the genes within the predictive signatures was speculated from machine-learning techniques to validate the predictive capacity (ref 22). Eukaryotic initiation factor-α kinase 4 (*EIF2AK4*) would be involved in programming professional antigen presenting cells (DCs) to stimulate $CD8^+$ T cell responses (Querec et al. 2009). A *TNFRSF17* gene signature was predicted to be correlated with neutralizing antibody responses. *TNFRSF17* encodes the receptor for the B-cell growth factor BLyS-BAFF known to play a key role in the differentiation of plasma cells (Vincent et al. 2013).

The potential application of systems vaccinology in humans was further extended by studies on immunity to human influenza vaccines, the trivalent inactivated seasonal influenza vaccine (TIV) (Nakaya et al. 2011; Bucasas et al. 2011; Franco et al. 2013), and LAIV (FluMist) (Nakaya et al. 2011). TIV is the most common flu shot vaccine, which is a mixture of inactivated split H1N1, H3N2, and influenza B vaccines. To determine whether molecular signatures after YF-17D vaccination would be similar to other vaccines such as influenza, a recent study carried out a systems analysis of responses to TIV and LAIV in young healthy adults during three consecutive influenza seasons (Nakaya et al. 2011). The group of people who received TIV showed higher antibody titers and more plasmablasts compared to the group who received nasal spray of LAIV. As expected from the fact that replicating LAIV infects mucosal tissues of respiratory tracts, humans with a nasal spray of LAIV showed a robust type I IFN antiviral transcriptomic signatures. TIV-vaccinated humans also expressed some gene encoding type I interferons and related proteins as well as gene encoding proinflammatory mediators (Nakaya et al. 2011). In this study, genes that are involved in innate sensing of viruses and antiviral responses were highly expressed within 1 to 3 days after vaccination of humans (Nakaya et al. 2011). After 3 to 7 days of vaccination, the up-regulated genes (*TNRSF17*, *XBP-1*) were found to be involved in the differentiation of plasmablasts, which is likely to be correlated with the magnitude of the later hemagglutin titers (Nakaya et al. 2011). Other studies also demonstrated a plausible correlation between this 'plasmablast signature' and its capacity to predict antibody titers (Obermoser et al. 2013; Furman et al. 2013; Tsang et al. 2014).

Conclusions

In summary, recent advances in applying systems-level approaches to virus, vaccines, cells, organs, animals, and even to humans revealed extensive new information on gene expression and protein components, thus showing some promises in future. This new information extend our understanding in the pathogen-host cell interactions, host cellular responses, disease pathogenesis, host immune responses, and eventually new therapeutics and novel vaccine development. However, it should be reminded that microarray data may fail to provide convincing significance in our complex biological systems. In addition, the results of systems analysis need to be validated by experiments generating functional data such as protein techniques, gene perturbation, or deficient animal models. Finally, systems biology requires multidisciplinary and close collaborative experts including biologists, vaccinologists, immunologists, systems bioinformatics, computational specialists, and clinicians.

Authors' contributions
YL, YJK, YJJ, KHK, YMK, and SMK carried out collecting all information and arranging the Tables. SIK has collected the proteomic data. YL has contributed to drafting the manuscript and arranging the references. SMK finalized writing the manuscript with complete references. All authors read and approved the final manuscript.

Competing interests
The authors declare that they have no competing interests.

Acknowledgements
This work was partially supported by NIH/NIAID grants AI105170 (S.M.K.) and AI093772 (S.M.K.).

Author details
[1]Center for Inflammation, Immunity & Infection, Institute for Biomedical Sciences, Georgia State University, Atlanta, GA 30303, USA. [2]Division of Life Science, Korea Basic Science Institute, Daejeon 305-333, South Korea.

References

Abdel-Ghafar AN, Chotpitayasunondh T, Gao Z, Hayden FG, Nguyen DH, de Jong MD, Naghdaliyev A, Peiris JS, Shindo N, Soeroso S, Uyeki TM (2008) Update on avian influenza A (H5N1) virus infection in humans. N Engl J Med 358:261–273

Aldridge JR Jr, Moseley CE, Boltz DA, Negovetich NJ, Reynolds C, Franks J, Brown SA, Doherty PC, Webster RG, Thomas PG (2009) TNF/iNOS-producing dendritic cells are the necessary evil of lethal influenza virus infection. Proc Natl Acad Sci U S A 106:5306–5311

Allen IC, Scull MA, Moore CB, Holl EK, McElvania-TeKippe E, Taxman DJ, Guthrie EH, Pickles RJ, Ting JP (2009) The NLRP3 inflammasome mediates in vivo innate immunity to influenza A virus through recognition of viral RNA. Immunity 30:556–565

Arico E, Wang E, Tornesello ML, Tagliamonte M, Lewis GK, Marincola FM, Buonaguro FM, Buonaguro L (2005) Immature monocyte derived dendritic cells gene expression profile in response to virus-like particles stimulation. J Transl Med 3:45

Baskin CR, Bielefeldt-Ohmann H, Tumpey TM, Sabourin PJ, Long JP, Garcia-Sastre A, Tolnay AE, Albrecht R, Pyles JA, Olson PH, Aicher LD, Rosenzweig ER, Murali-Krishna K, Clark EA, Kotur MS, Fornek JL, Proll S, Palermo RE, Sabourin CL, Katze MG (2009) Early and sustained innate immune response defines pathology and death in nonhuman primates infected by highly pathogenic influenza virus. Proc Natl Acad Sci U S A 106:3455–3460

Bont L, Versteegh J, Swelsen WT, Heijnen CJ, Kavelaars A, Brus F, Draaisma JM, Pekelharing-Berghuis M, van Diemen-Steenvoorde RA, Kimpen JL (2002) Natural reinfection with respiratory syncytial virus does not boost virus-specific T-cell immunity. Pediatr Res 52:363–367

Boon AC, Finkelstein D, Zheng M, Liao G, Allard J, Klumpp K, Webster R, Peltz G, and Webby RJ (2011) H5N1 influenza virus pathogenesis in genetically diverse mice is mediated at the level of viral load. mBio 2(5).

Bortz E, Whitelegge JP, Jia Q, Zhou ZH, Stewart JP, Wu TT, Sun R (2003) Identification of proteins associated with murine gammaherpesvirus 68 virions. J Virol 77:13425–13432

Brincks EL, Katewa A, Kucaba TA, Griffith TS, Legge KL (2008) CD8 T cells utilize TRAIL to control influenza virus infection. J Immunol 181:4918–4925

Bucasas KL, Franco LM, Shaw CA, Bray MS, Wells JM, Nino D, Arden N, Quarles JM, Couch RB, Belmont JW (2011) Early patterns of gene expression correlate with the humoral immune response to influenza vaccination in humans. J Infect Dis 203:921–929

Buonaguro L, Tornesello ML, Tagliamonte M, Gallo RC, Wang LX, Kamin-Lewis R, Abdelwahab S, Lewis GK, Buonaguro FM (2006) Baculovirus-derived human immunodeficiency virus type 1 virus-like particles activate dendritic cells and induce ex vivo T-cell responses. J Virol 80:9134–9143

Buonaguro L, Monaco A, Arico E, Wang E, Tornesello ML, Lewis GK, Marincola FM, Buonaguro FM (2008) Gene expression profile of peripheral blood mononuclear cells in response to HIV-VLPs stimulation. BMC Bioinformatics 9(Suppl 2):S5

Cameron CM, Cameron MJ, Bermejo-Martin JF, Ran L, Xu L, Turner PV, Ran R, Danesh A, Fang Y, Chan PK, Mytle N, Sullivan TJ, Collins TL, Johnson MG, Medina JC, Rowe T, Kelvin DJ (2008) Gene expression analysis of host innate immune responses during Lethal H5N1 infection in ferrets. J Virol 82:11308–11317

Cantin R, Methot S, Tremblay MJ (2005) Plunder and stowaways: incorporation of cellular proteins by enveloped viruses. J Virol 79:6577–6587

Chang ST, Tchitchek N, Ghosh D, Benecke A, Katze MG (2011) A chemokine gene expression signature derived from meta-analysis predicts the pathogenicity of viral respiratory infections. BMC Syst Biol 5:202

Chertova E, Chertov O, Coren LV, Roser JD, Trubey CM, Bess JW Jr, Sowder RC 2nd, Barsov E, Hood BL, Fisher RJ, Nagashima K, Conrads TP, Veenstra TD, Lifson JD, Ott DE (2006) Proteomic and biochemical analysis of purified human immunodeficiency virus type 1 produced from infected monocyte-derived macrophages. J Virol 80:9039–9052

Chung CS, Chen CH, Ho MY, Huang CY, Liao CL, Chang W (2006) Vaccinia virus proteome: identification of proteins in vaccinia virus intracellular mature virion particles. J Virol 80:2127–2140

Cilloniz C, Pantin-Jackwood MJ, Ni C, Goodman AG, Peng X, Proll SC, Carter VS, Rosenzweig ER, Szretter KJ, Katz JM, Korth MJ, Swayne DE, Tumpey TM, Katze MG (2010) Lethal dissemination of H5N1 influenza virus is associated with dysregulation of inflammation and lipoxin signaling in a mouse model of infection. J Virol 84:7613–7624

Coffman RL, Sher A, Seder RA (2010) Vaccine adjuvants: putting innate immunity to work. Immunity 33:492–503

De-Jong MD, Simmons CP, Thanh TT, Hien VM, Smith GJ, Chau TN, Hoang DM, Chau NV, Khanh TH, Dong VC, Qui PT, Cam BV, Ha-do Q, Guan Y, Peiris JS, Chinh NT, Hien TT, Farrar J (2006) Fatal outcome of human influenza A (H5N1) is associated with high viral load and hypercytokinemia. Nat Med 12:1203–1207

Demirov DG, Ono A, Orenstein JM, Freed EO (2002) Overexpression of the N-terminal domain of TSG101 inhibits HIV-1 budding by blocking late domain function. Proc Natl Acad Sci U S A 99:955–960

Dove BK, Surtees R, Bean TJ, Munday D, Wise HM, Digard P, Carroll MW, Ajuh P, Barr JN, Hiscox JA (2012) A quantitative proteomic analysis of lung epithelial (A549) cells infected with 2009 pandemic influenza A virus using stable isotope labelling with amino acids in cell culture. Proteomics 12:1431–1436

Franco LM, Bucasas KL, Wells JM, Nino D, Wang X, Zapata GE, Arden N, Renwick A, Yu P, Quarles JM, Bray MS, Couch RB, Belmont JW, Shaw CA (2013) Integrative genomic analysis of the human immune response to influenza vaccination. eLife 2:e00299

Franke EK, Yuan HE, Luban J (1994) Specific incorporation of cyclophilin A into HIV-1 virions. Nature 372:359–362

Fries LF, Smith GE, Glenn GM (2013) A recombinant viruslike particle influenza A (H7N9) vaccine. N Engl J Med 369:2564–2566

Furman D, Jojic V, Kidd B, Shen-Orr S, Price J, Jarrell J, Tse T, Huang H, Lund P, Maecker HT, Utz PJ, Dekker CL, Koller D, Davis MM (2013) Apoptosis and other immune biomarkers predict influenza vaccine responsiveness. Mol Syst Biol 9:659

Garland SM, Hernandez-Avila M, Wheeler CM, Perez G, Harper DM, Leodolter S, Tang GW, Ferris DG, Steben M, Bryan J, Taddeo FJ, Railkar R, Esser MT, Sings HL, Nelson M, Boslego J, Sattler C, Barr E, Koutsky LA (2007) Quadrivalent vaccine against human papillomavirus to prevent anogenital diseases. N Engl J Med 356:1928–1943

Garofalo RP, Hintz KH, Hill V, Patti J, Ogra PL, Welliver RC Sr (2005) A comparison of epidemiologic and immunologic features of bronchiolitis caused by influenza virus and respiratory syncytial virus. J Med Virol 75:282–289

Gaucher D, Therrien R, Kettaf N, Angermann BR, Boucher G, Filali-Mouhim A, Moser JM, Mehta RS, Drake DR 3rd, Castro E, Akondy R, Rinfret A, Yassine-Diab B, Said EA, Chouikh Y, Cameron MJ, Clum R, Kelvin D, Somogyi R, Greller LD, Balderas RS, Wilkinson P, Pantaleo G, Tartaglia J, Haddad EK, Sekaly RP (2008) Yellow fever vaccine induces integrated multilineage and polyfunctional immune responses. J Exp Med 205:3119–3131

Gill MA, Long K, Kwon T, Muniz L, Mejias A, Connolly J, Roy L, Banchereau J, Ramilo O (2008) Differential recruitment of dendritic cells and monocytes to respiratory mucosal sites in children with influenza virus or respiratory syncytial virus infection. J Infect Dis 198:1667–1676

Glezen WP, Taber LH, Frank AL, Kasel JA (1986) Risk of primary infection and reinfection with respiratory syncytial virus. Am J Dis Child 140:543–546

Hall CB, Walsh EE, Long CE, Schnabel KC (1991) Immunity to and frequency of reinfection with respiratory syncytial virus. J Infect Dis 163:693–698

Ho SH, Martin F, Higginbottom A, Partridge LJ, Parthasarathy V, Moseley GW, Lopez P, Cheng-Mayer C, Monk PN (2006) Recombinant extracellular domains of tetraspanin proteins are potent inhibitors of the infection of macrophages by human immunodeficiency virus type 1. J Virol 80:6487–6496

Huang YC, Li Z, Hyseni X, Schmitt M, Devlin RB, Karoly ED, Soukup JM (2008) Identification of gene biomarkers for respiratory syncytial virus infection in a bronchial epithelial cell line. Genomic Med 2:113–125

Ibricevic A, Pekosz A, Walter MJ, Newby C, Battaile JT, Brown EG, Holtzman MJ, Brody SL (2006) Influenza virus receptor specificity and cell tropism in mouse and human airway epithelial cells. J Virol 80:7469–7480

Ioannidis I, McNally B, Willette M, Peeples ME, Chaussabel D, Durbin JE, Ramilo O, Mejias A, Flano E (2012) Plasticity and virus specificity of the airway epithelial cell immune response during respiratory virus infection. J Virol 86:5422–5436

Iwasaki A, Medzhitov R (2010) Regulation of adaptive immunity by the innate immune system. Science 327:291–295

Janssen R, Pennings J, Hodemaekers H, Buisman A, van Oosten M, de Rond L, Ozturk K, Dormans J, Kimman T, Hoebee B (2007) Host transcription profiles upon primary respiratory syncytial virus infection. J Virol 81:5958–5967

Johannsen E, Luftig M, Chase MR, Weicksel S, Cahir-McFarland E, Illanes D, Sarracino D, Kieff E (2004) Proteins of purified Epstein-Barr virus. Proc Natl Acad Sci U S A 101:16286–16291

Jolly C, Sattentau QJ (2007) Human immunodeficiency virus type 1 assembly, budding, and cell-cell spread in T cells take place in tetraspanin-enriched plasma membrane domains. J Virol 81:7873–7884

Kang SM, Pushko P, Bright RA, Smith G, Compans RW (2009a) Influenza virus-like particles as pandemic vaccines. Curr Top Microbiol Immunol 333:269–289

Kang SM, Song JM, Quan FS, Compans RW (2009b) Influenza vaccines based on virus-like particles. Virus Res 143:140–146

Kapikian AZ, Mitchell RH, Chanock RM, Shvedoff RA, Stewart CE (1969) An epidemiologic study of altered clinical reactivity to respiratory syncytial (RS) virus infection in children previously vaccinated with an inactivated RS virus vaccine. Am J Epidemiol 89:405–421

Kattenhorn LM, Mills R, Wagner M, Lomsadze A, Makeev V, Borodovsky M, Ploegh HL, Kessler BM (2004) Identification of proteins associated with murine cytomegalovirus virions. J Virol 78:11187–11197

Kawai T, Akira S (2007) TLR signaling. Semin Immunol 19:24–32

Khurana S, Wu J, Verma N, Verma S, Raghunandan R, Manischewitz J, King LR, Kpamegan E, Pincus S, Smith G, Glenn G, Golding H (2011) H5N1 virus-like particle vaccine elicits cross-reactive neutralizing antibodies that preferentially bind to the oligomeric form of influenza virus hemagglutinin in humans. J Virol 85:10945–10954

Kim HW, Canchola JG, Brandt CD, Pyles G, Chanock RM, Jensen K, Parrott RH (1969) Respiratory syncytial virus disease in infants despite prior administration of antigenic inactivated vaccine. Am J Epidemiol 89:422–434

Kim MC, Lee JS, Kwon YM, Eunju O, Lee YJ, Choi JG, Wang BZ, Compans RW, Kang SM (2013a) Multiple heterologous M2 extracellular domains presented on virus-like particles confer broader and stronger M2 immunity than live influenza A virus infection. Antivir Res 99:328–335

Kim MC, Song JM, Eunju O, Kwon YM, Lee YJ, Compans RW, Kang SM (2013b) Virus-like particles containing multiple M2 extracellular domains confer improved cross-protection against various subtypes of influenza virus. Mol Ther 21:485–492

Kim MC, Lee YN, Hwang HS, Lee YT, Ko EJ, Jung YJ, Cho MK, Kim YJ, Lee JS, Ha SH, Kang SM (2014) Influenza M2 virus-like particles confer a broader range of cross protection to the strain-specific pre-existing immunity. Vaccine 32:5824–5831

Klausberger M, Wilde M, Palmberger D, Hai R, Albrecht RA, Margine I, Hirsh A, Garcia-Sastre A, Grabherr R, Krammer F (2014) One-shot vaccination with an insect cell-derived low-dose influenza A H7 virus-like particle preparation protects mice against H7N9 challenge. Vaccine 32:355–362

Le Goffic R, Balloy V, Lagranderie M, Alexopoulou L, Escriou N, Flavell R, Chignard M, Si-Tahar M (2006) Detrimental contribution of the Toll-like receptor (TLR)3 to influenza A virus-induced acute pneumonia. PLoS Pathog 2:e53

Lin KL, Suzuki Y, Nakano H, Ramsburg E, Gunn MD (2008) CCR2+ monocyte-derived dendritic cells and exudate macrophages produce influenza-induced pulmonary immune pathology and mortality. J Immunol 180:2562–2572

Lopez-Macias, C (2012) Virus-like particle (VLP)-based vaccines for pandemic influenza: Performance of a VLP vaccine during the 2009 influenza pandemic. Human Vaccines Immunotherapeutics 8(3):411–414.

Lopez-Macias C, Ferat-Osorio E, Tenorio-Calvo A, Isibasi A, Talavera J, Arteaga-Ruiz O, Arriaga-Pizano L, Hickman SP, Allende M, Lenhard K, Pincus S, Connolly K, Raghunandan R, Smith G, Glenn G (2011) Safety and immunogenicity of a virus-like particle pandemic influenza A (H1N1) 2009 vaccine in a blinded, randomized, placebo-controlled trial of adults in Mexico. Vaccine 29:7826–7834

Mangeat B, Turelli P, Caron G, Friedli M, Perrin L, Trono D (2003) Broad antiretroviral defence by human APOBEC3G through lethal editing of nascent reverse transcripts. Nature 424:99–103

Mariani R, Chen D, Schrofelbauer B, Navarro F, Konig R, Bollman B, Munk C, Nymark-McMahon H, Landau NR (2003) Species-specific exclusion of APOBEC3G from HIV-1 virions by Vif. Cell 114:21–31

Martinez I, Lombardia L, Garcia-Barreno B, Dominguez O, Melero JA (2007) Distinct gene subsets are induced at different time points after human respiratory syncytial virus infection of A549 cells. J Gen Virol 88:570–581

Mejias A, Dimo B, Suarez NM, Garcia C, Suarez-Arrabal MC, Jartti T, Blankenship D, Jordan-Villegas A, Ardura MI, Xu Z, Banchereau J, Chaussabel D, Ramilo O (2013) Whole blood gene expression profiles to assess pathogenesis and disease severity in infants with respiratory syncytial virus infection. PLoS Med 10:e1001549

Munday DC, Emmott E, Surtees R, Lardeau CH, Wu W, Duprex WP, Dove BK, Barr JN, Hiscox JA (2010) Quantitative proteomic analysis of A549 cells infected with human respiratory syncytial virus. Mol Cell Proteomics 9:2438–2459

Nakaya HI, Wrammert J, Lee EK, Racioppi L, Marie-Kunze S, Haining WN, Means AR, Kasturi SP, Khan N, Li GM, McCausland M, Kanchan V, Kokko KE, Li S, Elbein R, Mehta AK, Aderem A, Subbarao K, Ahmed R, Pulendran B (2011) Systems biology of vaccination for seasonal influenza in humans. Nat Immunol 12:786–795

Nokes DJ, Okiro EA, Ngama M, Ochola R, White LJ, Scott PD, English M, Cane PA, Medley GF (2008) Respiratory syncytial virus infection and disease in infants and young children observed from birth in Kilifi District, Kenya. Clin Infect Dis 46:50–57

Obermoser G, Presnell S, Domico K, Xu H, Wang Y, Anguiano E, Thompson-Snipes L, Ranganathan R, Zeitner B, Bjork A, Anderson D, Speake C, Ruchaud E, Skinner J, Alsina L, Sharma M, Dutartre H, Cepika A, Israelsson E, Nguyen P, Nguyen QA, Harrod AC, Zurawski SM, Pascual V, Ueno H, Nepom GT, Quinn C, Blankenship D, Palucka K, Banchereau J, Chaussabel D (2013) Systems scale interactive exploration reveals quantitative and qualitative differences in response to influenza and pneumococcal vaccines. Immunity 38:831–844

Peiris JS, Yu WC, Leung CW, Cheung CY, Ng WF, Nicholls JM, Ng TK, Chan KH, Lai ST, Lim WL, Yuen KY, Guan Y (2004) Re-emergence of fatal human influenza A subtype H5N1 disease. Lancet 363:617–619

Peiris JS, Cheung CY, Leung CY, Nicholls JM (2009) Innate immune responses to influenza A H5N1: friend or foe? Trends Immunol 30:574–584

Pennings JL, Schuurhof A, Hodemaekers HM, Buisman A, de Rond LC, Widjojoatmodjo MN, Luytjes W, Kimpen JL, Bont L, Janssen R (2011) Systemic signature of the lung response to respiratory syncytial virus infection. PLoS One 6:e21461

Piedra PA (2003) Clinical experience with respiratory syncytial virus vaccines. Pediatr Infect Dis J 22:S94–99

Pulendran B (2009) Learning immunology from the yellow fever vaccine: innate immunity to systems vaccinology. Nat Rev Immunol 9:741–747

Pulendran B, Ahmed R (2006) Translating innate immunity into immunological memory: implications for vaccine development. Cell 124:849–863

Pulendran B, Oh JZ, Nakaya HI, Ravindran R, Kazmin DA (2013) Immunity to viruses: learning from successful human vaccines. Immunol Rev 255:243–255

Querec T, Bennouna S, Alkan S, Laouar Y, Gorden K, Flavell R, Akira S, Ahmed R, Pulendran B (2006) Yellow fever vaccine YF-17D activates multiple dendritic cell subsets via TLR2, 7, 8, and 9 to stimulate polyvalent immunity. J Exp Med 203:413–424

Querec TD, Akondy RS, Lee EK, Cao W, Nakaya HI, Teuwen D, Pirani A, Gernert K, Deng J, Marzolf B, Kennedy K, Wu H, Bennouna S, Oluoch H, Miller J, Vencio RZ, Mulligan M, Aderem A, Ahmed R, Pulendran B (2009) Systems biology approach predicts immunogenicity of the yellow fever vaccine in humans. Nat Immunol 10:116–125

Radhakrishnan A, Yeo D, Brown G, Myaing MZ, Iyer LR, Fleck R, Tan BH, Aitken J, Sanmun D, Tang K, Yarwood A, Brink J, Sugrue RJ (2010) Protein analysis of purified respiratory syncytial virus particles reveals an important role for heat shock protein 90 in virus particle assembly. Mol Cell Proteomics 9:1829–1848

Rhorer J, Ambrose CS, Dickinson S, Hamilton H, Oleka NA, Malinoski FJ, Wittes J (2009) Efficacy of live attenuated influenza vaccine in children: A meta-analysis of nine randomized clinical trials. Vaccine 27:1101–1110

Roy P, Noad R (2009) Virus-like particles as a vaccine delivery system: myths and facts. Adv Exp Med Biol 655:145–158

Sanders CJ, Doherty PC, Thomas PG (2011) Respiratory epithelial cells in innate immunity to influenza virus infection. Cell Tissue Res 343:13–21

Saphire AC, Gallay PA, Bark SJ (2006) Proteomic analysis of human immunodeficiency virus using liquid chromatography/tandem mass spectrometry effectively distinguishes specific incorporated host proteins. J Proteome Res 5:530–538

Scott PD, Ochola R, Ngama M, Okiro EA, James Nokes D, Medley GF, Cane PA (2006) Molecular analysis of respiratory syncytial virus reinfections in infants from coastal Kenya. J Infect Dis 193:59–67

Segovia J, Sabbah A, Mgbemena V, Tsai SY, Chang TH, Berton MT, Morris IR, Allen IC, Ting JP, Bose S (2012) TLR2/MyD88/NF-kappaB pathway, reactive oxygen species, potassium efflux activates NLRP3/ASC inflammasome during respiratory syncytial virus infection. PLoS One 7:e29695

Shaw ML, Stone KL, Colangelo CM, Gulcicek EE, Palese P (2008) Cellular proteins in influenza virus particles. PLoS Pathog 4:e1000085

Shinya K, Gao Y, Cilloniz C, Suzuki Y, Fujie M, Deng G, Zhu Q, Fan S, Makino A, Muramoto Y, Fukuyama S, Tamura D, Noda T, Eisfeld AJ, Katze MG, Chen H, Kawaoka Y (2012) Integrated clinical, pathologic, virologic, and transcriptomic analysis of H5N1 influenza virus-induced viral pneumonia in the rhesus macaque. J Virol 86:6055–6066

Smith GE, Flyer DC, Raghunandan R, Liu Y, Wei Z, Wu Y, Kpamegan E, Courbron D, Fries LF 3rd, Glenn GM (2013) Development of influenza H7N9 virus like particle (VLP) vaccine: homologous A/Anhui/1/2013 (H7N9) protection and heterologous A/chicken/Jalisco/CPA1/2012 (H7N3) cross-protection in vaccinated mice challenged with H7N9 virus. Vaccine 31:4305–4313

Song H, Wittman V, Byers A, Tapia T, Zhou B, Warren W, Heaton P, Connolly K (2010) In vitro stimulation of human influenza-specific CD8+ T cells by dendritic cells pulsed with an influenza virus-like particle (VLP) vaccine. Vaccine 28:5524–5532

Song JM, Choi CW, Kwon SO, Compans RW, Kang SM, Kim SI (2011) Proteomic characterization of influenza H5N1 virus-like particles and their protective immunogenicity. J Proteome Res 10:3450–3459

Steinman RM (2008) Dendritic cells in vivo: a key target for a new vaccine science. Immunity 29:319–324

Tate MD, Deng YM, Jones JE, Anderson GP, Brooks AG, Reading PC (2009) Neutrophils ameliorate lung injury and the development of severe disease during influenza infection. J Immunol 183:7441–7450

Tate MD, Brooks AG, Reading PC, Mintern JD (2012) Neutrophils sustain effective CD8(+) T-cell responses in the respiratory tract following influenza infection. Immunol Cell Biol 90:197–205

Thomas PG, Dash P, Aldridge JR Jr, Ellebedy AH, Reynolds C, Funk AJ, Martin WJ, Lamkanfi M, Webby RJ, Boyd KL, Doherty PC, Kanneganti TD (2009) The intracellular sensor NLRP3 mediates key innate and healing responses to influenza A virus via the regulation of caspase-1. Immunity 30:566–575

Tisoncik JR, Korth MJ, Simmons CP, Farrar J, Martin TR, Katze MG (2012) Into the eye of the cytokine storm. Microbiol Mol Biol Rev 76:16–32

To KF, Chan PK, Chan KF, Lee WK, Lam WY, Wong KF, Tang NL, Tsang DN, Sung RY, Buckley TA, Tam JS, Cheng AF (2001) Pathology of fatal human infection associated with avian influenza A H5N1 virus. J Med Virol 63:242–246

Tripp RA, Mejias A, Ramilo O (2013) Host gene expression and respiratory syncytial virus infection. Curr Top Microbiol Immunol 372:193–209

Tsang JS, Schwartzberg PL, Kotliarov Y, Biancotto A, Xie Z, Germain RN, Wang E, Olnes MJ, Narayanan M, Golding H, Moir S, Dickler HB, Perl S, Cheung F (2014) Global analyses of human immune variation reveal baseline predictors of postvaccination responses. Cell 157:499–513

Tumpey TM, Garcia-Sastre A, Taubenberger JK, Palese P, Swayne DE, Pantin-Jackwood MJ, Schultz-Cherry S, Solorzano A, Van Rooijen N, Katz JM, Basler CF (2005) Pathogenicity of influenza viruses with genes from the 1918 pandemic virus: functional roles of alveolar macrophages and neutrophils in limiting virus replication and mortality in mice. J Virol 79:14933–14944

Vester D, Rapp E, Kluge S, Genzel Y, Reichl U (2010) Virus-host cell interactions in vaccine production cell lines infected with different human influenza A virus variants: a proteomic approach. J Proteome 73:1656–1669

Vincent FB, Saulep-Easton D, Figgett WA, Fairfax KA, Mackay F (2013) The BAFF/APRIL system: emerging functions beyond B cell biology and autoimmunity. Cytokine Growth Factor Rev 24:203–215

Welliver TP, Garofalo RP, Hosakote Y, Hintz KH, Avendano L, Sanchez K, Velozo L, Jafri H, Chavez-Bueno S, Ogra PL, McKinney L, Reed JL, Welliver RC Sr (2007) Severe human lower respiratory tract illness caused by respiratory syncytial virus and influenza virus is characterized by the absence of pulmonary cytotoxic lymphocyte responses. J Infect Dis 195:1126–1136

Wu CY, Yeh YC, Yang YC, Chou C, Liu MT, Wu HS, Chan JT, Hsiao PW (2010) Mammalian expression of virus-like particles for advanced mimicry of authentic influenza virus. PLoS One 5:e9784

Yang F, Wang F, Guo Y, Zhou Q, Wang Y, Yin Y, Sun S (2008) Enhanced capacity of antigen presentation of HBc-VLP-pulsed RAW264.7 cells revealed by proteomics analysis. J Proteome Res 7:4898–4903

Zhu FX, Chong JM, Wu L, Yuan Y (2005) Virion proteins of Kaposi's sarcoma-associated herpesvirus. J Virol 79:800–811

7

Removal of traces of mercury from a carrier gas for analytical purpose

Cristina MI Theil[1], Luis FH Niencheski[2], Gilberto Fillmann[2] and Marcio R Milani[1*]

Abstract

Background: The analysis of mercury by cold vapor requires a gas, usually argon or helium, to transport elementary mercury to the gold trap or directly to the detector. When analyzing mercury in environmental matrices, a gas with a metal concentration as low as a few picograms per cubic meter is needed. Different sorbents have been used to purify the gas for a long time, but little information is available about them, mainly considering the analytical purpose. This paper presents results of the absorption capacity for solids and hypochlorite solutions that usually are used as mercury sorbents, giving technical information to the analyst to decide the best gas cleaning process to be used.

Findings: The absorption capacities of different sorbents were tested using atomic fluorescence spectrometry. Among the tested solids, platinum presented the highest absorption capacity (13.04 pg Hg per gram of Pt). Interaction between sodium hypochlorite, sodium chloride, and EDTA in the absorption capacity was investigated by a 2^3 factorial design. Results showed a significant interaction between hypochlorite and chloride.

Conclusions: A solution of 1.26 mmol L^{-1} sodium hypochlorite, 0.48 mol L^{-1} sodium chloride, and 0.6 mmol L^{-1} EDTA shows the highest absorption capacity (167.3 pg Hg) among the tested compositions. That solution has eliminated even traces of mercury from gases, resulting in a carrier free of mercury, what cannot be achieved using the solid sorbents tested, despite the use of solutions which is more tedious than the use of solids to clean gas. Anyway, the hypochlorite solution shows to be a good option to clean gases that have to be used in the analysis of mercury in samples with very low concentration.

Keywords: Environmental sample; Sodium hypochlorite; Absorption capacity

Introduction

The determination of the concentration of metals in environmental samples has drawn the attention of researchers over the years, and the chemical speciation analyses of metals have received particular attention (Borges et al. 2012; Wotter et al. 2011). The impact of mercury in the environment concerning its toxicity, mobility, bioaccumulation, and biomagnification is responsible for a continuous interest to investigate the fate of this metal (Santos et al. 2008; Stergarsek et al. 2013; Emili et al. 2012). The development of analytical methods and lower limits of quantification have allowed increasing the understanding of biogeochemical processes involving mercury and other metals. Among the most popular analytical methods to analyze mercury is the cold vapor atomic fluorescence spectrometry (CVAFS) (Santos et al. 2008; Leermakers et al. 2005), whose main advantage is separation of the analyte from the matrix by purging the volatile mercury with an inert gas to the detector, with a net result of high sensitivity and selectivity and lower limit of quantification. However, the success of the determination by CVAFS is closely related to the quality of the gas used to purge the analyte, since it is often necessary to make determinations of a few picograms of mercury (Liang et al. 1996).

Concentrations of dissolved gaseous mercury in surface water or dimethylmercury in environmental matrices, for example, are extremely low, and therefore, the need for extremely low blank values is essential (Emili et al. 2012; Hines et al. 2012). The care with cleaning materials, purification of reagents, and nature of the materials to be used in the mercury analyses are perfectly defined (MESL 1997).

* Correspondence: marcmilafurg@gmail.com
[1]Escola de Química e Alimentos, Federal University of Rio Grande – FURG, Av. Italia, km 8, 96201-900, Rio Grande, RS, Brazil
Full list of author information is available at the end of the article

However, if the mercury concentration in the gas is not low enough, one should purify it.

The gas purification can be achieved by different ways. Solid sorbents, as traps containing particles of gold, platinum, or active carbon, usually are used by the analyst due to their operational facility (Liang et al. 1996). Solutions also can be used to remove mercury from gaseous streams. The mercury can be oxidized by hypochlorite solutions, producing Hg^{2+} (Sizeneva et al. 2005), but the importance of the physical states during the reaction must be remembered. Thus, the dissolution of one drop of metallic mercury in a hypochlorite solution is a heterogeneous reaction which takes place on the surface, depending on the area and the surface state. Sizeneva et al. (2005) established that the reaction of mercury dissolution in hypochlorite solution is first order. If mercury is considered to be in the gaseous state reacting with the hypochlorite solution, then the reaction is second order (Zhao and Rochelle 1999).

This paper presents a study comparing the capacity of well-known sorbents to remove gaseous mercury and the mercury concentration obtained after a helium stream passed through those sorbents.

Materials and methods

The tests were carried out with helium gas 5.0 in which the mercury concentration was determined by atomic fluorescence spectrometry after absorption on a gold-coated sand trap (MESL 1997; Horvat et al. 1993) to be equal to 6.35 ng m^{-3} ($n = 5$). The concentration of a commercial sodium hypochlorite (0.21 mol L^{-1}) was determined by iodometric titration (Lagowski 1995). Sodium chloride and EDTA were pro-analysis (Merck, Whitehouse Station, NJ, USA). Ultrapure water (resistivity 18.2 MΩ cm, Milli-Q, Millipore Co., Billerica, MA, USA) was used to prepare all the solutions.

Figure 1 shows a schematic of the system used to perform the experiments. Briefly, the gas flow was set at

200 mL min^{-1} using a flowmeter (FM) (model PMR5-010005, Cole Parmer, Vernon Hills, IL, USA). The gas flows through the eliminator of mercury (EHg), which can be a column containing a solid or a bubbler with a specific solution. A septum (SI) was placed in line for injecting the gaseous mercury standard. The mercury is absorbed on a gold-coated sand trap (GT) (Tekran Instruments Corporation, Toronto, Ontario, Canada) during sometime and, soon after, is desorbed under temperature control by a desorption system (DS) (QMS multifunction controller, Slovenia). A fluorescence detector (DT) (Brooks Rand model III, Brooks Rand Labs, Seattle, WA, USA) records the intensity of the analytical signal (Mercury Guru Software version 4.7).

In the experiments which aim to compare the performance of solid sorbents, glass columns of 124-mm length and 9-mm diameter containing the solid to be evaluated were used and placed in the EHg position (Figure 1). The gas flows through the system for 3 min, and mercury not adsorbed by EHg is the pre-concentrate; at the end of the experiment, the signal is recorded. The experiment is repeated five times for each solid, and the means of absorption capacities are compared using Duncan's post hoc test at a significance level of 5%. The absorption capacity (AC) is calculated as follows:

$$AC = W1 - W2,$$

where W1 is the total mass of mercury that flows through the system during the experiment time, calculated by multiplying the time of the experiment, mercury concentration in the gas, and the gas flow rate; W2 is the mass of mercury not retained in EHg (W2) calculated using the peak height.

Four different solid sorbents are tested, namely (a) coal taken from a chemical cartridge mask of a multi-gas personal protection with a bed height of 65 mm, here referred to as C1; (b) coal taken from a chemical cartridge to be used against mercury vapors or chlorine gas from an individual protective mask with a bed height of 60 mm, here referred to as C2; (c) metallic gold particles with an average diameter of 0.1 mm and bed height of 15 mm, referred to as A1; and (d) platinum wire coil with a bed height of 10 mm, referred to as A2. The solids are packed into columns, with quartz wool used to seal the ends of the column.

When solutions are evaluated to sorb mercury, the system is assembled as shown in Figure 1, and a 150-mL glass flask is used, containing 50 mL of the solution to be tested, in EHg position. The gas flows through the system during a period (TPC) determining the pre-concentration of mercury. Then, the metal is desorbed by heating, and the fluorescence is measured. If the TPC is low and no peak appears, the procedure has to be repeated until a peak is recorded. The total time is recorded, i.e., the total volume of gas, and thus, the total mass of mercury (W1) that flows

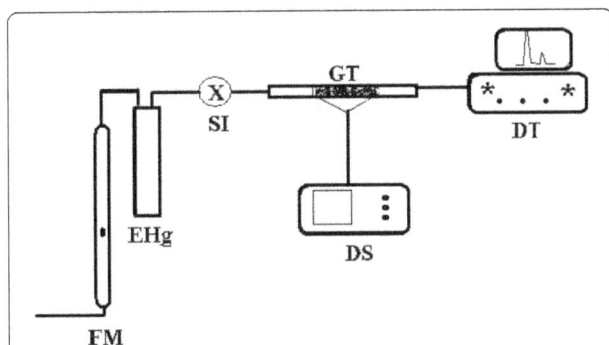

Figure 1 Schematic setup of the system for the determination of gaseous mercury. FM, flowmeter; EHg, eliminator of mercury; SI, septum; GT, gold-coated sand trap; DS, desorption system; DT, fluorescence detector.

Table 1 Performance of sorbents for removing mercury from a gas mixture

Sorbent	AC	s	Duncan
A1 + A2	3.38	0.465	A
A2	3.34	0.444	A
C1	2.93	0.508	AB
A1	2.52	0.360	B
C2	1.91	0.435	C

Gas flow = 200 mL min^{-1}; mercury concentration in the gas mixture = 6.35 pg L^{-1}; experiment time = 3 min. A1 and A2 columns contain Au and Pt, respectively. C1 and C2 columns contain activated carbon for general use and specifically for Hg, respectively. Absorptive capacity (AC) is expressed in pg Hg. Standard deviation (s) is for five replicates. Duncan test results to a level of 5%.

through the system during the experiment time can be calculated, as well as the mass of Hg that EHg does not retain during the experiment. In the following, AC is calculated.

A 2^3 factorial design is used to evaluate the influence of the composition of the absorbent solution on the absorption capacity. All possible combinations between three factors ($N = 3$) and two levels of concentration ($k = 2$) are evaluated, i.e., a full factorial design is run. Three replicates for each combination ($n = 3$) are run. The following factors and levels are evaluated: (a) chloride concentration (0.48 and 0.96 mol L^{-1}), (b) hypochlorite concentration (0.63 and 1.26 mmol L^{-1}), and (c) EDTA concentration (0.30 and 0.60 mmol L^{-1}). The analysis of variance is used to identify the significance and the interaction between the factors for a significance level of 5%.

The mass of mercury is calculated comparing the height of the peak of the samples with the height of the peak from the standard. The peak of the standard is recorded after an injection of an aliquot of a gas mixture containing a mercury concentration is well established, as described elsewhere (Gardfeldt et al. 2002). A 50-μL syringe gas tight

with removable needle (model 1705, Hamilton, Reno, NV, USA) is used to transfer an aliquot of 10 μL of gas mixture kept at 6.5°C in a thermostatic bath (model MQBTC99-20, Microquimica, Florianópolis, Brazil), corresponding to a mass of 40.4 pg Hg. The aliquot is injected into the system through the septum in the SI (Figure 1). The analyte is preconcentrated on a gold-coated sand trap, desorbed by heating, and the analytical signal corresponding to 40.4 pg Hg is then recorded.

Results and discussion
Solid sorbents

The mean absorption capacity ($n = 5$) for each solid and Duncan test for 5% significance level are shown in Table 1. The Duncan test shows that there is no significant difference in AC when using a pair of columns (A1 + A2) or only a platinum column (A2). Moreover, a similar performance is obtained using the column C1, although it shows an AC slightly below the previous ranges. The column containing gold particles (A1) provides an AC which is lower than A2, but its performance is comparable with C1. Although the coal used in column C2 is marketed specifically for removing gaseous mercury, this material had the lowest AC, i.e., it has the worst performance for the conditions tested. In summary, the results show that when using a platinum column, one can expect to remove 88.7% of the mercury, while using column C2 results in the removal of 50.1% of the mercury present.

Absorbent solutions

The ability of a hypochlorite solution to remove gaseous mercury is evaluated by measuring the AC for different sodium hypochlorite solutions (0.42, 0.84, 1.26, 1.68, and 2.10 mmol L^{-1}). The results are shown in Figure 2, and a direct relationship between AC and hypochlorite concentration, up to 1.68 mmol L^{-1}, is observed; after that, a

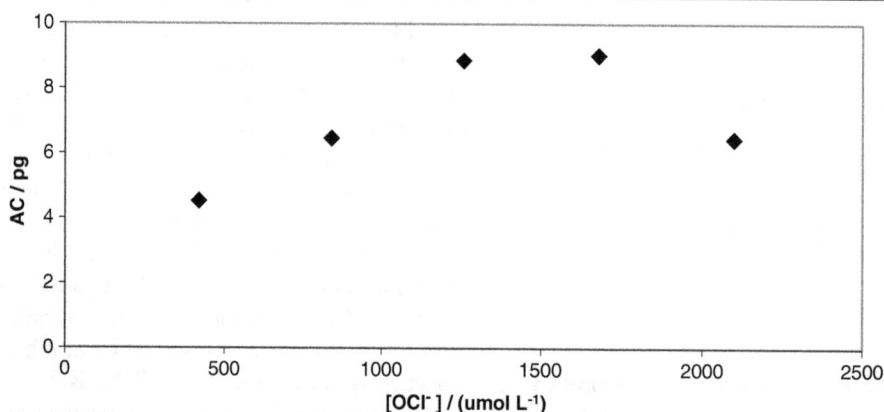

Figure 2 Influence of hypochlorite on absorption capacity. Variation of the AC of gaseous Hg due to the increase of the concentration of the aqueous solution of sodium hypochlorite [OCl$^-$].

Table 2 Results of ANOVA

Causes of variation	df	SS	MS	F
OCl^-	1	38,286.4	38,286.4	1,380.0
EDTA	1	2,668.1	2,668.1	96.2
Cl^-	1	379.0	379.0	13.7
$OCl^- \times$ EDTA	1	4,873.7	4,873.7	175.7
$OCl^- \times Cl^-$	1	1,401.7	1,401.7	50.5
EDTA $\times Cl^-$	1	3,407.5	3,407.5	122.8
$OCl^- \times$ EDTA $\times Cl^-$	1	1,746.6	1,746.6	63.0
Treatment	7	52,763.0	7,537.6	271.7
Residual	16	44.9	27.7	
Total	23	53,206.9		

Value F for a significance level of 5% = 2.66. *df*, degrees of freedom; *SS*, sum of squares; *MS*, mean square; *F*, Fisher F test value.

reduction in AC is observed. The distribution of the concentrations of OCl^-, HOCl, and Cl_2 essentially depends on the pH and the chloride concentration (see reactions below). Since in this experiment no buffer solution is used, then the increased concentration of hypochlorite causes the pH to increase and a decrease of the chlorine concentration. Thus, for hypochlorite concentrations higher than 2.10 mmol L^{-1}, an effective reduction of chlorine concentration must have taken place, leading to the consequent reduction of AC, as seen in Figure 2.

A study shows that the removal of gaseous mercury for the hypochlorite solution depends mainly on the presence of Cl_2 (Zhao and Rochelle 1999). However, the concentration of chlorine in the hypochlorite solution is defined by the following equilibria:

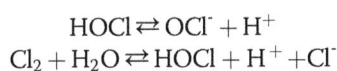

$$HOCl \rightleftarrows OCl^- + H^+$$
$$Cl_2 + H_2O \rightleftarrows HOCl + H^+ + Cl^-$$

The mutual influence of the concentrations of hypochlorite and chloride on mercury removal is experimentally investigated by a 2^3 factorial design with the following factors: concentration of hypochlorite, chloride, and EDTA, investigated at two levels of concentration.

The results in Figure 2 are used to define the hypochlorite concentrations to be tested, being chosen as 0.63 and 1.26 mmol L^{-1}. The highest hypochlorite concentration is chosen based on the maximum AC seen in Figure 2, and the lowest concentration is chosen as 50% of that value.

Zhao and Rochelle (1999) have demonstrated that, when using 1.0 mol L^{-1} NaCl in the presence of sodium hypochlorite, the concentration of mercury in the gas phase is stabilized for their experiments. However, when using 0.1 mol L^{-1} NaCl, those authors proved that the mercury concentration gradually increased in the gas phase (Zhao and Rochelle 1999). Thus, to perform the tests, 0.48 and 0.96 mol L^{-1} chloride concentrations are chosen.

It is assumed that the use of EDTA can increase AC because EDTA ensures the stabilization of Hg^{2+} formed by redox reaction, since the logKest HgEDTA ranges from 21.5 to 23.5. The selected concentrations are 0.30 and 0.60 mmol L^{-1} to perform the tests, ensuring large excess of this reagent with respect to the concentration of Hg^{2+}.

Table 2 shows the results of analysis of variance, and it can be seen that the F values for the main factors (OCl^-, Cl^-, and EDTA) are significant at the 5% level, but the interactions between these factors are also significant in the same level. So, it is not recommended to compare directly the means of the factors (Vieira and Hoffmann 1989). For this reason, the influence of OCl^- and Cl^- inside the EDTA level is evaluated, which is shown in Figure 3a,b. It is clear that increasing the hypochlorite concentration also increases AC regardless of the concentrations of chloride or EDTA. Increasing the chloride concentration will increase AC only for the lower concentration of EDTA (0.30 mmol L^{-1}).

Conclusions

The use of a column containing a platinum coil proves to be more suitable for removing gaseous mercury, among the solid sorbents tested. The use of a solution of 1.26 mmol L^{-1} OCl^-, 0.48 mol L^{-1} Cl^-, and 0.6 mmol L^{-1} EDTA has an absorption capacity of 167 pg for gaseous Hg. This solution can purify gases that shall be used to analyze mercury in environmental samples, but the higher the mercury concentration in the gas, the more frequent

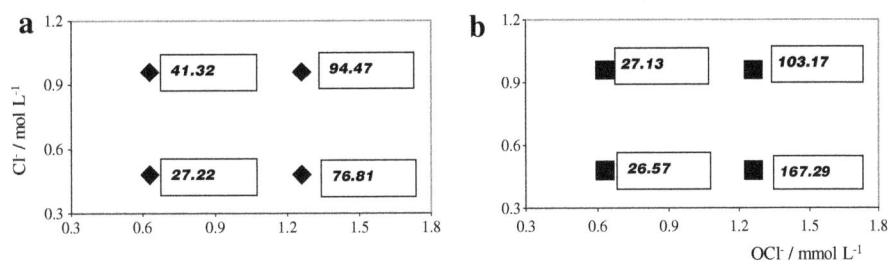

Figure 3 Representation of the interaction between hypochlorite, chloride and EDTA concentration. The interaction $OCl^- \times Cl^-$ for EDTA **(a)** 0.30 mmol L^{-1} (diamonds) and **(b)** 0.60 mmol L^{-1} (squares) is considered. The numerical values inside the graph represent the AC (pg Hg) for each condition tested.

the substitution of the solution. It also should be pointed out that it is a very easy way to remove the gaseous mercury and to get good blanks, even when a high-purity gas is not available.

Competing interests
The authors declare that they have no competing interests.

Authors' contributions
CMIT carried out the laboratorial experiments. LFHN drafted the manuscript. GF participated in the design of the study and performed the statistical analysis. MRM conceived and coordinated the study. All authors read and approved the final manuscript.

Author details
[1]Escola de Química e Alimentos, Federal University of Rio Grande – FURG, Av. Italia, km 8, 96201-900, Rio Grande, RS, Brazil. [2]Instituto de Oceanografia, Federal University of Rio Grande – FURG, Av. Italia, km 8, 96201-900, Rio Grande, RS, Brazil.

References
Borges AR, Niencheski LFH, Milani ICB, Milani MR (2012) Optimisation and application of the voltammetric technique for speciation of chromium in the Patos Lagoon Estuary-Brazil. Envir Moni Assess 184:5553–5562. doi:10.1007/s10661-011-2361-7

Emili A, Acquavita A, Koron N, Covelli S, Faganell J, Horvat M, Zizek S, Fajon V (2012) Benthic flux measurements of Hg species in a northern Adriatic lagoon environment (Marano and Grado Lagoon, Italy). Est Coast Schelf Sci 113:71–85. doi:10.1016/j.ecss.2012.05.018

Gardfeldt K, Horvat M, Sommar J, Kotnik J, Fajon V, Wangberg I, Lindqvist O (2002) Comparison of procedures for measurements of dissolved gaseous mercury in seawater performed on a Mediterranean cruise. Anal Bioanal Chem 374:1002–1008. doi:10.1007/s00216-002-1592-4

Hines ME, Poitras EM, Covelli S, Faganeli J, Emili A, Zizek S, Horvat M (2012) Mercury methylation and demethylation in Hg-contaminated lagoon sediments (Marano and Grado Lagoon, Italy). Est Coast Schelf Sci 113:85–95. doi:10.1016/j.ecss.2011.12.021

Horvat M, Lian L, Bloom NS (1993) Comparison of distillation with other current isolation methods for the determination of methyl mercury compounds in low level environmental samples. Part II. Water Anal Chim Acta 282:153–168. doi:0003-2670/93/$06.00

Lagowski JJ (1995) Experiences in chemistry. McGraw-Hill, New York

Leermakers M, Baeyens W, Quevauviller P, Horvat M (2005) Mercury in environmental samples: speciation, artifacts and validation. Trends Anal Chem 24:383–393. doi:10.1016/j.trac.2004.01.001

Liang L, Horvat M, Danilchik P (1996) A novel analytical method for determination of picogram levels of total mercury in gasoline and other petroleum based products. Sci Total Environ 87:57–64. doi:0048-9697(96)05129-7

Marine Environmental Studies Laboratory - MESL (1997) International Atomic Energy Agency. Standard Operating Procedures, Monaco

Santos EJ, Hermann AB, Frescura VLA, Sturgeon RE, Curtius AJ (2008) A novel approach to cold vapor generation for the determination of mercury in biological samples. J Braz Chem Soc 19:929–934

Sizeneva IP, Val'tsifer VA, Strel'nikov VN (2005) A study of mercury dissolution in aqueous solutions of sodium hypochlorite. Russ J Appl Chem 78:546–548. doi:1070-4272/05/7804-0546

Stergarsek A, Horvat M, Frkal P, Guevara SR, Kocjancic R (2013) Removal of Hg0 in wet FGD by catalytic oxidation with air – a contribution to the development of a process chemical model. Fuel 107:183–191. doi:10.1016/j.fuel.2012.08.001

Vieira S, Hoffmann R (1989) Estatística Experimental. Atlas SA, São Paulo

Wotter SET, Niencheski LFH, Milani MR (2011) Chemical speciation and dissolved iron in the pore water of Patos Lagoon sediments - Brazil. P Electrochim Acta 29:155–163. doi:10.4152/pea.201103155

Zhao LL, Rochelle GT (1999) Mercury absorption in aqueous hypochlorite. Chem Eng Sci 54:655–662. doi:0009-2509(98)00263-2

Simultaneous determination of β-alanine betaine and trimethylamine in bacterial culture and plant samples by capillary electrophoresis

Isam A Mohamed Ahmed[1,2*], Ailijiang Maimaiti[2], Nobuhiro Mori[3], Norikazu Yamanaka[2] and Takeshi Taniguchi[2]

Abstract

Background: 3-*N*-trimethylaminopropionic acid (β-alanine betaine) and trimethylamine (TMA) are important nitrogenous compounds that perform fundamental roles in biological pathways throughout all kingdoms of life; however, yet their simultaneous determination method is hardly reported.

Methods: Capillary electrophoresis method for the simultaneous determination of TMA and β-alanine betaine in microbial culture and plant samples was developed. To increase the sensitivity, TMA and β-alanine betaine in the samples were first derivatized with bromophenacyl bromide and then analyzed by capillary electrophoresis under low pH.

Results: The derivatization was found to be practically useful for the elimination of interfering substances from plant and microbial extracts, as well as giving well resolved peaks for the analytes (β-alanine betaine esters and TMA salt). Analytical features of the developed method showed its respectable performance in terms of linearity ($r^2 > 0.99$), precision (relative standard deviation (RSD) < 5%), and detection limits (0.01 mM).

Conclusion: The developed method allows the quantitative determination of TMA and β-alanine betaine in complex biological samples and assists to study biosynthetic and degradation pathways of these important compounds.

Keywords: β-alanine betaine; Capillary electrophoresis; Microbial culture; Plant leaves; Trimethylamine

Background

In nature, many plants, bacteria, and marine algae accumulate quaternary ammonium compounds in response to various environmental stresses such as flooding, freezing, heating, drought, and salinity (Rhodes and Hanson 1993; Gorham 1995). These compounds form a structurally heterogeneous class of compounds with a unifying character of a polar and fully methyl substituted nitrogen atom, creating a permanent positive charge on the N moiety (Rhodes and Hanson 1993). Of them, glycine betaine, 3-*N*-trimethylaminopropionic acid (β-alanine betaine), and proline betaine are known to be the most effective osmoprotectants and are widely distributed in the biosphere (Yancey 2005). Although glycine betaine is extensively accumulated in many plants in response to various environmental stresses, various members of the highly stress-tolerant plant family Plumbaginaceae accumulate β-alanine betaine instead of glycine betaine (Hanson et al. 1994). β-Alanine betaine synthesis is not controlled by choline availability, because it is derived from β-alanine by three-step methylation (Rathinasabapathi et al. 2000). Distinct from glycine betaine synthesis, β-alanine betaine synthesis does not require oxygen, and therefore, it was proposed to be suitable for osmoprotection under saline and hypoxic conditions (Hanson et al. 1994; Rathinasabapathi et al. 2000). Consequently, β-Alanine betaine appears to be effective over a broader ecological spectrum than glycine betaine (Rhodes and Hanson 1993).

In soil microorganisms, our recent reports revealed that β-alanine betaine was accumulated as an intermediate metabolite in the degradation pathway of homocholine by members of the genera *Arthrobacter*, *Rhodococcus*, and *Pseudomonas* (Mohamed Ahmed et al. 2009a, b; Mohamed

* Correspondence: isamnawa@yahoo.com
[1]Department of Food Science and Technology, Faculty of Agriculture, University of Khartoum, Shambat 13314, Sudan
[2]Arid Land Research Center, Tottori University, 1390 Hamasaka, Tottori 680-0001, Japan
Full list of author information is available at the end of the article

Ahmed et al. 2010; Mohamed Ahmed et al. 2014). The potential role of β-alanine betaine in plants' and microorganisms' tolerance to salinity and hypoxia makes its synthetic pathway an interesting target for metabolic engineering. However, the estimation methods of this interesting metabolite are still scarce.

Trimethylamine (TMA) is a volatile low molecular weight tertiary aliphatic amine that has been recognized widely in many animal and plant tissues and is one of the degradation products of nitrogenous organic material such as quaternary ammonium compounds such as choline and homocholine (Craciun and Balskus 2012; Mohamed Ahmed et al. 2010). Commonly, the amount of TMA is a useful indicator of spoilage in fresh and lightly preserved seafood as it increases during the breakdown of seafood, such as fish and shrimp (Dalgaard 2006; Ghaly et al. 2010). In medical diagnosis, an increase in the concentration of TMA in the breath of patients can be used as a sign of viremic disease (Siminhoff et al. 1977). Therefore, detection of TMA is of high interests in many fields such environmental protection, food industry, and medical diagnosis. However, one of the challenging aspects of the analysis of β-alanine betaine and TMA lies in their lack of useful chromospheres, and their chemical structures have permanently charged groups that prevent gas chromatographic separation in their intact forms. In the past, analyses of β-alanine betaine and TMA relied on qualitative or semi-quantitative colorimetric tests that employed either thin-layer chromatography and Dragendorff's reagent or reaction with picric acid to form colored complex (Blunden et al. 1981; Grieve and Grattan 1983). Since these methods are limited in their sensitivity, selectivity and quantitative accuracy, and ability to assay betaines and TMA in one sample, knowledge of the identities and absolute concentrations of β-alanine betaine and TMA in biological materials remained inadequate. Recently, capillary electrophoresis has been applied in many different fields because of its extremely high resolution, its speed, and its applicability to a wide range of molecules whether they are charged or uncharged, or of low or high molecular weight (Shintani and Polonsky 1997). In the present work, capillary electrophoresis method under low pH (Nishimura et al. 2001; Zhang et al. 2002) was effectively improved for simultaneous determination of β-alanine betaine and TMA in both plant leaves and microbial culture samples.

Methods
Materials
β-alanine betaine was synthesized by *N*-methylation of dimethylaminopropionic acid (Tokyo Kasei Kogyo Co. Ltd, Tokyo, Japan) with methyl iodide as described previously (Mohamed Ahmed et al. 2010). Briefly, 4 ml of methyl iodide was added to a suspension of dimethylaminopropionic

acid (1 g, 6.5 mM) and $KHCO_3$ (1.3 g, 13 mM) in 20 ml of methanol. The mixture was stirred overnight at room temperature and then decanted. Thereafter, the liquid phase was concentrated, and the residue was extracted using 15 ml of mixed solvent (acetonitrile/methanol = 10:1, *v/v*). The combined extracts were dried under a nitrogen stream to give β-alanine betaine as a colorless powder (1.2 g, 63.2%). The structure and purity of β-alanine betaine were confirmed using proton nuclear magnetic resonance (^{1}H NMR) and capillary electrophoresis. Unless otherwise specified, all other reagents were of analytical grade and were from either Wako (Wako Pure Chemical Industries Ltd, Tokyo, Japan) or Sigma (St. Louis, MO, USA).

Extract preparation from microbial samples
Homocholine-degrading strains were isolated from the soil samples obtained from different locations at Tottori University and around Tottori City, Japan. The bacterial strains were cultivated for 24 h at 30°C on 75 ml of basal homocholine liquid media containing 20 mM homocholine as a sole source of carbon, nitrogen, and energy. The cells were harvested at the exponential phase by centrifugation at 10,000 × *g* for 20 min at 4°C. The supernatant was collected and preserved at −20°C until used for detection of β-alanine betaine and TMA. The harvested bacterial cells were washed three times with saline solution (8.5 g/l KCl), and re-suspended in 50 mM potassium phosphate buffer (pH 7.5). The resting cell reaction was started by the addition of homocholine (20 mM) to the cell suspension. The suspension was incubated on a shaker at 120 rpm and 30°C. At appropriate time intervals (30 min, 1 h, 2 h, 3 h, and 6 h), aliquots of the cell suspension were withdrawn and boiled for 3 to 5 min to stop the reaction. These extracts were preserved at −20°C until used for sample derivatization.

Extract preparation from plant samples
Plant (*Limonium suffruticosum*, *Phragmites australis*, and *Elaeagnus oxycarpa*) leaf samples, at productive stage, were collected from an area around Aiding Lake in the Turpan Basin, Xinjiang, China, in August 2010. The area of the study site is about 10,000 m^2 (100 m × 100 m), and three plots (10 m × 10 m) were established randomly. The samples were collected from five plants of each species and carefully washed with water. The samples were dried in oven at 85°C for 48 h, ground to fine power, and then brought to Arid Land Research Center, Tottori University, Japan, for analysis. For extract preparation, about 100 mg of powdered samples were added to 1.5 ml water, mixed in a plastic tube, incubated at 75°C for 20 min, and then centrifuged at 15,000 × *g* for 10 min. These samples were preserved at −20°C until used for sample derivatization.

Simultaneous determination of β-alanine betaine and trimethylamine in bacterial culture and plant samples by...

51

Sample derivatization

One of the challenging aspects of analysis of β-alanine betaine and TMA lies in their lack of useful chromophores and thus could not be detected in ultraviolet-visible (UV/vis) light range. To overcome this limitation, the samples were derivatized with 4-bromphencyl bromide before analysis with capillary electrophoresis. Esterification was carried out following the methods of Nishimura et al. (2001) with some modifications. Briefly, 0.1 ml of the sample extract and/or authentic standards of β-alanine betaine and TMA were placed in a microtube and mixed with 0.05 ml of buffer solution (100 mM KH_2PO_4/distilled water/acetonitrile = 1:1:4). To the mixture, 0.3 ml of 4-bromophenacyl bromide (20 mg/ml in acetonitrile) was added. The tube was capped and heated at 90°C for 90 min. The reaction mixture was evaporated to dryness with a centrifugal evaporator (CVE-200D; Tokyo Rikakikai, Tokyo, Japan). The residue was dissolved in 300 μl of 50 mM sodium phosphate buffer (pH 3.0), mixed well, and centrifuged at 10,000 × g for 20 min at 4°C. The supernatants, which contained ester and salt of the metabolites β-alanine betaine and TMA, were filtered using 45-μm filter (Millex Millipore, Billerica, MA, USA) to remove the micro-particles that might block the flow through the capillary tube. The filtered samples were then analyzed by capillary electrophoresis.

Capillary electrophoresis analysis

Capillary electrophoresis analysis was conducted using a capillary electrophoresis system model Photal CAPI-3300 (Otsuka Electronics. Co. Ltd., Osaka, Japan) equipped with a fused silica capillary of 75-μm i.d. with a total length of 80 cm (effective length of 68 cm). Before starting the analysis, the capillary was conditioned with 0.1 M NaOH for 5 min followed by conditioning with distilled water for 3 min and electrolyte buffer for 3 min (50 mM sodium phosphate buffer, pH 3.0). Between each run, the capillary was flushed with distilled water (1 min) and electrolyte buffer (3 min). The temperature of the capillary was set at 25°C and then the samples and/or the authentic standards (β-alanine betaine and TMA) were injected hydrostatically (25 mm, 60 s). During the run and in order to avoid sample carry-over into the electrophoresis buffer, the capillary was dipped twice in distilled water and washing buffer (same electrophoresis buffer that set in other tubes). The applied potential was 20 kV, and the peaks of TMA-salt and β-alanine betaine-ester were monitored at 262 nm.

Statistical analyses

Statistical analyses were performed with the SPSS v. 18.0 software (SPSS Inc., Chicago, IL, USA). One-factor ANOVA was performed to identify statistically significant differences among treatments, followed by Tukey's HSD test ($P \leq 0.05$).

Results and discussion

Sample derivatization

In the literature, it has been established that 4-bromophenacyl bromide reagent reacts with quaternary ammonium compounds and can accurately be used for their quantification (Gorham et al. 1982). We have adapted and modified this derivatization method for the determination of β-alanine betaine and TMA in plant and microbial samples by capillary electrophoresis. Since β-alanine betaine and TMA lack of useful chromophores that lead to the inability for detection at the UV/vis range, in the current work, they were derivatized with 4-bromophenacyl bromide to form β-alanine betaine-ester and TMA-salt (Figure 1). These reaction products showed a maximum absorption at 262 nm, which was within the range 214 to 266 nm reported previously for various betaine esters (Gorham et al. 1982; Zhang et al. 2002). Determination of betaine as the 4-bromophenacyl ester has previously been reported to be exceptionally sensitive and specific (Gorham et al. 1982). Generally, absolute acetonitrile is usually used as solvent of 4-bromophenacyl for reaction with betaines (Gorham et al. 1982) giving a rapid esterification reaction at neutral or slightly alkaline pH levels. Nevertheless, in the current work, the reaction was carried out in less than 80% acetonitrile solution that gave an average yield of 4-bromophenacyl esters and salts of more than 70%. To avoid the chemical decomposition of β-alanine betaine through C-N bond cleavage that may lead to the production of trimethylamine and acrylate under alkaline condition (Gorham et al. 1982; Zhang et al. 2002), the derivatization condition was optimized and carried out at slightly acidic conditions (pH 5.6) in the reaction mixture that contain potassium dihydrogen phosphate/distilled water/acetonitrile (1:1:4 v/v). Interestingly, these conditions resulted in efficient detection and quantification of β-alanine betaine in its intact form without decomposition as shown from the derivatized authentic standard of β-alanine betaine (Figure 2). In the meanwhile, TMA was also effectively detected and quantified after derivatization under the same conditions. Thus, the derivatization protocol of the current study could strikingly be used for simultaneous determination of both β-alanine betaine and TMA in plant, food, and microbial samples. In many previous reports, various betaines were esterified with 4-bromophenacyl bromide in the presence of a potassium bicarbonate/potassium dihydrogen phosphate/acetonitrile (1:1:4 v/v) (Zhang et al. 1997; Nishimura et al. 2001; Zhang et al. 2002). Under these conditions, peak of trimethylamine salt derived from the breakdown of β-alanine betaine was observed in the capillary electropherogram as confirmed by using authentic trimethylamine (Nishimura et al. 2001; Zhang et al. 2002). During derivatization with 4-bromophenacyl bromide, the

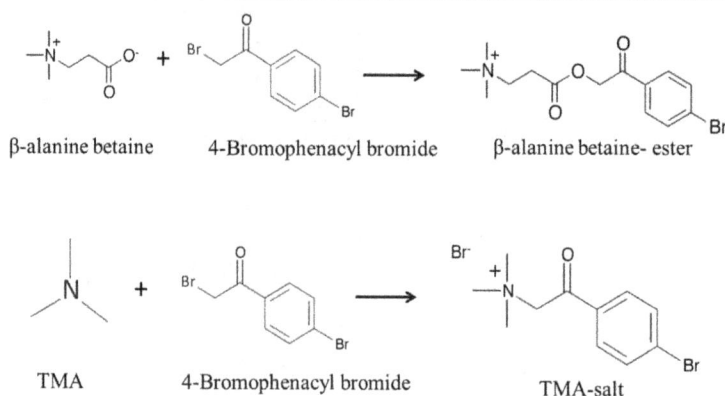

Figure 1 Reaction of β-alanine betaine and trimethylamine (TMA) with 4-bromophenacyl bromide and formation of β-alanine betaine-ester and TMA-salt.

compound dimethylsulfoniopropionate (DMSP), which structurally resembles β-alanine betaine, was also degraded to dimethylsulfide (DMS) and acrylate under an alkaline condition (Gorham et al. 1982; Zhang et al. 2005). Zhang et al. (2005) also reported that they kept the decomposition of DMSP as minimum by lowering the pH of the esterification reaction mixture to 4.0 by using 18-Crown-6 instead of potassium carbonate. In the current study, the decomposition of β-alanine betaine and consequently the production of the trimethylamine were completely avoided by omitting potassium bicarbonate from the derivatization buffer.

Repeatability, linearity, and detection limit of TMA and β-alanine betaine

The repeatability of the proposed method was explored by five consecutive runs of separate authentic standards (2.5 mM) of TMA and β-alanine betaine. The relative standard deviation (RSD) for migration time of TMA and β-alanine betaine was 0.09% and 0.16%, respectively.

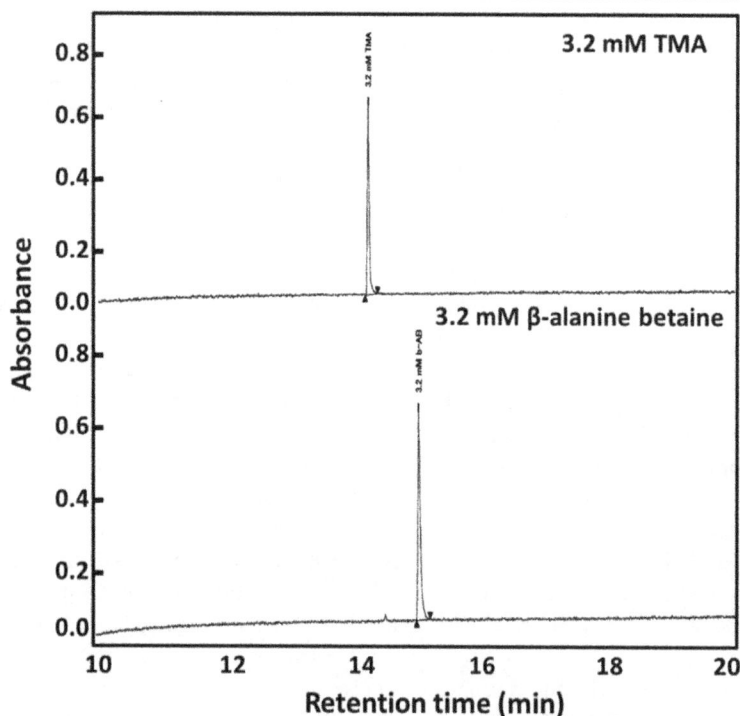

Figure 2 Authentic standard of TMA salt and β-alanine betaine ester detected by capillary electrophoresis. About 3.2 mM of β-alanine betaine and TMA were esterified with 4-bromophenacyl bromide as described in the 'Methods' section. The samples were injected hydrostatically (25 mm, 60 s) at 25°C. The applied potential was 20 kV, and the peaks were monitored at 262 nm.

While, the RSD for peak area of the analytes TMA and β-alanine betaine was 4.24% and 2.57%, respectively. These values were in general agreement with those reported previously for TMA and glycine betaine when analyzed by capillary electrophoresis (Timm and Jørgensen 2002; Zhang et al. 2002). However, lower percentage of RSD of glycine betaine, choline, and TMA analyzed by ion exchange chromatography with non-suppressed conductivity detection method was recently reported (Zhang and Zhu 2007). To determine the day-to-day repeatability of the analysis, the samples were analyzed by capillary electrophoresis for three consecutive days, performed as sequence of five runs each day. The RSD of the migration time was 2.45% and 1.96% for TMA and β-alanine betaine, respectively, whereas that of the peak area was 3.49% and 3.74% for TMA and β-alanine betaine, respectively. In their analysis of day-to-day repeatability of capillary electrophoresis method for estimation of various amines, Timm and Jørgensen (2002) reported a RSD of less than 3% for migration time and less than 10% for

peak area. They stated that these figures are acceptable in many applications; however, the percentage of RSD could be improved by including an internal standard within the samples. Linearity was investigated using the stock solution containing either TMA or β-alanine betaine, which was serially diluted. Then, eight concentrations (0.05 to 32 mM) of the analytes were analyzed in triplicate, and the calibration curves were constructed by plotting the peak area versus the concentration (mM) of each analyte. The results showed that the suggested procedure produced highly linear calibration curves (Figure 3) with the correlation coefficients of 0.9933 and 0.9997 for TMA and β-alanine betaine, respectively. The linearity range of β-alanine betaine in the current study is greater than the range 0.05 to 5.0 mM reported previously for many betaines analyzed by capillary electrophoresis (Zhang et al. 2002). Moreover, the linearity range of TMA in the current study is also greater than the range 0.25 to 10.0 µg/ml reported previously for the analysis of TMA in water samples by liquid chromatography

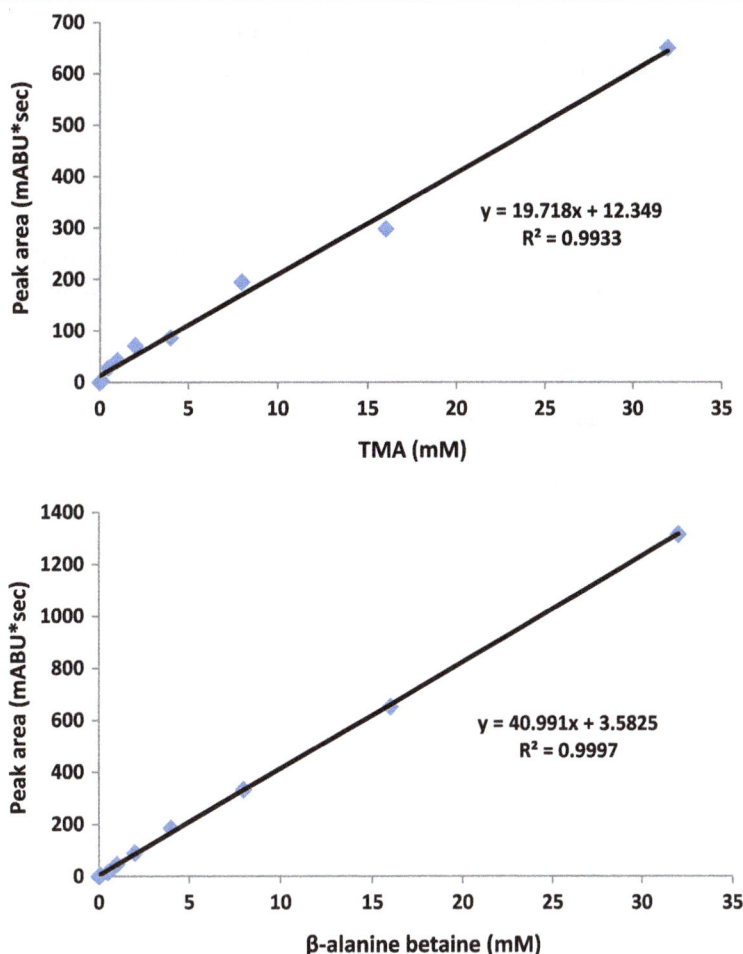

Figure 3 Linearity curves of β-alanine betaine and trimethylamine as estimated by the developed capillary electrophoresis method. Trimethylamine (upper panel); β-alanine betaine (lower panel).

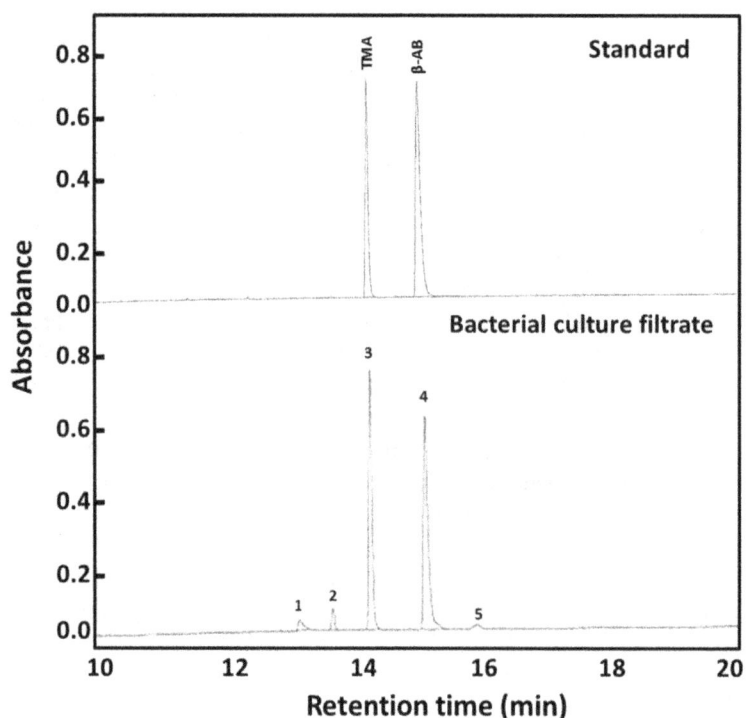

Figure 4 Capillary electrophoresis chromatogram of the intermediate metabolites of homocholine biodegradation by soil microorganism. The culture filtrate samples were esterified with 4-bromophenacyl bromide as described in the 'Methods' section. The samples were injected hydrostatically (25 mm, 60 s) at 25°C. The applied potential was 20 kV, and the peaks were monitored at 262 nm. The chromatogram showed the generated metabolite peaks from the degraded homocholine by the isolated strains alongside with authentic standard peaks of trimethylamine (TMA) and β-alanine betaine (β-AB).

(Chafer-Pericas et al. 2004). The good linearity range of the current methods is a good indication for its applicability to accurately estimate these analytes in various samples including plants, microbial, food, and clinical samples. The limit of detection was calculated as the concentration that produced a signal-to-noise ratio of 3 and was estimated by analyzing solution of decreasing concentration of TMA and β-alanine betaine until this

Table 1 Trimethylamine and β-alanine betaine content in bacterial culture and plant samples

Samples	β-alanine betaine	Trimethylamine
Bacterial culture (mmol/l)		
Arthrobacter sp. strain E5	5.52	9.22
Pseudomonas sp. strain A9	4.62	41.48
Rhodococcus sp. strain A2	3.89	30.93
Rhodococcus sp. strain A4	4.38	23.96
Plant (μmol/g DW)		
Limonium suffruticosum	65.97	28.93
Phragmites australis	64.44	26.96
Elaeagnus oxycarpa (200 mM NaCl)	16.29	n.d.

DW, dry weight; n.d., not detected.

ratio was observed. The limit of detection of both compounds was found to be 0.01 mM (100 μM). Similarly, the detection limit of glycine betaine estimated by low pH capillary electrophoresis method was reported to be 0.01 mM (Nishimura et al. 2001; Zhang et al. 2002). Moreover, slightly lower detection limit (0.005 mM) for DMSP-ester by low pH capillary electrophoresis has also been reported (Zhang et al. 2005). On the other hand, a lower detection limit (50 ng/ml) of TMA in water samples analyzed by liquid chromatography has also been reported (Chafer-Pericas et al. 2004). However, the detection limit of many amines including trimethylamine as analyzed by capillary electrophoresis with indirect UV-detection mode was reported to be 0.01 mM (Timm and Jørgensen 2002). Generally, the detection limits of both compounds by the proposed method in the current study well agreed with those of the previously reported detection methods for individual compounds. Strikingly, in the current method, both compounds could simultaneously be determined even if they exist in relatively lower concentrations. Although, the proposed method may be suitable for most applications concerning the determination of TMA and β-alanine betaine in plants, microbial, environmental, and food samples, methods with extremely lower detection limits for these analytes might still be required.

Application of the method to microbial and plant samples

To illustrate an application of the developed method, three samples of either microbial cultures or dry cell reaction mixtures or plant leaves were prepared as described in the 'Methods' section. The intermediate metabolites in the microbial culture filtrate of different microorganisms grown on homocholine as the sole source of carbon and nitrogen was analyzed by the developed method. During the consumption of homocholine by the growing cell cultures of the isolated strains of the genera *Arthrobacter*, *Pseudomonas*, and *Rhodococcus* (Mohamed Ahmed et al. 2009a, b; Mohamed Ahmed et al. 2010), there were a concurrent formation and accumulation of some soluble metabolites identified as trimethylamine (peak 3, TMA) and β-alanine betaine (peak 4, β-AB) as detected by capillary electrophoresis method (Figure 4). Under the optimized derivatization conditions of the developed method, the TMA-salt and β-alanine betaine-ester were successfully separated and detected with very clear and sharp peaks using a UV detector at 262 nm. The amounts of these metabolites were successfully estimated (Table 1) by calculating the area of each peak. It can be seen that the quantity

of TMA in all microbial culture samples was higher compared to the quantity of β-alanine betaine. This is an ideal phenomenon because these strains were observed to cleave C/N pond of β-alanine betaine and rapidly use the resulted carbon chain as source of carbon with the release of TMA as a major metabolite (Mohamed Ahmed et al. 2009a, b; Mohamed Ahmed et al. 2010; Mohamed Ahmed et al. 2014). To check the accuracy of this method, the TMA concentration in the culture filtrate and intact cell reaction products of the isolated strains was analyzed by using picric acid-based colorimetric method (Dyer 1945). The results showed no significant differences from that obtained by the developed capillary electrophoresis method. To expand the applicability of the developed method, the plant leaves that were obtained from Xinjiang, China, in August 2010 were also prepared and analyzed by this method as described in the 'Methods' section. The results again showed clear peaks of TMA and β-alanine betaine in the leaf samples of many of plants growing on saline soil in this area (Figure 5). These osmotolerant plants accumulated sufficient amounts of nitrogenous compounds to cope with these osmotic

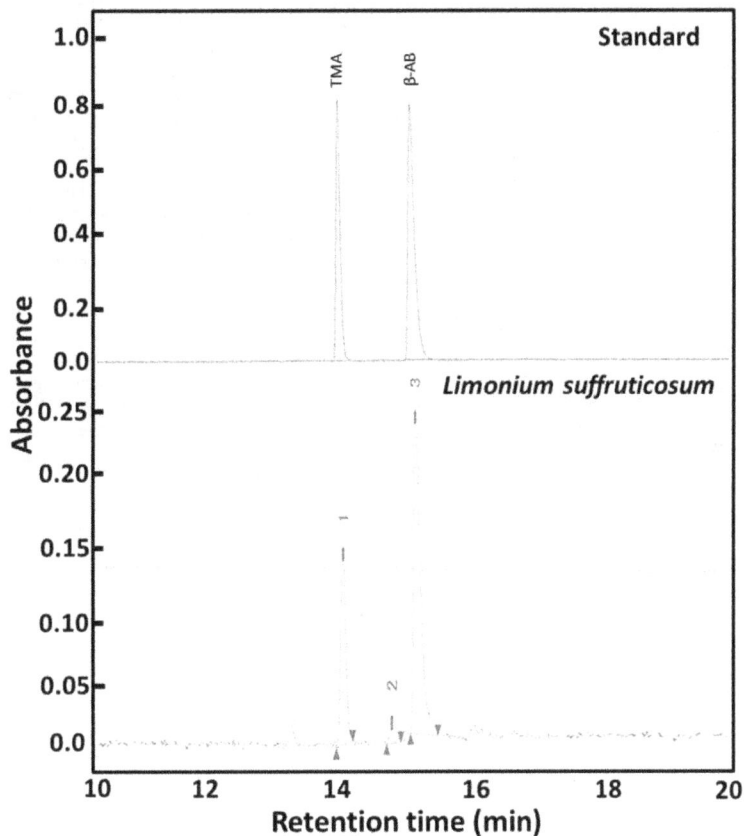

Figure 5 Capillary electrophoresis chromatogram of the osmolytes in the leaves of *Limonium suffruticosum* grown in saline soil.
Plant leaf samples were esterified with 4-bromophenacyl bromide as described in the 'Methods' section. The samples were injected hydrostatically (25 mm, 60 s) at 25°C. The applied potential was 20 kV, and the peaks were monitored at 262 nm. The chromatogram showed the accumulated osmolytes in the plant samples alongside with authentic standard peaks of trimethylamine (TMA) and β-alanine betaine (β-AB).

stresses. Of these nitrogenous compounds, considerable amounts of both TMA and β-alanine betaine were quantitatively estimated using the developed method (Table 1). Similarly, the concentration of β-alanine betaine in many plants of the family Plumbaginaceae was found in the range of 1 to 147 µg/g DW, which was estimated by either TLC and autoradiography (Rathinasabapathi et al. 2000) or ^{1}H NMR (Baysalfurtana et al. 2013). Collectively, the above findings clearly demonstrated the suitability of the developed method for the simultaneous detection and quantification of these analytes in microbial culture and plant samples. Although it is not tested in the current study, the developed method could efficiently be used for the estimation of TMA and β-alanine betaine in both food and feed samples.

Conclusion

A capillary electrophoresis method for the simultaneous determination of TMA and β-alanine betaine was developed. The method described here has generally wide detection range suitable for analysis of TMA and β-alanine betaine in microbial and plant samples. The advantages of the current method are its low cost, low detection limit, simple operation, rapid, and high sensitivity.

Competing interests
The authors declare that they have no competing interests.

Authors' contributions
IAMA has performed most of the experimental and analytical work and prepared the draft of the manuscript. AM has performed the plant-related parts of the experimental work with the direct help from IAMA. The guidelines and supervision of this work was provided by NM, NY, and TT. All authors read and approved the final manuscript.

Acknowledgements
Financial assistance from the Ministry of Education, Culture, Sports, Science, and Technology of Japan in the form of a scholarship for the first and second authors is gratefully acknowledged.

Author details
[1]Department of Food Science and Technology, Faculty of Agriculture, University of Khartoum, Shambat 13314, Sudan. [2]Arid Land Research Center, Tottori University, 1390 Hamasaka, Tottori 680-0001, Japan. [3]School of Agricultural, Biological, and Environmental Sciences, Faculty of Agriculture, Tottori University, Koyama, Tottori 680-8553, Japan.

References
Baysalfurtana G, Duman H, Tipirdamaz R (2013) Seasonal changes of inorganic and organic osmolyte content in three endemic Limonium species of Lake Tuz (Turkey). Turk J Bot 37:455–463

Blunden G, El Barouni MM, Gordon SM, McLean WFH, Rogers DJ (1981) Extractions, purification and characterization of Dragendorff-positive compounds from some British marine algae. Botanica Marina 24:451–456

Chafer-Pericas C, Herraez-Hernandez R, Campins-Falco P (2004) Liquid chromatographic determination of trimethylamine in water. J Chromatogr A 1023:27–31

Craciun S, Balskus EP (2012) Microbial conversion of choline to trimethylamine requires a glycyl radical enzyme. Proc Natl Acad Sci USA 109:21307–21312

Dalgaard P (2006) Microbiology of marine muscle foods. In: Hui YH (ed) Handbook of food science, technology and engineering. CRC, Boca Raton, pp. 53-1–53-20

Dyer WJ (1945) Amines in fish muscles 1. Colorimetric determination of trimethylamine as the picrate salt. J Fish Res Board Can 6(5):351–358

Ghaly AE, Dave D, Budge S, Brooks MS (2010) Fish spoilage mechanisms and preservation techniques: review. Amer J Appl Sci 7:859–877

Gorham J (1995) Betaines in higher plants—biosynthesis and role in stress metabolisms. In: Wallsgrove RM (ed) Amino acids and their derivatives in higher plants. Cambridge University Press, Cambridge, England, pp 173–203

Gorham J, McDonnell E, Wyn Jones RG (1982) Determination of betaines as ultraviolet-absorbing esters. Anal Chim Acta 138:277–283

Grieve CM, Grattan SR (1983) Rapid assay for determination of water soluble quaternary ammonium compounds. Plant Soil 70:303–307

Hanson AD, Rathinasabapathi B, Rivoal J, Burnet M, Dillon MO, Gage DA (1994) Osmoprotective compounds in the Plumbaginaceae: a natural experiment in metabolic engineering of stress tolerance. Proc Natl Acad Sci USA 91:306–310

Mohamed Ahmed IA, Jiro A, Ichiyanagi T, Sakuno E, Mori N (2009a) Isolation and characterization of 3-N-trimethylamino-1-propanol degrading Rhodococcus sp. strain A2. FEMS Microbiol Lett 296(2):219–225

Mohamed Ahmed IA, Jiro A, Ichiyanagi T, Sakuno E, Mori N (2009b) Isolation and characterization of 3-N-trimethylamino-1-propanol degrading Arthrobacter sp. strain E5. Res J Microbiol 4(2):49–58

Mohamed Ahmed IA, Jiro A, Ichiyanagi T, Sakuno E, Mori N (2010) Isolation and characterization of homocholine degrading Pseudomonas sp. strain A9 and B9b. World J Microbiol Biotechnol 26(8):1455–1464

Mohamed Ahmed IA, Eltayeb MM, Arima J, Mori N, Yamanaka N, Taniguchi T (2014) Screening for enzymatic activities in the degradation pathway of homocholine by soil microorganisms. Australian J Bas Appl Sci 8(2):222–233

Nishimura N, Zhang J, Abo M, Okubo A, Yamazaki S (2001) Application of capillary electrophoresis to the simultaneous determination of betaines in plants. Anal Sci J 17:103–106

Rathinasabapathi B, Sigua C, Ho J, Gage DA (2000) Osmoprotectant β-alanine betaine synthesis in the Plumbaginaceae: S-adenosyl-L-methionine dependent N methylation of β-alanine to its betaine via N-methyl and N, N-dimethyl β-alanines. Physiol Plant 109:225–231

Rhodes D, Hanson AD (1993) Quaternary ammonium and tertiary sulfonium compounds in higher plants. Annu Rev Plant Physiol Plant Mol Biol 44:357–384

Shintani H, Polonsky J (1997) Handbook of capillary electrophoresis applications. Blackie Academic & Professional, London

Siminhoff ML, Burke JF, Saukkonen JJ, Ordinario AT, Doty R, Dunn S (1977) Biochemical profile of uremic breath. The New Engl J Med 297:132–135

Timm M, Jørgensen BM (2002) Simultaneous determination of ammonia, dimethylamine, trimethylamine and trimethylamine-N-oxide in fish extracts by capillary electrophoresis with indirect UV-detection. Food Chem 76(4):509–518

Yancey PH (2005) Organic osmolytes as compatible, metabolic and counteracting cytoprotectants in high osmolarity and other stresses. J Exp Biol 208:2819–2830

Zhang J, Zhu Y (2007) Determination of betaine, choline and trimethylamine in feed additive by ion-exchange liquid chromatography/non-suppressed conductivity detection. J Chromatogr A 1170:114–117

Zhang J, Okubo A, Yamazaki S (1997) Determination of betaines in plants by low-pH capillary electrophoresis as their phenacyl esters. Bunseki Kagaku (in Japanese) 46:275–279

Zhang J, Nishimura N, Okubo A, Yamazaki S (2002) Development of an analytical method for the determination of betaines in higher plants by capillary electrophoresis at low pH. Phytochem Anal 13:189–194

Zhang J, Nagahama T, Abo M, Okubo A, Yamazaki S (2005) Capillary electrophoretic analysis of dimethylsulfoniopropionate in sugarcane and marine algal extracts. Talanta 66:244–248

9

Identification of a *postmortem* redistribution factor (*F*) for forensic toxicology

Iain M McIntyre

Abstract

Background: *Postmortem* redistribution (PMR) refers to the changes that may occur in drug concentrations after death. Consequently, *postmortem* concentrations in blood may not always replicate the *antemortem* drug levels. Literature supports the model describing drugs with a liver (L) concentration to peripheral blood (P) concentration ratio less than 5 (L/kg) being prone to little or no PMR. Conversely, drugs with a L/P ratio greater than 20 to 30 (L/kg) have propensity for substantial PMR.

Findings: Expanding upon this prior work, the current paper presents the concept of a *postmortem* redistribution factor (*F*) for a drug, which characterizes the direct relationship between *postmortem* peripheral blood and the corresponding *antemortem* whole blood concentration.

Conclusions: Development of the concept of a "*postmortem* redistribution factor" will provide a more definitive and authoritative drug ranking, and possibly, numerical interpretation of PMR for forensic toxicologists.

Keywords: *Postmortem* redistribution factor; Peripheral blood; Liver; *Antemortem*; Concentration; Ratio

Findings

Introduction

A potentially significant issue complicating interpretation of *postmortem* drug concentrations results from the phenomenon referred to as *postmortem* redistribution (PMR). *Postmortem* drug concentrations in the blood may not always straightforwardly parallel *antemortem* drug concentrations in the blood due to the movement of the drugs after death. Accordingly, some authors have argued a cautious approach in interpreting *postmortem* concentrations, and others have taken a far more pessimistic and even cynical perspective. The mechanisms involved in PMR are both complicated and poorly understood. However, *postmortem* drug concentrations in the blood may follow some commonly accepted trends that aid with interpretation. Generally, the characteristics of the drug itself can be used to predict if a drug is subject to PMR. Substantial changes in blood drug concentrations are predicted for basic, lipophilic drugs with a high volume of distribution (>3 L/kg) (Prouty & Anderson 1990). When PMR occurs, blood specimens drawn from the central body cavity and heart generally exhibit higher drug concentrations *postmortem* than specimens drawn from peripheral areas, most commonly the femoral region. Diffusion of drugs from organ tissues into the blood may explain the observed phenomenon.

Previous attempts to assess and account for PMR have utilized *postmortem* blood specimens collected from at least two areas of the body at autopsy, a peripheral area and a central area (often the heart), so that a comparison could be made. The resulting *postmortem* blood ratio was considered to reflect a drug's potential for PMR (Prouty & Anderson 1990; Dalpe-Scott et al. 1995). Recent work, however, has described ambiguities with this approach (McIntyre et al. 2012).

The collection, analysis, and comparison of *antemortem* blood specimens are obviously helpful in assisting with the interpretation of *postmortem* blood drug concentrations, but relevant specimens are only rarely available. In a set of case studies of six drugs, concentrations in the *postmortem* femoral blood specimens exceeded the *antemortem* concentrations in five of the drugs studied, suggesting that even peripheral blood exhibited redistribution (Cook et al. 2000). The potential for redistribution of other drugs in *postmortem* peripheral blood has also been documented (Gerostamoulos et al. 2012).

Correspondence: iain.mcintyre@sdcounty.ca.gov
Forensic Toxicology Laboratory Manager, County of San Diego Medical Examiner's Office, 5570 Overland Ave., Suite 101, San Diego, CA 92123, USA

The liver (L) to peripheral blood (P) ratio has been proposed as a more dependable marker for PMR, with ratios less than 5 (L/kg) indicating little to no propensity towards PMR, and ratios exceeding 20 to 30 (L/kg) indicative of a propensity for substantial PMR (McIntyre et al. 2012). A number of reports elaborating on, and supporting, this model have now been published (McIntyre & Mallett 2012; McIntyre & Meyer Escott 2012; McIntyre & Anderson 2012; McIntyre et al. 2013a; McIntyre et al. 2013b). Furthermore, a direct correlation between the *postmortem* peripheral blood and corresponding *antemortem* concentration - by consideration of the L/P ratio - has been expressed (McIntyre et al. 2013c). The report, describing methamphetamine cases, found that the *postmortem* peripheral blood concentrations were approximately 1.5 times higher than the corresponding concentrations attained in whole blood specimens collected before death. Given that the L/P ratios for methamphetamine had been confirmed to be approximately 6 (L/kg), it was then projected that drugs exhibiting L/P ratios between 5 and 10 (L/kg) would theoretically yield *postmortem* peripheral blood concentrations up to twice the corresponding *antemortem* concentrations - a measure of PMR potential. It was further hypothesized that L/P ratios ranging from 10 to 20 (L/kg) would demonstrate greater potential for PMR with *postmortem* peripheral blood concentrations 2 to 3 times that of the corresponding *antemortem* levels and consequently even higher L/P ratios indicative of even greater potential for PMR.

The current document sets out to expound upon this L/P model and its resultant implications by proposing the concept of a *postmortem* redistribution factor (F) for a drug. The *postmortem* redistribution factor has been defined as a factor that characterizes the direct relationship between a drug's *postmortem* peripheral blood and the corresponding *antemortem* (AM) whole blood concentration.

Hypothesis

Equation 1 presents the proposed relationship between the *antemortem* whole blood concentration of a compound and the corresponding *postmortem* peripheral blood concentration:

$$AM = P/F \tag{1}$$

where AM = *antemortem* whole blood concentration, P = *postmortem* peripheral blood concentration, and F = *postmortem* redistribution factor.

Rearrangement of Equation 1 gives

$$F = P/AM \tag{2}$$

Thus, an example of an experimental (or actual) F could be determined for a drug where both the *postmortem* peripheral blood and *antemortem* whole blood drug concentrations have been determined in the same individual (assuming an insignificant delay between the collection of the *antemortem* blood and the time of death).

Discussion

Considering the methamphetamine data (McIntyre et al. 2013c), an experimental (actual) F for methamphetamine of 1.5 is predicted - *postmortem* peripheral blood concentrations being 1.5 times (on average) greater than the corresponding *antemortem* concentrations.

A related approach to assess potential for PMR has also recently been described (Launiainen & Ojanpera 2013). This study presented data for 129 drugs comparing *postmortem* femoral blood concentrations to therapeutic plasma concentrations to describe drugs' propensity for PMR. This study analyzed a large number of cases where median *postmortem* drug concentrations were compared with estimations of the therapeutic concentrations. These authors projected a similar ratio for methamphetamine of 1.8. Although these data represent a practical attempt to describe PMR, it is conceivable that the determination of an F value from analytically determined *postmortem* data (such as the unique drug L/P ratio) may well produce more consistently accurate estimates.

The principal goal of these endeavors was to attempt to develop a ranking of drugs and indicate their propensity for and, subsequently, their potential extent of PMR. Until now, most efforts in interpretation have simply described PMR by an aphorism, ranging from 'the drug has not been found to exhibit PMR' to 'the drug is subject to PMR.' Such descriptions have never been particularly useful in the interpretation of *postmortem* drug concentrations, especially in relation to deducing what the drug concentration may have been at the time of death. The development of the concept of a systematically based *postmortem* redistribution factor will provide a more definitive and authoritative ranking and possibly numerical interpretation of PMR.

Competing interests
The author declares that there are no competing interests.

References

Cook J, Braithwaite RA, Hale KA (2000) Estimating antemortem drug concentrations from postmortem blood samples: the influence of postmortem redistribution. J Clin Path 53:282–285

Dalpe-Scott M, Degouffe M, Garbutt D, Drost M (1995) A comparison of drug concentrations in postmortem cardiac and peripheral blood in 320 cases. Can Soc For Sci J 28:113–121

Gerostamoulos D, Beyer J, Staikos V, Tayler P, Woodford N, Drummer OH (2012) The effect of the postmortem interval on the redistribution of drugs: a comparison of mortuary admission and autopsy blood specimens. Forensic Sci Med Path 8:373–379

Launiainen T, Ojanpera I (2013) Drug concentrations in post-mortem femoral blood compared with therapeutic concentrations in plasma. Drug Test Anal. doi:10.1002/dta.1507

McIntyre IM, Anderson DT (2012) Postmortem fentanyl concentrations: a review. J Forensic Res 3:157. doi:10.4172/2157-7145.1000157

McIntyre IM, Mallett P (2012) Sertraline concentrations and postmortem redistribution. Forensic Sci Int 223:349–352

McIntyre IM, Meyer Escott C (2012) Postmortem drug redistribution. J Forensic Res 3:e108. doi:10.4172/2157-7145.1000e108

McIntyre IM, Sherrard J, Lucas J (2012) Postmortem carisoprodol and meprobamate concentrations in blood and liver: lack of significant distribution. J Anal Tox 36:177–181

McIntyre IM, Mallett P, Trochta A, Morhaime J (2013a) Hydroxyzine distribution in postmortem cases and potential for redistribution. Forensic Sci Int 231:28–33

McIntyre IM, Gary RD, Estrada J, Nelson CL (2013b) Antemortem and postmortem fentanyl concentrations: a case report. Int J Legal Med. http://dx.doi.org/10.1007/s00414-013-0897-5

McIntyre IM, Nelson CL, Schaber B, Hamm CE (2013c) Antemortem and postmortem methamphetamine blood concentrations: three case reports. J Anal Tox 37(6):386–389

Prouty RW, Anderson WH (1990) The forensic science implications of site and temporal influences on postmortem blood-drug concentrations. J Forensic Sci 35:243–270

10

A new and sensitive reaction rate method for spectrophotometric determination of trace amounts of thiourea in different water samples based on an induction period

Mansour Arab Chamjangali[*], Gadamali Bagherian, Nasser Goudarzi and Shima Mehrjoo-Irani

Abstract

Background: Thiourea (TU) has various industrial, agricultural and analytical applications. TU has been labeled as having carcinogenic activity. Hypothyroidism was induced in animals by using TU. A simple and sensitive spectrophotometric reaction rate method was proposed for the determination of trace amounts of TU.

Method: The method is based on the inhibitory effect of TU on the rate of meta cresol purple (MCP) with bromate in the presence of bromide. The reaction progress was followed by monitoring the absorbance of MCP at 525 nm.

Results: The effects of different variables on the sensitivity of the proposed method were studied and optimized. Under optimum conditions a linear relationship between induction period time and TU concentration was found in the concentration range of 0.10 – 6.0 µg mL^{-1} of TU. The detection limit (3σ) of 0.020 µg mL^{-1} was found. The relative standard deviations for six replicate determinations of 0.10, 2.0 and 5.0 µg mL^{-1} of TU were 2.3%, 1.8% and 1.1%, respectively.

Conclusion: In this study a new reaction system was proposed for the kinetic spectrophotometric determination of TU in water samples. The new method not only benefit from high selectivity and sensitivity, but also it has the advantage of fast and simple operation.

Keywords: Thiourea (TU); Kinetic; Spectrophotometry; Meta cresol purple (MCP); Induction period

Background

Thiourea (TU) has various industrial, agricultural and analytical applications. This material is widely used in photography as a fixing agent and also removes stains from negative. In agriculture, it is used as fungicides, herbicides and rodenticides (Pérez-Ruiz et al. 1995) and also to decrease the content of nitrifying bacteria in soil (Smyth and Osteryoung 1977). TU is also used for induction of early ripening in several fruits (de Oliveira et al. 2004). In analytical chemistry TU is used as a spectrophotometric reagent for determination of several metals (HE et al. 1999). TU is also used as a reagent for copper electrolytes refinery (Akeneev et al. 2005). Compounds of TU are often added to citrus fruits as a fungicide during cold storage. TU has been labeled as having carcinogenic activity. Hypothyroidism was induced in animals by using TU (Sokkar et al. 2000; Bhide et al. 2001). Therefore, determination of TU at trace levels is of interest. There are some problems with determination of TU in waste water. These problems are due to the existence of a large number of organic components with relatively high concentrations, non-extractable nature of TU with traditional organic solvents and poor volatility of TU, which is not easily analyzed by gas chromatography (Toyoda et al. 1979). However, various methods have been proposed for the determination of TU such as titrimetry with iodine (Amin 1985; Pillai and Indrasenan 1980) or N-bromosuccinimdie (Sarwar and Thibert 1968), Raman spectroscopy (Bowley et al. 1986), spectrophotometry (Abd El-Kader et al. 1984; Hutchinson and Boltz 1958), polarography (Trojánek and Kopanica 1985), voltammetry

* Correspondence: marab@shahroodut.ac.ir
Department of Chemistry, Shahrood University, Shahrood, P.O. Box 36155–316, Iran

(Stará and Kopanica 1984), liquid chromatography-mass spectrometry (Xiao-Lan et al. 2009), high performance liquid chromatography (Rethmeier et al. 2001), ion selective electrode potentiometry (Radić and Komljenović 1991), FTIR (Kargosha et al. 2001), and tandem mass spectrometry (Raffaelli et al. 1997).

One of the useful methods which have been widely applied to trace determination of analytes is kinetic spectrophotometric method in which the main required equipment is a spectrophotometer. Based on the literature survey only a few numbers of indicator reactions for the kinetic determination of TU by spectrophotometric method have been published (Abbasi et al. 2009; Abbasi et al. 2010). All of these methods are based on the catalytic effect of the TU on the certain indicators reactions. To the best of our knowledge, there is no report on the use of induction period effect of TU for its kinetic determination and so this is the first report on the kinetic determination of TU based on induction period on the MCP-bromate-sulfuric acid as a novel reaction system.

In the present report a new sensitive and selective kinetic spectrophotometric method is proposed for determination of TU based on the induction period associated with TU on the catalytic oxidation of MCP by bromate. TU acts as an inhibitor on the catalytic effect of bromide ion. The reaction induction period at 525 nm is proportional to the TU concentration.

Method

Reagents and solutions

A 1000 µg mL^{-1} stock standard solution of TU was prepared by dissolving 0.1020 g TU (Merck) in distilled water and diluting it to 100 mL. In order to prepare the working solutions, appropriate dilution of the stock standard solution was carried out. For preparing 0.013 M potassium bromide solution in 100 mL volumetric flask, 0.1544 g of KBr (Merck) was dissolved in distilled water and diluted to the mark. A 100 mL 5.2×10^{-4} M MCP solution was prepared by dissolving 0.0200 g of MCP (Merck) in 25 mL ethanol and diluted with distilled water in a 100 mL calibrated flask. In order to prepare a 100 mL bromate ion solution (0.030 M), 0.5010 g of KBrO$_3$ (Merck) was dissolved in distielld water. Sulfuric acid solution (0.30 M) was prepared by diluting a known volume of concentrated solution (Merck) and standardized against sodium carbonate. All reagents used in this study were of analytical grade and double distilled water was used to prepare sample solutions.

Apparatus

Absorption-time graphs at a fixed wavelength and absorption spectra were recorded on a Shimadzu UV-160 Spectrophotometer with a pair of 1.0 cm quartz cell. In order to control the temperature of the reaction a water bath thermostat (n-BIOTEK, INC, model NB-301) was used in this study. A stopwatch was also applied to record the time of the reactions.

General procedure for determination of TU

The reagent solutions and water were kept at 25°C in the thermostatic water bath for 30 min. An appropriate volume of sample or standard solutions were transferred to a 10.0 mL standard flask, then 1.0 mL of 0.30 M sulfuric acid, 1.0 mL of 5.2×10^{-4} M MCP and 1.0 mL of 0.013 M potassium bromide solution were added sequentially and the mixture was then diluted to ca. 8 mL. After mixing, 1.0 mL of 0.030 M KBrO$_3$ was added and diluted to the mark with doubly distilled water. The stopped clock was started and after transferring ca 2 mL of this reaction mixture to spectrophotometer cell. The change in the absorbance at 525 nm was recorded against water for the first 15–450 s reaction time interval. The same procedure was applied for the measurement of the blank solution (without TU). In order to construct a calibration graph reaction induction period (t_{ip}) was plotted against TU concentration in a series of standard working solutions.

Results and discussion

MCP has been widely used in analytical chemistry in acid–base titrations, the kinetic determination of inorganic substance such as bromide (Ensafi et al. 2004a), nitrite and nitrate (Ensafi et al. 2004b) and organic analyts such as piroxicam (Arab Chamjangali et al. 2012) and phenylhydrazine (Arab Chamjangali et al. 2009). Results of experiments revealed that, the reaction rate between MCP and sodium bromate in solutions with low pH values at ambient temperature is not fast. When bromide ions are present in the solution, they reacts with bromate ion to produce Br$_2$, which cause an increase in the reaction rate. A catalytic cycle is shaped when MCP reacts with the produced Br$_2$ and convert to a colorless ingredient. Figure 1a shows the result of decrement in MCP absorption at $\lambda_{max} = 525$ nm with time. It was found that TU ion has an inhibition effect on the catalytic reaction due to possible reaction with Br$_2$ and/or bromate. Therefore, as can be seen in Figure 1b, in the presence of TU the reaction rate is depressed and an induction period was appeared. Further studies were carried out by recording the absorbance time profile of MCP-bromate-bromide reaction system at different concentration of TU. The results (Figure 2) show that the reaction induction period, which is appeared at the initiation of the reaction, is proportional to the TU concentration.

Optimization of variables

In order to find the optimum conditions, the effects of different variables on the absorbance changes of the catalyzed (ΔA_c) and inhibited (ΔA_i) reaction were studied

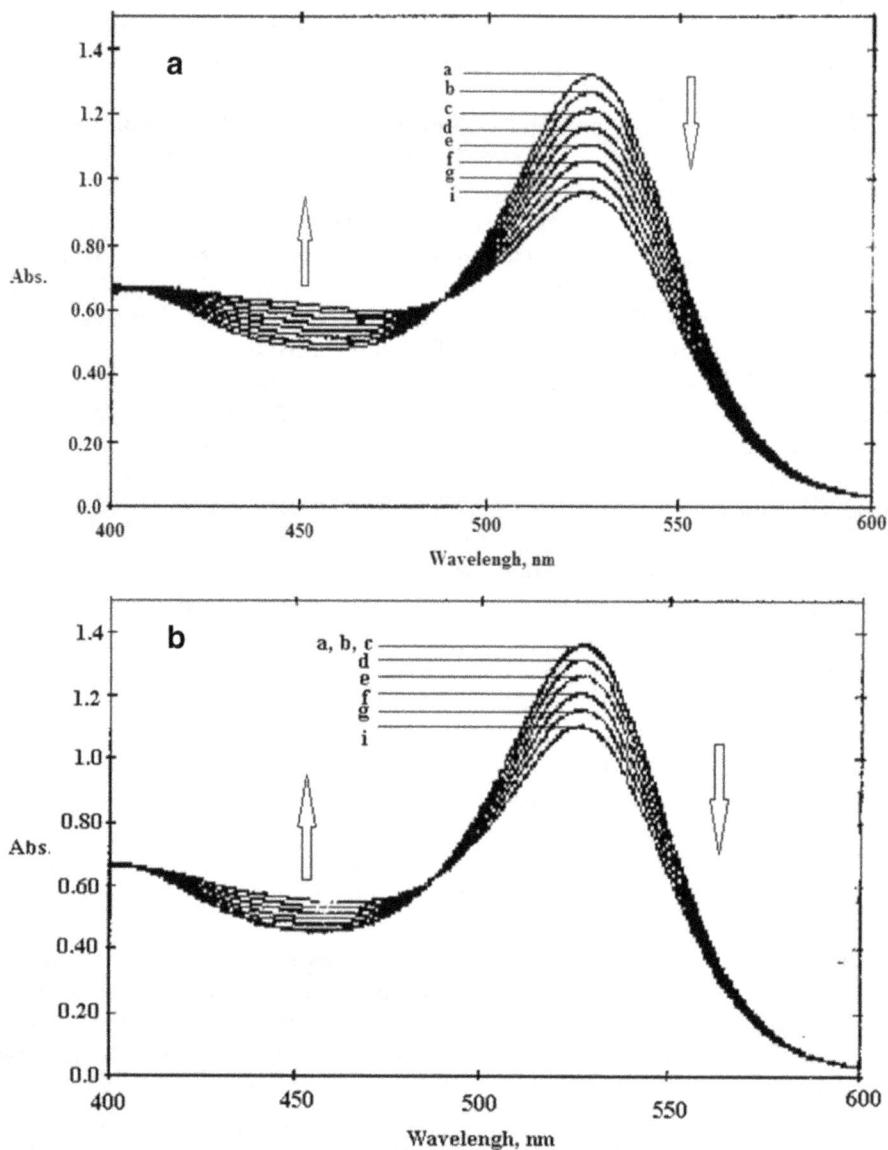

Figure 1 Absorption spectra of reaction system with scan time intervals of 30 s (a to i). (a) In the absence of TU and **(b)** in the presence of 0.50 µg mL^{-1} TU. Conditions: H$_2$SO$_4$, 0.030 M; MCP, 5.2×10^{-5}; KBr, 1.3×10^{-3} M; KBrO$_3$, 3.0×10^{-3} and temperature of 25°C.

during a fixed time of 15–115 s. One-at-a time optimization procedure was applied in this study and the difference between absorbance changes of catalyzed and inhibited reaction ($\Delta A = \Delta A_c - \Delta A_i$) was determined and used as an analytical signal.

Based on primary studies, acidic media is the best media for observation of reaction induction period caused by TU. Thus, different acids such as sulfuric, hydrochloric and nitric acid with the same concentration were tested to find the best type of reaction medium. According to the results, sulfuric acid showed higher sensitivity. So, sulfuric acid was chosen as the best reaction medium.

In order to find out the influence of sulfuric acid concentration on the sensitivity, the effect of sulfuric acid was investigated in the concentration range of 0.010-0.045 M. Based on the obtained results (Figure 3), ΔA_i and ΔA_c increase with increasing sulfuric acid concentration. This is due to this fact that the oxidation ability of bromate ions increases with increasing in the hydronium ion concentration. It was obvious that the maximum difference between changes in absorbance of the catalyzed and inhibited reaction was reached at 0.030 M. So, for further studies, a sulfuric acid concentration of 0.030 M, was used.

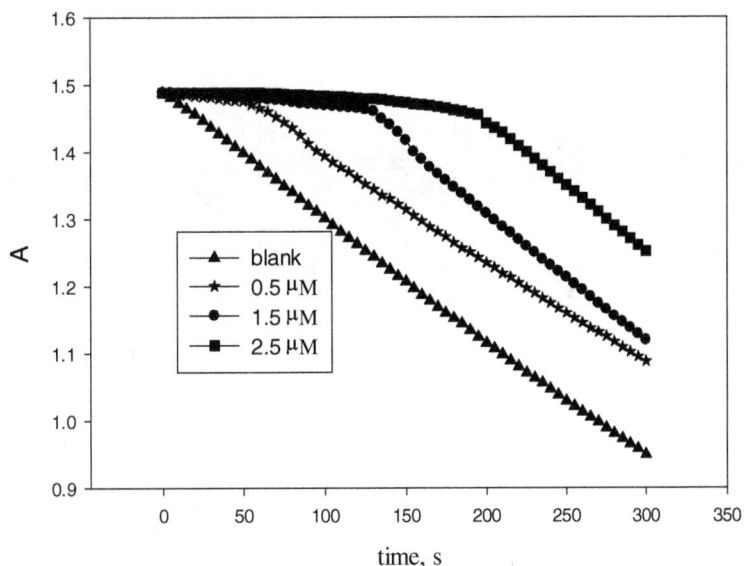

Figure 2 Absorbance-time profile. Conditions: same as Figure 1.

The dependence of sensitivity of the method on the potassium bromide concentration was studied in the range of 5.0×10^{-4} to 2.0×10^{-3} M bromide under the optimum concentration of sulfuric acid and potassium bromate. Figure 4 reveals that both ΔA_i and ΔA_c increase with increasing KBr concentration and their difference ($\Delta A_c - \Delta A_i$) reaches its maximum value at 1.3×10^{-3} M. The increases in ΔA_c and ΔA_i with increase in bromide ion concentration can be justified considering the catalytic nature of the reaction in the existence of bromide ion as the catalyst. However regarding to the obtained results, KBr concentration of 1.3×10^{-3} M was used as the best concentration for further studies.

The other factor which affects the analytical signal (sensitivity), is the concentration of potassium bromate (Figure 5). Result revealed that by increasing bromate concentration, ΔA_c and ΔA_i increased because of augmentation in oxidation capability of bromate with concentration. Moreover, the analytical signal and so sensitivity, increased and reached the highest value at 3.0×10^{-3} M and thus, the concentration of 3.0×10^{-3} was used as the optimum concentration for bromate.

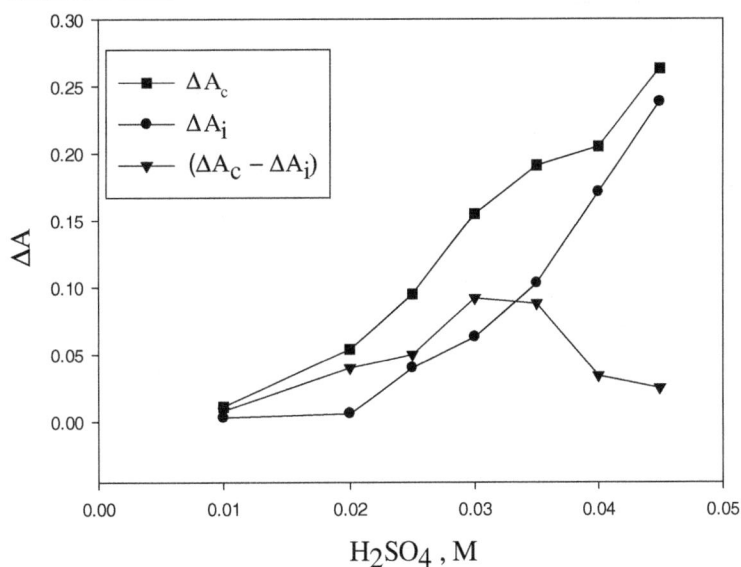

Figure 3 Effect of H$_2$SO$_4$ concentration. Conditions: TU 0.50 µg mL^{-1}; MCP, 5.2×10^{-5} M; KBr, 1.3×10^{-3} M, KBrO$_3$, 3.0×10^{-3} and temperature of 25°C.

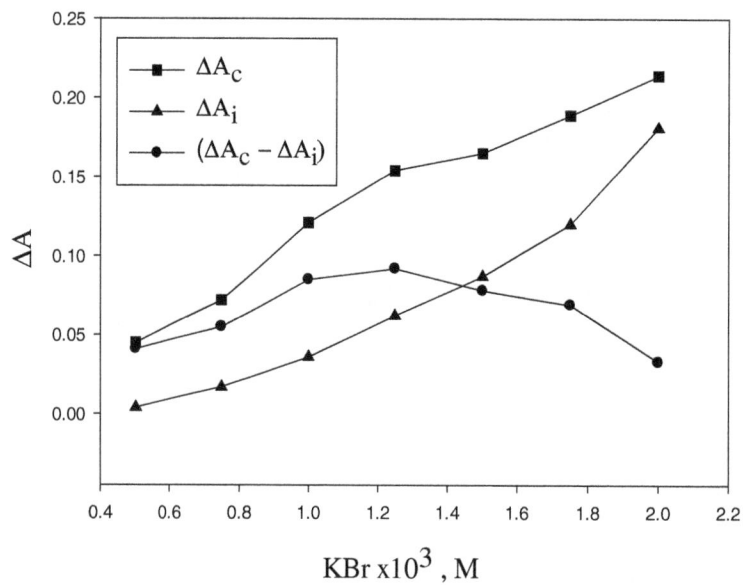

Figure 4 Effect of KBr concentration. Conditions: TU 0.5.0 μg mL^{-1}; H$_2$SO$_4$, 0.030 M; MCP, 5.2 × 10^{-5} M; KBrO$_3$, 3.0 × 10^{-3} and temperature of 25°C.

The sensitivity increased slightly as the concentration of MCP increased from 1.0×10^{-5} to 5.7×10^{-5} M and then it deceased. For this reason, 5.2×10^{-5} M of MCP was selected for the recommended procedure.

The other parameter which has a severe effect on the rate of both catalyzed and inhibited reactions in this study is the temperature. So, by applying the optimized concentration of reagents, the effect of temperature was studied in the range of 5 – 40°C. The results showed that the analytical signal was increased with increasing temperature. However 25°C (about room temperature) was used during this study.

In order to study the impact of ionic strength on the reaction induction period (analytical signal used in construction calibration curve) potassium nitrate (2.0 M) was used under the aforementioned optimum values.

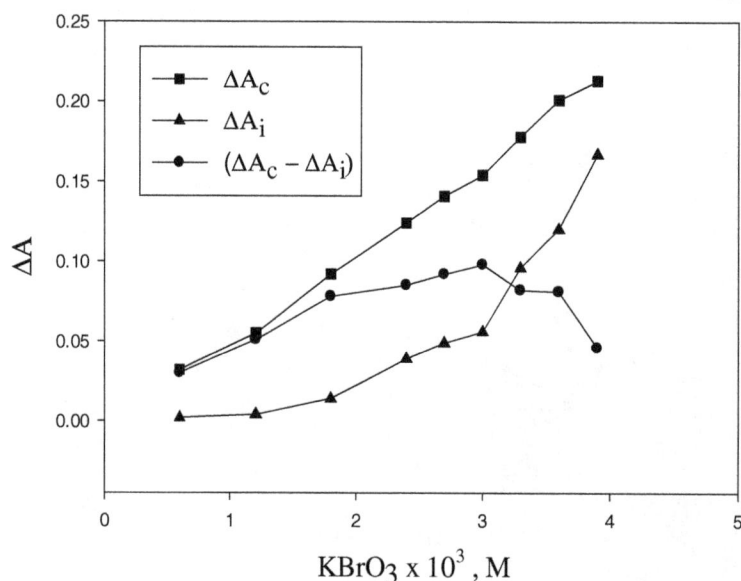

Figure 5 Effect of KBrO$_3$ concentration. Conditions: TU 0.50 μg mL^{-1}; H$_2$SO$_4$, 0.030 M; MCP, 5.2 × 10^{-5} M; KBr, 1.30 × 10^{-3} M and temperature of 25°C.

Table 1 Interferences for the determination of TU (0.50 μg mL^{-1})

Species	Tolerance limit (W$_{Species}$/W$_{TU}$)
Co^{+2}, Na$^+$, Ca^{+2}, Al^{+3}, K$^+$, NO$_3^-$, S$_2$O$_8^{-2}$, Urea , Glucose	1000
Cu^{+2}	800
Cd^{+2}, Zn^{+2}, Mn^{+2}, Mg^{+2}, Fe^{+3}	500
Citrate, Acetate, Ba^{+2}	300
EDTA, Pb^{+2}, Oxalate	200
Formate, Salicylate, PO$_4^{3-}$	100
CO$_3^{2-}$	80
CN$^-$	65
Ni^{+2}	40
Cl$^-$, Tartaric acid	20
Cr$_2$O$_7^{2-}$, Ag$^+$	5
Hg^{+2}	2

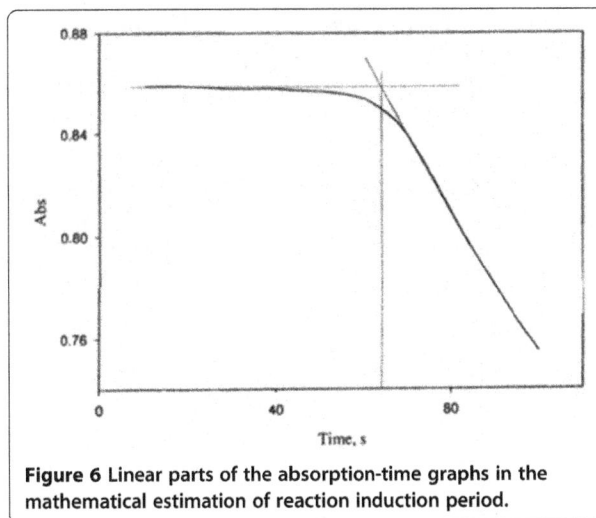

Figure 6 Linear parts of the absorption-time graphs in the mathematical estimation of reaction induction period.

Based on the obtained results, the induction period was not affected by the ionic strength up to 0.32 M (maximum value tested) of sodium nitrate.

Selectivity

The effect of some foreign species on the determination of 0.50 μg mL^{-1} TU were examined. The extreme tolerable concentration was taken as a concentration of foreign species that produce a change in the induction period more than ± 5%. Table 1 shows the achieved results. Most cations and anions did not interfere even if they were present in amounts of 1000-fold greater than TU. The inhibitory effect of Hg^{2+} was observed. The interference effect of mercury (II) ion was decreased (up to 150-fold) by precipitation using 0.010 M sodium hydroxide solution. The results (Table 1) demonstrate that this method has good selectivity.

Calibration and analytical parameters

The optimum condition of 0.030 M sulfuric acid, 3.0 × 10^{-3} M potassium bromate, 1.3 × 10^{-3} M potassium bromide, and 5.2 × 10^{-5} M MCP at a temperature of 25°C was applied to plot a calibration graph. The reaction induction period under optimum condition was linear to TU concentration in the ranges of 0.10 – 6.0 μg mL^{-1} with a regression equation of t_{ip} = 67.1 (±0.6) C_{TU} + 20.7 (±0.5) (r = 0.9990, n = 14). Where, C_{TU} is the TU concentration in μg mL^{-1} and t$_{ip}$ is the reaction induction period in seconds. The induction period was measured by extrapolation as illustrated in Figure 6.

The experimental 3σ detection limit is 20 ng mL^{-1}. For investigation of the accuracy and precision of the proposed method, standard solutions of 0.10, 2.0 and 5.0 μg mL^{-1} of TU were studied using suggested

procedure. Each concentration was tested six times and the relative standard deviations (RSD) of 2.3%, 1.8% and 1.1% were achieved respectively. The analytical parameters of reported methods for determination of TU are summarized in Table 2. It was clear that the detection limit of the proposed method is better than or comparable with othe reported methods (Toyoda et al. 1979, Abbasi et al. 2009). Some of these methods have better detection limit than the proposed method (Abbasi et al. 2010, Raffaelli et al. 1997), but suffer from the expensive equipments (Raffaelli et al. 1997)or time consuming (Abbasi et al. 2010).

Real sample analysis

The proposed method was effectively used for analysis of different spiked water samples (mineral and springer water). In analysis of water samples the procedure described in the experimental section was applied and the

Table 2 The analytical parameters of reported methods for determination of TU

No	Method	Linear range (μg mL^{-1})	Detection limit (μg mL^{-1})	Ref
1	Gas–liquid chromatography	NR	0.08	Toyoda et al. 1979
2	Kinetic catalytic spectrophotometry	0.03-10	0.02	Abbasi et al. 2009
3	Kinetic catalytic spectrophotometry	0.01-12	0.008	Abbasi et al. 2010
4	Tandem Mass Spectrometry	0.01-5.0	0.001	Raffaelli et al. 1997
5	Kinetic induction base spectrophotometry	0.10 – 6.0	0.02	This work

Table 3 Results for determination of TU in different spiked water samples

Sample	Add ($\mu g\ mL^{-1}$)	Found ($\mu g\ mL^{-1}$)	Recovery%	RSD% (n = 3)
Abshar Spring	-----	ND	-----	----
	0.40	0.41	102	1.7
	1.00	1.06	106	2.1
	2.00	2.03	102	2.4
	5.50	5.47	99	1.3
Mireral Water	-----	ND	-----	-----
	0.20	0.21	105	1.6
	0.50	0.52	104	2.9
	1.50	1.52	101	2.2
	3.00	3.01	101	2.3
	5.00	5.03	101	1.8

concentration of TU was calculated using constructed calibration graph. The results are shown in Table 3. The validity of the proposed method in the analysis of real samples is evident from the calculated recoveries. The student's t- test at 95% confidence level did not show any systematic error in the proposed method and thus confirms its reliability.

Conclusions

In this study a new reaction system was proposed for the kinetic spectrophotometric determination of TU in water samples. The new method not only benefit from high selectivity and sensitivity, but also it has the advantage of fast and simple operation. Besides, using cheap reagents and simple instrumentations in which minimum maintenance is required are the other reasons for favorably of the proposed method.

Competing interests
The authors declare that they have no competing interests.

Authors' contributions
All authors read and approved the final manuscript.

Acknowledgment
The authors are thankful to the Shahrood University Research Council for the support of this work.

References

Abbasi S, Khani H, Gholivand MB, Naghipour A, Farmany A, Abbasi F (2009)) A kinetic method for the determination of thiourea by its catalytic effect in micellar media. Spectrochim. Acta. Adv Physiol Educ 72:327–331

Abbasi S, Khani H, Hosseinzadeh L, Safari Z (2010) Determination of thiourea in fruit juice by a kinetic spectrophotometric method. J Hazard Mater 174:257–262

Abd El-Kader AK, Ahmed YZ, El-mottalb MA (1984) Spectrophotometric determination of thiosulfate, thiocyanate and thiourea ions by using sodium Nitroprusside as a complexing agent. Anal Lett 17:2259–2266

Akeneev YA, Zakharova EA, Slepchenko GB, Pikula NP (2005) Voltammetric determination of thiourea in copper refinery electrolytes. J Anal Chem 60:514–517

Amin D (1985) Determination of thiourea, phenylthiourea and allythiourea with iodine. Analyst 110:215–216

Arab Chamjangali M, Bagherian G, Ameri S (2009) A new induction period based reaction rate method for determination trace amounts of phenylhydrazine in water samples. J Hazard Mater 166:701–705

Arab Chamjangali M, Bagherian G, Mehrjoo-Irani S (2012) Determination of Piroxicam in different pharmaceutical products by a simple kinetic procedure based on an induction period effect. Anal Chem Lett 2:44–55

Bhide SV, Deshmukh BT, Talvelkar BA, Nagvekar AS (2001) Effect of induced hypothyroidism on the blood biochemical constituents in goats. Indian Vet J 78:205–208

Bowley HJ, Crathorne EA, Gerrard DL (1986) Quantitative determination of thiourea in aqueous solution in the presence of sulphur dioxide by Raman spectroscopy. Analyst 111:539–542

de Oliveira AN, de Santana H, Zaia CTBV, Zaia DAM (2004) A study of reaction between quinones and thiourea: determination of thiourea in orange juice. J Food Comp Anal 17:165–177

Ensafi AA, Rezaei B, Nouroozi S (2004a) Highly selective spectrophotometric flow-injection determination of trace amounts of bromide by catalytic effect on the oxidation of m-cresolsulfonephthalein by periodate. Spectrochim. Acta. Adv Physiol Educ 60:2053–2057

Ensafi AA, Rezaei B, Nouroozi S (2004b) Simultaneous spectrophotometric determination of nitrite and nitrate by flow injection analysis. Anal Sci 20:1749–1753

HE Z, Wu F, MENG H, LING L, YUAN L, LUO Q, ZENG Y (1999) Chemiluminescence determination of thiourea using tris (2, 2'-bipyridyl) ruthenium (II)-KMnO₄ system. Anal Sci 15:381–383

Hutchinson K, Boltz DF (1958) Spectrophotometric determination of nitrite and thiourea. Anal Chem 30:54–56

Kargosha K, Khanmohammadi M, Ghadiri M (2001) Fourier transform infrared spectrometric determination of thiourea in the presence of sulfur dioxide in aqueous solution. Anal Chim Acta 437:139–143

Pérez-Ruiz T, Martínez-Lozano C, Tomás V, Casajús R (1995) Flow injection fluorimetric determination of thiourea. Talanta 42:391–394

Pillai CPK, Indrasenan P (1980) Iodamine-T as an oxidemetric titrant in aqueous medium. Talanta 27:751–753

Radić N, Komljenović J (1991) Potentiometric determination of mercury(II) and thiourea in strong acid solution using an ion-selective electrode with AgI-based membrane hydrophobised by PTFE. Fresenius J Anal Chem 341:592–596

Raffaelli A, Pucci S, Lazzaroni R, Salvadori P (1997) Rapid determination of thiourea in waste water by atmospheric pressure chemical ionization tandem mass spectrometry using selected-reaction monitoring. Rapid Commun Mass Spectrom 11:259–264

Rethmeier J, Neumann G, Stumpf C, Rabenstein A, Vogt C (2001) Determination of low thiourea concentrations in industrial process water and natural samples using reversed-phase high-performance liquid chromatography. J Chromatogr A 934:129–134

Sarwar M, Thibert RJ (1968) Titrimetric determination of thiourea and thioacetamide using N-bromosuccinimide. Anal Lett 1:381–384

Smyth MR, Osteryoung JG (1977) Determination of some thiourea-containing pesticides by pulse voltammetric methods of analysis. Anal Chem 49:2310–2314

Sokkar SM, Soror AH, Ahmed YF, Ezzo OH, Hamouda MA (2000) Pathological and biochemical studies on experimental hypothyroidism in growing lambs. J. Vet. Med., Ser. Biogeosciences 47:641–652

Stará V, Kopanica M (1984) Adsorptive stripping voltammetric determination of thiourea and thiourea derivatives. Anal Chim Acta 159:105–110

Toyoda M, Ogawa S, Ito Y, Iwaida M (1979) Gas–liquid chromatographic determination of thiourea in citrus peels. J Assoc Off Anal Chem 62:1146–1149

Trojánek A, Kopanica M (1985) Thin-layer polarographic detector for the high-performance liquid chromatographic detection of thiourea derivatives. J Chromatogr 328:127–133

Xiao-Lan H, Hui-Tai L, Hui-Qin W, Wen-Rui C, Zhi-Fei Z, Fang H, Xiao-Shan L, Zhi-Xin Z (2009) Determination of thiourea in noodle and rice flour by liquid chromatography-mass spectrometry. Chinese J Anal Chem 37:1531–1534

11

Radio analytical technique in characterization of nuclear grade ion exchangers Duolite ARA-9366B and Purolite NRW-5010

Pravin U Singare

Abstract

Background: Ion exchange is one of the widely used techniques in nuclear industries for treatment of liquid radioactive waste. Regular efforts are being made in order to develop new ion exchange resins and their subsequent characterization so as to bring about efficient industrial performance. Among the different characterization techniques, radioactive tracer technique is one of the sensitive analytical techniques, mainly because of its non-destructive nature, high detection sensitivity, capability of *in-situ* detection, and physico-chemical compatibility with the material under study. The present work was therefore performed to demonstrate the application of the radioactive tracer technique in performance evaluation of two closely related nuclear grade anion exchange resins Duolite ARA-9366B and Purolite NRW-5010.

Methods: The short-lived radioisotope ^{131}I and ^{82}Br were used in the present experimental work to trace the kinetics of iodide and bromide ion-isotopic exchange reactions. The radioactivity was measured at various time intervals using γ-ray spectrometer having well type NaI(Tl) scintillation detector of Nucleonix make. From the radioactivity measured at various time intervals, the values of specific reaction rate (min^{-1}), amount of ion exchanged (mmol), and initial rate of ion exchange (mmol/min) were calculated.

Results: It was observed that for iodide ion-isotopic exchange reaction under identical experimental conditions of 30.0°C, 1.000 g of ion exchange resins and 0.001 mol/L labeled iodide ion solution, the above values were calculated as 0.246, 0.155, and 0.038, respectively for Purolite NRW-5010 resin, which was higher than the respective values of 0.201, 0.139, and 0.028 obtained for Duolite ARA-9366B resins. The identical trend was observed for the two resins during bromide ion-isotopic exchange reaction.

Conclusions: The overall results indicate that under identical experimental conditions, Purolite NRW-5010 resins exhibit superior performance over Duolite ARA-9366B resins. The same technique can be extended further for performance evaluation of different nuclear as well as non-nuclear grade ion exchange resins.

Keywords: Nuclear grade resins; Duolite ARA-9366B; Purolite NRW-5010; Characterization; Reaction kinetics; Ion-isotopic exchange reactions

Background

There are a number of liquid processes and waste streams at nuclear facilities (i.e., nuclear power plants, fuel reprocessing plants, nuclear research centers, etc.) that require treatment for process chemistry control reasons and/or the removal of radioactive contaminants (IAEA 2002). The treatment processes may be required for reactor primary coolants, the cleanup of spent fuel pools, liquid radioactive waste management systems, etc. One of the most common treatment methods for such aqueous streams is the ion exchange, which is a well-developed technique that has been employed for many years in both the nuclear industry and in other industries. In spite of its advanced stage of development, various aspects of ion exchange technology are being studied in many countries to improve its efficiency and economy in its application to radioactive waste management. Organic ion exchange resins have been developed over a much longer period of time than the selective inorganic ion exchangers that

Correspondence: pravinsingare@gmail.com
Department of Chemistry, Bhavan's College, Munshi Nagar, Andheri (West), Mumbai 400 058, India

have recently become available in commercial quantities and can now meet the demands of the nuclear industry. The organic ion exchange resins are very effective at transferring the radioactive content of a large volume of liquid into a small volume of solid and have proved to be reliable and effective for the control of both the chemistry and radiochemistry of water coolant systems at nuclear power plants and also for processing some liquid radioactive waste (Samanta et al. 1992; Samanta et al. 1993; Samanta et al. 1995; Kulkarni et al. 1996; Bray et al. 1990). In a number of cases, for specific physical and chemical reasons, organic resins cannot be replaced by commercially available inorganic ion exchangers. Also, organic ion exchange resins are used globally (Tomoi et al. 1997; Zhu et al. 2009; Kumaresan et al. 2006). The selection of an appropriate ion exchange material for treatment of liquid waste is possible on the basis of information provided by the manufacturer. However, since the selection of the appropriate ion-exchange material depends on the needs of the system, it is expected that the data obtained from the actual experimental trials will prove to be more helpful in the characterization and subsequent selection of the resins.

Although there are many alternative methods available for characterization of ion exchange resins (de Villiers et al. 1964; Harland 1994; Zeng et al. 1996; Patel et al. 2004; Liu et al. 2005; Masram et al. 2010), radioactive tracer isotopic technique is one of the sensitive analytical techniques (Sood et al. 2004; IAEA 2004). By monitoring the radioactivity continuously, the migration of the tracer and in turn, of the bulk matter under investigation, can be followed. As a result, radioisotopes have become a useful tool and almost every branch of industry, which uses them (Sood et al. 2004; IAEA 2004), and every radiotracer methodology, is described extensively in the literature (Clark et al. 2011; Dagadu et al. 2012; Koron et al. 2012). Considering the extensive technological application of radioactive tracers, in the present investigation, attempts are made to apply the same technique to study the kinetics of ion-isotopic exchange reactions in nuclear grade anion exchange resins Duolite ARA-9366B and Purolite NRW-5010. It is expected that the kinetics data obtained here will not only be used in characterization of these resins but also in standardization of the process parameters for their efficient application.

Methods

Conditioning of ion exchange resins

Duolite ARA-9366B (by Auchtel Products Ltd., Mumbai, India) and Purolite NRW-5010 (by Purolite International India Private Limited, Pune, India) are type I strong base, quaternary ammonium, nuclear grade anion exchange resins in hydroxide form. Details regarding the properties of the resins used are given in Table 1. These resins were converted separately in to iodide/bromide form by treatment with 10% (w/v) KI/KBr solution in a conditioning column which is adjusted at the flow rate as 1 mL/min. The resins were then washed with double distilled water, until the washings were free from iodide/bromide ions as tested by $AgNO_3$ solution. These resins in bromide and iodide form were then dried separately over P_2O_5 in desiccators at room temperature.

Radioactive tracer isotopes

The radioisotope ^{131}I and ^{82}Br used in the present experimental work was obtained from Board of Radiation and Isotope Technology (BRIT), Mumbai, India. Details regarding the isotopes used in the present experimental work are given in Table 2.

Study on kinetics of iodide ion-isotopic exchange reaction

In a stoppered bottle 250 mL (V) of 0.001 mol/L iodide ion solution was labeled with diluted ^{131}I radioactive solution using a micro syringe, such that 1.0 mL of labeled solution has a radioactivity of around 15,000 counts per minute (cpm) when measured with γ-ray spectrometer having NaI(Tl) scintillation detector. The γ-ray spectrometer used was Type GR 612 supplied by Nucleonix Systems Pvt. Ltd., India, having a well-type scintillation detector of Nucleonix make. Since only about 50 to 100 μL of the radioactive iodide ion solution was required for labeling the solution, its concentration will remain unchanged, which was further confirmed by potentiometer titration against $AgNO_3$ solution. The above-labeled solution of known initial activity (A_i) was kept in a thermostat adjusted to 30.0°C. The swelled and conditioned dry ion exchange resins in iodide form weighing exactly 1.000 g (m) were transferred quickly into this labeled solution which was vigorously stirred by using a mechanical stirrer and the activity in counts per minute of 1.0 mL of solution was measured. The solution was transferred back to the same bottle containing labeled solution after measuring activity. The iodide ion-isotopic exchange reaction can be represented as:

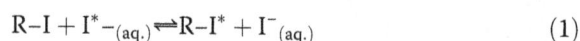

$$R\text{-}I + I^{*-}_{(aq.)} \rightleftharpoons R\text{-}I^{*} + I^{-}_{(aq.)} \qquad (1)$$

Here, R-I represents ion exchange resin in iodide form; $I^{*-}_{(aq.)}$ represents aqueous iodide ion solution labeled with ^{131}I radiotracer isotope.

The activity of solution was measured at a fixed interval of every 2.0 min. The final activity (A_f) of the solution was also measured after 3 h which was sufficient time to attain the equilibrium (Singare et al. 2012; Lokhande et al. 2007; Lokhande et al. 2008a; Lokhande and Singare 2008; Lokhande et al. 2006). The activity measured at various time intervals was corrected for background counts.

Table 1 Properties of ion exchange resins

Ion exchange resin	Matrix	Functional group	Particle size (mm)	Moisture content (%)	Operating pH	Maximum operating temperature (°C)	Total exchange capacity (mEq/mL)
Duolite ARA-9366B	Cross-linked polystyrene divinylbenzene	$-N^+R_3$	0.3 to 1.2	55	0 to 14	60	1.0
Purolite NRW-5010	Cross-linked polystyrene divinylbenzene	$-N^+R_3$	0.43 to 1.20	60	0 to 14	60	0.8

Similar experiments were carried out by equilibrating separately 1.000 g of ion exchange resin in iodide form with labeled iodide ion solution of four different concentrations ranging up to 0.004 mol/L at a constant temperature of 30.0°C. The same experimental sets were repeated for higher temperatures up to 45.0°C.

Study on kinetics of bromide ion-isotopic exchange reaction

The experiment was also performed to study the kinetics of bromide ion-isotopic exchange reaction by equilibrating 1.000 g of ion exchange resin in bromide form with labeled bromide ion solution in the same concentration and temperature range as above. The labeling of bromide ion solution was done by using ^{82}Br as a radioactive tracer isotope for which the same procedure as explained above was followed. The bromide ion-isotopic exchange reaction can be represented as:

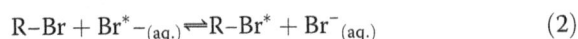

$$R-Br + Br^{*-}_{(aq.)} \rightleftharpoons R-Br^{*} + Br^{-}_{(aq.)} \qquad (2)$$

Here, R – Br represents ion exchange resin in bromide form; $Br^{*-}_{(aq.)}$ represents aqueous bromide ion solution labeled with ^{82}Br radiotracer isotope.

Results and discussion

Comparative study of ion-isotopic exchange reactions

In the present investigation, it was observed that due to the rapid ion-isotopic exchange reaction taking place, the activity of solution decreases rapidly initially, then due to the slow exchange, the activity of the solution decreases slowly and finally remains nearly constant. Preliminary studies show that the above exchange reactions are of first order (Singare and Lokhande 2012; Lokhande and Singare 2007, 2008; Lokhande et al. 2006, 2008a). Therefore, logarithm of activity when plotted against time gives a composite curve in which the activity initially decreases sharply and thereafter very slowly giving nearly straight line (Figure 1). Thus, it is evident that rapid and slow ion-

isotopic exchange reactions are occurring simultaneously (Singare and Lokhande 2012; Lokhande and Singare 2007, 2008; Lokhande et al. 2006, 2008a). For both the ion-isotopic exchange reactions, using the two resins, the ion exchange process is film diffusion controlled during an initial short period and the tendency toward particle diffusion control increases as exchange continues (Lokhande et al. 2008b). Presumably, the rapid exchange process taking place during the initial short period was film diffusion controlled; while the slow exchange process near the end was particle diffusion controlled. The specific reaction rate for the rapid process was obtained by resolving the composite curve as is done to determine the decay constants of radioactive isotopes in a mixture. The rapid process is completed in a short time, but the slow process continues for a much longer time. At the late stage, the slow exchange is the only reaction that takes place, and the logarithm of activity (cpm) curve against time is a straight line. Now, the straight line was extrapolated back to zero time. The extrapolated portion represents the contribution of slow process to the total activity which now includes rapid process also. The activity, due to slow process, was subtracted from the total activity at various time intervals. The difference gives the activity due to the rapid process only. From the activity exchanged due to rapid process at various time intervals, the specific reaction rates (k) of rapid ion-isotopic exchange reaction were calculated. The amount of iodide/bromide ions exchanged (mmol) on the resin were obtained from the initial and final activities of solution and the amount of exchangeable ions in 250 mL of solution. From the amount of ions exchanged on the resin (mmol) and the specific reaction rates (min^{-1}), the initial rate of ion exchanged (mmol/min) was calculated.

Because of the larger solvated size of bromide ions as compared to that of iodide ions, it was observed that the exchange of bromide ions occurs at the slower rate than that of iodide ions. Hence, under identical experimental conditions, the values of specific reaction rate (min^{-1}),

Table 2 Properties of ^{131}I and ^{82}Br tracer isotopes (Sood et al. 2004)

Isotopes	Half-life	Radioactivity/MBq	γ-energy (MeV)	Chemical form	Physical form
^{131}I	8.04 days	185	0.36	Iodide*	Aqueous
^{82}Br	36 h	185	0.55	Bromide**	Aqueous

*Sodium iodide in dilute sodium sulfite. **Ammonium bromide in dilute ammonium hydroxide.

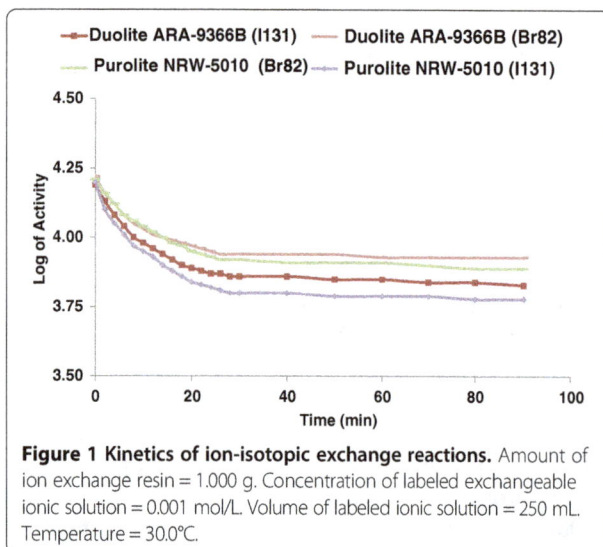

Figure 1 Kinetics of ion-isotopic exchange reactions. Amount of ion exchange resin = 1.000 g. Concentration of labeled exchangeable ionic solution = 0.001 mol/L. Volume of labeled ionic solution = 250 mL. Temperature = 30.0°C.

amount of ion exchanged (mmol) and initial rate of ion exchange (mmol/min) are calculated to be lower for bromide ion-isotopic exchange reaction than that for iodide ion-isotopic exchange reaction as summarized in Tables 3 and 4. For both bromide and iodide ion-isotopic exchange reactions, the value of specific reaction rate increases with increase in the concentration of iodide and bromide ions in solution from 0.001 mol/L to 0.004 mol/L (Table 3). However, with rise in temperature from 30.0°C to 45.0°C, the specific reaction rate was observed to decrease (Table 4). Thus, in case of Purolite NRW-5010 at 30.0°C, when the concentration of iodide and bromide ions in solution increases from 0.001 mol/L to 0.004 mol/L, the specific reaction rate value for iodide ion-isotopic exchange increases from 0.246 to 0.290 min^{-1}, while for bromide ion-isotopic exchange, the value increases from 0.203 to 0.245 min^{-1}. Similarly, in the case of Duolite ARA-9366B, the value for iodide ion-isotopic exchange increases from 0.201 to 0.235 min^{-1}, while for bromide ion-isotopic exchange the value increases from 0.165 to 0.186 min^{-1}. However, when the concentration of iodide and bromide ions in solution is kept constant at 0.001 mol/L and temperature is raised from 30.0°C to 45.0°C; in the case of Purolite NRW-5010, the specific reaction rate value for iodide ion-isotopic exchange decreases from 0.246 to 0.219 min^{-1}, while for bromide ion-isotopic exchange, the value decreases from 0.203 to 0.178 min^{-1}. Similarly, in case of Duolite ARA-9366B, the specific reaction rate value for iodide ion-isotopic exchange decreases from 0.201 to 0.156 min^{-1}, while for bromide ion-isotopic exchange, the value decreases from 0.165 to 0.125 min^{-1}. From the results, it appears that iodide ion exchange has the faster rate as compared to that of bromide ions which was related to the extent of solvation (Tables 3 and 4).

Comparative study of anion exchange resins

From the Tables 3 and 4, it is observed that for iodide ion-isotopic exchange reaction using Purolite NRW-5010 resin, the values of specific reaction rate (min^{-1}), amount of iodide ion exchanged (mmol), and initial rates of iodide ion exchange (mmol/min) were 0.246, 0.155, and 0.038, respectively, which were higher than 0.201, 0.139, and 0.028, respectively, as that obtained using Duolite ARA-9366B resins under identical experimental conditions of 30.0°C, 1.000 g of ion exchange resins, and 0.001 mol/L labeled iodide ion solution. The identical trend was observed for the two resins during bromide ion-isotopic exchange reaction.

From Table 3, it is observed that using Purolite NRW-5010 resins at a constant temperature of 30.0°C, as the concentration of labeled iodide ion solution, increases from 0.001 to 0.004 mol/L, the percentage of iodide ions exchanged also increases from 61.80% to 68.90% while using Duolite ARA-9366B resins, under identical experimental conditions, the percentage of iodide ions exchanged increases from 55.40% to 62.90%. Similarly, in the case of bromide ion-isotopic exchange reaction, the percentage of bromide ions exchanged increases from 52.20% to 55.70% using Purolite NRW-5010 resin, while for Duolite ARA-9366B resin, it increases from 48.30% to 52.80%. The effect of ionic concentration on the percentage of ions exchanged is graphically represented in Figure 2.

From Table 4, it is observed that using Purolite NRW-5010 resins, for 0.001 mol/L labeled iodide ion solution, as the temperature increases from 30.0°C to 45.0°C, the percentage of iodide ions exchanged decreases from 61.80% to 59.80% while using Duolite ARA-9366B resins, the percentage of iodide ions exchanged decreases from 55.40% to 50.10%. Similarly, in case of bromide ion-isotopic exchange reaction, the percentage of bromide ions exchanged decreases from 52.20% to 45.40% using Purolite NRW-5010 resin, while for Duolite ARA-9366B resin, it decreases from 48.30% to 40.60%. The effect of temperature on percentage of ions exchanged is graphically represented in Figure 3.

The overall results indicate that under identical experimental conditions, as compared to Duolite ARA-9366B resins, Purolite NRW-5010 resins shows higher percentage of ions exchanged. Thus, Purolite NRW-5010 resins show superior performance over Duolite ARA-9366B resins under identical operational parameters.

Statistical correlations

The results of the present investigation show a strong negative correlation between the amount of ions exchanged and temperature of exchanging medium (Figures 4 and 5). In the case of iodide ion-isotopic exchange reactions, the values of r calculated for Purolite NRW-5010 and Duolite

Table 3 Concentration effect on ion-isotopic exchange reactions

Concentration of ionic solution (mol/L)	Amount of ions in 250 mL solution (mmol)	Reaction (1)						Reaction (2)					
		Duolite ARA-9366B			Purolite NRW-5010			Duolite ARA-9366B			Purolite NRW-5010		
		Specific reaction rate of rapid process (min⁻¹)	Amount of iodide ion exchanged (mmol)	Initial rate of iodide ion exchange (mmol/min)	Specific reaction rate of rapid process (min⁻¹)	Amount of iodide ion exchanged (mmol)	Initial rate of iodide ion exchanged (mmol/min)	Specific reaction rate of rapid process (min⁻¹)	Amount of bromide ion exchanged (mmol)	Initial rate of bromide ion exchange (mmol/min)	Specific reaction rate of rapid process (min⁻¹)	Amount of bromide ion exchanged (mmol)	Initial rate of bromide ion exchange (mmol/min)
0.001	0.250	0.201	0.139 $A_i = 15{,}560$ cpm $A_f = 6{,}909$ cpm	0.028	0.246	0.155 $A_i = 16{,}010$ cpm $A_f = 6{,}084$ cpm	0.038	0.165	0.121 $A_i = 16{,}532$ cpm $A_f = 8{,}531$ cpm	0.020	0.203	0.131 $A_i = 16{,}229$ cpm $A_f = 7{,}725$ cpm	0.026
0.002	0.500	0.218	0.293 $A_i = 16{,}009$ cpm $A_f = 6{,}628$ cpm	0.064	0.253	0.322 $A_i = 16{,}564$ cpm $A_f = 5{,}897$ cpm	0.081	0.173	0.253 $A_i = 16{,}176$ cpm $A_f = 7{,}991$ cpm	0.044	0.215	0.269 $A_i = 15{,}632$ cpm $A_f = 7{,}222$ cpm	0.058
0.003	0.750	0.227	0.451 $A_i = 15{,}543$ cpm $A_f = 6{,}196$ cpm	0.102	0.274	0.499 $A_i = 15{,}889$ cpm $A_f = 5{,}318$ cpm	0.137	0.180	0.388 $A_i = 15{,}579$ cpm $A_f = 7{,}519$ cpm	0.070	0.228	0.409 $A_i = 15{,}999$ cpm $A_f = 7{,}274$ cpm	0.093
0.004	1.000	0.235	0.629 $A_i = 15{,}877$ cpm $A_f = 5{,}890$ cpm	0.148	0.290	0.689 $A_i = 15{,}895$ cpm $A_f = 4{,}943$ cpm	0.200	0.186	0.528 $A_i = 15{,}953$ cpm $A_f = 7{,}530$ cpm	0.098	0.245	0.557 $A_i = 16{,}005$ cpm $A_f = 7{,}090$ cpm	0.136

Amount of ion exchange resin = 1.000 g; volume of labeled ionic solution = 250 mL; temperature = 30.0°C.

Table 4 Temperature effect on ion-isotopic exchange reactions

Temperature (°C)	Reaction (1)						Reaction (2)					
	Duolite ARA-9366B			Purolite NRW-5010			Duolite ARA-9366B			Purolite NRW-5010		
	Specific reaction rate of rapid process (min^{-1})	Amount of iodide ion exchanged (mmol)	Initial rate of iodide ion exchange (mmol/min)	Specific reaction rate of rapid process (min^{-1})	Amount of iodide ion exchanged (mmol)	Initial rate of iodide ion exchange (mmol/min)	Specific reaction rate of rapid process (min^{-1})	Amount of bromide ion exchanged (mmol)	Initial rate of bromide ion exchange (mmol/min)	Specific reaction rate of rapid process (min^{-1})	Amount of bromide ion exchanged (mmol)	Initial rate of bromide ion exchange (mmol/min)
30.0	0.201	0.139 $A_i = 15{,}560$ cpm $A_f = 6{,}909$ cpm	0.028	0.246	0.155 $A_i = 16{,}010$ cpm $A_f = 6{,}084$ cpm	0.038	0.165	0.121 $A_i = 16{,}532$ cpm $A_f = 8{,}531$ cpm	0.020	0.203	0.131 $A_i = 16{,}229$ cpm $A_f = 7{,}725$ cpm	0.026
35.0	0.188	0.133 $A_i = 15{,}779$ cpm $A_f = 7{,}385$ cpm	0.025	0.235	0.153 $A_i = 15{,}043$ cpm $A_f = 5{,}837$ cpm	0.036	0.148	0.113 $A_i = 15{,}976$ cpm $A_f = 8{,}755$ cpm	0.017	0.194	0.125 $A_i = 16{,}101$ cpm $A_f = 8{,}050$ cpm	0.024
40.0	0.170	0.129 $A_i = 15{,}834$ cpm $A_f = 7{,}664$ cpm	0.022	0.228	0.150 $A_i = 15{,}756$ cpm $A_f = 6{,}302$ cpm	0.034	0.137	0.106 $A_i = 15{,}473$ cpm $A_f = 8{,}912$ cpm	0.014	0.185	0.120 $A_i = 15{,}998$ cpm $A_f = 8{,}319$ cpm	0.022
45.0	0.156	0.125 $A_i = 15{,}601$ cpm $A_f = 7{,}800$ cpm	0.020	0.219	0.150 $A_i = 15{,}539$ cpm $A_f = 6{,}216$ cpm	0.033	0.125	0.102 $A_i = 16{,}001$ cpm $A_f = 9{,}473$ cpm	0.013	0.178	0.114 $A_i = 15{,}678$ cpm $A_f = 8{,}529$ cpm	0.020

Amount of ion exchange resin = 1.000 g. Concentration of labeled exchangeable ionic solution = 0.001 mol/L. Volume of labeled ionic solution = 250 mL. Amount of exchangeable ions in 250 mL labeled solution = 0.250 mmol.

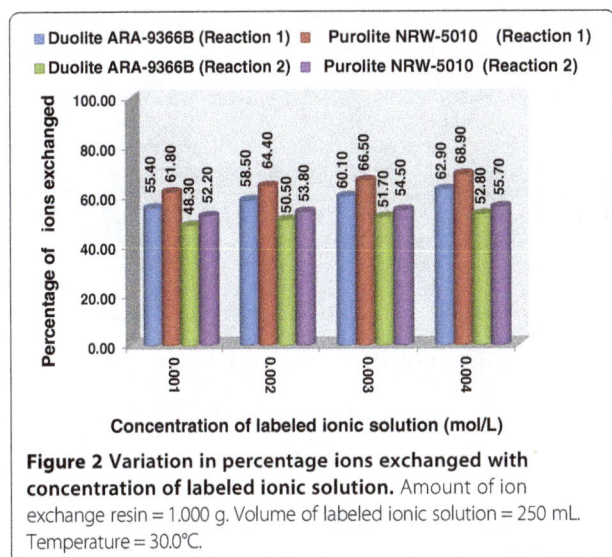

Figure 2 Variation in percentage ions exchanged with concentration of labeled ionic solution. Amount of ion exchange resin = 1.000 g. Volume of labeled ionic solution = 250 mL. Temperature = 30.0°C.

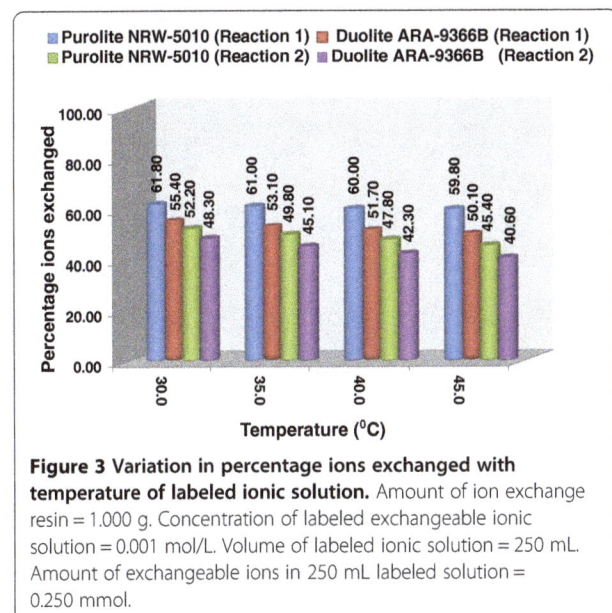

Figure 4 Correlation between temperature of exchanging medium and amount of iodide ion exchanged. Amount of ion exchange resin = 1.000 g. Concentration of labeled exchangeable ionic solution = 0.001 mol/L. Volume of labeled ionic solution = 250 mL. Amount of exchangeable ions in 250 mL labeled solution = 0.250 mmol. Correlation coefficient (r) for Duolite ARA-9366B = −0.9944. Correlation coefficient (r) for Purolite NRW-5010 = −0.9487.

ARA-9366B resins were −0.9487 and −0.9944, respectively. Similarly, in the case of bromide ion-isotopic exchange reactions, the r values calculated were −0.9994 and −0.9899, respectively, for both the resins.

Conclusion

Extensive work reported in the literature on synthesis of new ion exchange materials and their characterization itself is an indication that these resins are the wave of the present research and the material of new generation. Although there are a number of techniques reported in the literature on characterization of ion exchange resins,

the radioisotopic tracer technique has emerged as one of the effective non-destructive analytical techniques. The present research paper therefore is a successful demonstration on the application of radiotracer isotopes in the characterization of nuclear grade ion exchange resins Duolite ARA-9366B and Purolite NRW-5010. The results

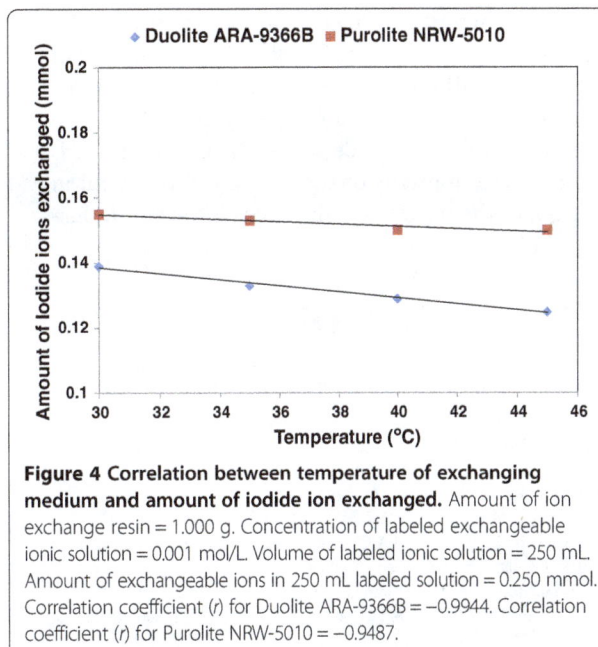

Figure 3 Variation in percentage ions exchanged with temperature of labeled ionic solution. Amount of ion exchange resin = 1.000 g. Concentration of labeled exchangeable ionic solution = 0.001 mol/L. Volume of labeled ionic solution = 250 mL. Amount of exchangeable ions in 250 mL labeled solution = 0.250 mmol.

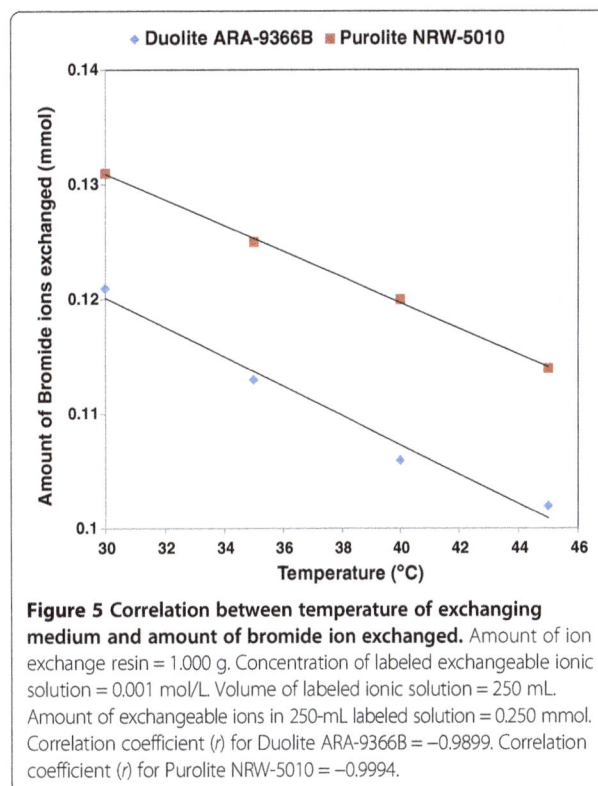

Figure 5 Correlation between temperature of exchanging medium and amount of bromide ion exchanged. Amount of ion exchange resin = 1.000 g. Concentration of labeled exchangeable ionic solution = 0.001 mol/L. Volume of labeled ionic solution = 250 mL. Amount of exchangeable ions in 250-mL labeled solution = 0.250 mmol. Correlation coefficient (r) for Duolite ARA-9366B = −0.9899. Correlation coefficient (r) for Purolite NRW-5010 = −0.9994.

of the present investigation may help to optimize the operational process parameters for efficient performance of these resins in various technical applications. The same technique can also be extended further for characterization of different non-nuclear grade ion exchange resins. It is anticipated that the results of such work will be useful in deciding about the selection of the resins in various industrial applications.

Competing interests

The author declares that there is no competing interests.

References

Bray LA, Elovich RJ, Carson KJ (1990) Cesium recovery using Savannah River laboratory resorcinol-formaldehyde ion exchange resin. Pacific Northwest Lab, Richland, WA, Rep PNL-7273

Clark MW, Harrison JJ, Payne TE (2011) The pH-dependence and reversibility of uranium and thorium binding on a modified bauxite refinery residue using isotopic exchange techniques. J Colloid Interface Sci 356(2):699–705

Dagadu CPK, Akaho EHK, Danso KA, Stegowski Z, Furman L (2012) Radiotracer investigation in gold leaching tanks. Appl Radiat Isot 70(1):156–161

de Villiers JP, Parrish JR (1964) Rapid characterization of ion-exchange resins by NMR. J Polym Sci Part A: Gen Papers 2(3):1331–1340

Harland CE (1994) Ion exchange, 2nd edn. RSC Publishing, UK, pp 49–89. doi:10.1039/9781847551184-00049, eISBN: 978-1-84755-118-4. ISBN 978-0-85186-484-6

International Atomic Energy Agency (IAEA) Vienna (2002) Application of ion exchange processes for the treatment of radioactive waste and management of spent ion exchangers. Technical reports series no. 408. International Atomic Energy Agency (IAEA), Vienna, Austria

International Atomic Energy Agency (IAEA) Vienna (2004) Radiotracer applications in industry - a guidebook. Technical reports series no.423. International Atomic Energy Agency (IAEA), Vienna, Austria

Koron N, Bratkic A, Ribeiro Guevara S, Vahcic M, Horvat M (2012) Mercury methylation and reduction potentials in marine water: an improved methodology using [197]Hg radiotracer. Appl Radiat Isot 70(1):46–50

Kulkarni Y, Samanta SK, Bakre SY, Raj K, Kumra MS (1996) Process for treatment of intermediate level radioactive waste based on radionuclide separation. Waste management'96 (Proc Int Symp Tucson, AZ, 1996). Arizona Board of Regents, Phoenix, AZ (CD-ROM)

Kumaresan R, Sabharwal KN, Srinivasan TG, Vasudeva Rao PR, Dhekane G (2006) Evaluation of new anion exchange resins for plutonium processing. Solvent Extr Ion Exc 24(4):589–602

Liu H, Zhang S, Nie S, Zhao X, Sun X, Yang X, Pan W (2005) Preparation and characterization of a novel pH-sensitive ion exchange resin. Chem Pharm Bull (Tokyo) 53(6):631–633

Lokhande RS, Singare PU (2007) Comparative study on ion-isotopic exchange reaction kinetics by application of tracer technique. Radiochim Acta 95 (03):173–176

Lokhande RS, Singare PU (2008) Comparative study on iodide and bromide ion-isotopic exchange reactions by application of radioactive tracer technique. J Porous Mater 15(03):253–258

Lokhande RS, Singare PU, Dole MH (2006) Comparative study on bromide and iodide ion-isotopic exchange reactions using strongly basic anion exchange resin duolite A-113. J Nucl Radiochem Sci 7(2):29–32

Lokhande RS, Singare PU, Patil VV (2008a) Application of radioactive tracer technique to study the kinetics and mechanism of reversible ion-isotopic exchange reaction using strongly basic anion exchange resin indion -850. Radiochemistry 50(6):638–641

Lokhande RS, Singare PU, Prabhavalkar TS (2008b) The application of the radioactive tracer technique to study the kinetics of bromide isotope exchange reaction with the participation of strongly basic anion exchange resin indion FF-IP. Russ J Phys Chem A 82(9):1589–1595

Masram DT, Kariya KP, Bhave NS (2010) A novel resin sef: synthesis, characterization and ion-exchange properties. Appl Sci Segment 1(1):APS/1513

Patel SA, Shah BS, Patel RM, Patel PM (2004) Synthesis, characterization and ion exchange properties of acrylic copolymers derived from 8-quinolinyl methacrylate. Iran Polym J 13(6):445–453

Samanta SK, Ramaswamy M, Misra BM (1992) Studies on cesium uptake by phenolic resins. Sep Sci Technol 27:255–267

Samanta SK, Ramaswamy M, Sen P, Varadarajan N, Singh RK (1993) Removal of radiocesium from alkaline IL waste. Natl symp on management of radioactive and toxic wastes (SMART-93). Kalpakkam, Bhabha Atomic Research Centre, Bombay, pp 56–58

Samanta SK, Theyyunni TK, Misra BM (1995) Column behavior of a resorcinol-formaldehyde polycondensate resin for radiocesium removal from simulated solution. J Nucl Sci Technol 32:425–429

Singare PU, Lokhande RS (2012) Studies on ion-isotopic exchange reactions using nuclear grade ion exchange resins. Ionics 18(4):351–357

Sood DD, Reddy AVR, Ramamoorthy N (2004) Applications of radioisotopes in agriculture and industry. In: Fundamentals of radiochemistry. Indian Association of Nuclear Chemists and Allied Scientists (IANCAS), Mumbai, India. pp 289–297

Tomoi M, Yamaguchi K, Ando R, Kantake Y, Aosaki Y, Kubota H (1997) Synthesis and thermal stability of novel anion exchange resins with spacer chains. J Appl Poly Sci 64(6):1161–1167

Zeng X, Murray GM (1996) Synthesis and characterization of site-selective ion-exchange resins templated for lead (II) ion. Sep Sci Technol 31(17):2403–2418

Zhu L, Liu Y, Chen J (2009) Synthesis of N-methylimidazolium functionalized strongly basic anion exchange resins for adsorption of Cr (VI). Ind Eng Chem Res 48(7):3261–3267

12

ROCK inhibitor, Y-27632, reduces FBS-induced structural alteration in organ-cultured mesenteric artery

Yang Hoon Huh[1,2], Hee-Seok Kweon[1*] and Toshio Kitazawa[2,3*]

Abstract

Background: Chronic treatment with fetal bovine serum (FBS) causes gradual vasoconstriction, vascular wall thickening, and contractility reduction in organ-cultured vascular tissues. We have previously demonstrated that Rho-associated kinase (ROCK) inhibitors prevent the functional alterations of small arteries in response to the FBS treatment. Here, we tested a further hypothesis that the chronic inhibition of ROCK has a protective effect on FBS-induced structural alterations.

Methods: To verify the new hypothesis, the rabbit mesenteric arterial rings were cultured in FBS-supplemented culture medium with or without Y-27632, a reversible ROCK inhibitor and then western blot, immunohistochemistry, apoptosis assay, and electron microscopy were performed using organ-cultured arterial rings.

Results: Chronic treatment with Y-27632 maintained the arterial diameter by preventing FBS-induced gradual arterial constriction during organ culture. Y-27632 also reduced the apoptosis and the loss of contractile myosin and actin filaments of smooth muscle cells. In addition, Y-27632 protected the morphological integrity between the endothelial cell layer and smooth muscle cell layer by preventing endothelial cell detachment and platelet endothelial cell adhesion molecule (PECAM) expression decrement.

Conclusions: Chronic ROCK inhibition provides protective effects against FBS-stimulated structural in addition to functional alterations of vascular smooth muscle cells and endothelial cells. These results strongly suggest that the RhoA/ROCK signaling is crucial for maintaining the structural and functional phenotypes of vasculature, and hence, chronic ROCK inhibition may provide protective effects on excessive growth factor-related vascular diseases including hypertension and atherosclerosis.

Keywords: Cardiovascular disease; Artery; Fetal bovine serum; Organ culture; RhoA/ROCK signaling

Background

Sustained vasoconstriction and vascular remodeling play important roles in the pathogenesis of vascular diseases such as hypertension, vasospasm, and atherosclerosis (Schwartz 1997; Noma et al. 2006). Recent studies provided better understanding of the pathophysiological mechanisms involved in vascular dysfunction. However, cardiovascular disease is the number one cause of death throughout

the world (World Health Organization 2011). Further studies are needed in order to clarify the mechanism that causes vasoconstriction and vascular remodeling-induced cardiovascular disease.

Ras homolog gene family member A (RhoA) acts as a molecular switch that regulates smooth muscle contraction (Brown et al. 2006), and Rho-associated kinase (ROCK), a RhoA effector, contributes to various physiological functions in blood vessels including endothelial and smooth muscle cells (Noma et al. 2006). Therefore, the RhoA/ROCK signaling pathway is proposed to play a pivotal role in diverse cellular functions, including expression and activity of endothelial nitric oxide synthase,

* Correspondence: hskweon@kbsi.re.kr; kitazawa.toshio@gmail.com
[1]Division of Electron Microscopic Research, Korea Basic Science Institute, 169-148 Gwahangno, Yuseong-gu, 305-806, Daejeon, Republic of Korea
[2]Boston Biomedical Research Institute, 64 Grove St, 02472, Watertown, MA, USA
Full list of author information is available at the end of the article

vascular smooth muscle contraction, actin cytoskeleton organization, cell adhesion and motility, proliferation, and hypertrophy in smooth muscle cells (Noma et al. 2006; Somlyo and Somlyo 2003; Gerthoffer 2007). Although deletion of either ROCK1 or 2 causes, respectively, embryonic and postnatal death in mouse models, abnormal upregulation of the RhoA/ROCK signaling pathway induces various vascular dysfunctions (Schwartz 1997; Félétou and Vanhoutte 2006; Cicek et al. 2013), which suggests that control of RhoA/ROCK signaling pathway or RhoA and/or ROCK expression itself is a useful therapeutic target in cardiovascular diseases (Noma et al. 2006; Cicek et al. 2013; Shimokawa and Takeshita 2005; Loirand et al. 2006).

To address the underlying mechanisms by which the upregulation of RhoA/ROCK signaling causes abnormal contractility of the vascular smooth muscle and following vascular disease, various studies were carried out by the use of an animal model and isolated or cultured cell model. In an animal model, the beneficial effect of ROCK-inhibiting compounds including Y-27632 has been reported on various cardiovascular diseases such as hypertension, coronary and cerebral vasospasm, and pulmonary arterial hypertension (Uehata et al. 1997; Shimokawa and Rashid 2007). *In vitro* cell culture models also have been used extensively to elucidate the molecular mechanism governed by RhoA/ROCK signaling with finely tuned culture conditions (Chen et al. 2010; Singh et al. 2011; Pagiatakis et al. 2012). However, both approaches have particular limitations. For *in vivo* animal models, the underlying mechanisms on the effects of ROCK inhibitors are difficult to identify because disease conditions are difficult to control and many factors can be involved. In the case of *in vitro* cell culture models, it is difficult to regenerate the contractile machinery in smooth muscle cells and the smooth muscle cell-endothelial cell interactions. In addition, smooth muscle cells in culture rapidly transform their phenotypes from contractile to synthetic (Somlyo and Somlyo 2003; Owens et al. 2004; Woodsome et al. 2006). Thus, organ culture of the vasculature has been used as an alternative method that can complement the limitations of animal- or cultured cell-based studies for investigating the chronic treatment effects of various inhibitors in arterial tissues (Ozaki and Karaki 2002; Thorne and Paul 2003; Huh et al. 2011).

Recently, using an organ culture model, we have reported that the ROCK inhibitor Y-27632 preserved the small mesenteric arterial rings from 1-week fetal bovine serum (FBS)-supplemented culture-induced functional abnormalities, such as gradual arterial constriction, wall thickening, reduced contractile activity, downregulation of endothelial nitric oxide synthase (eNOS), reduction of endothelium-induced relaxation, and increased ROCK-specific myosin phosphatase targeting subunit 1 (MYPT1)

Thr 853 phosphorylation (Huh et al. 2011). In this report, we further expanded the previous functional study to a structural study. We tested whether inhibition of ROCK mitigates the effect of the growth stimulant FBS on the structure of differentiated smooth muscle and endothelial cells in peripheral arteries, which primarily regulate blood pressure. Here, we report that the ROCK inhibitor Y-27632 prevents the detachment and loss of the endothelial cell layer, the disappearance of contractile thick and thin filaments, and apoptosis and necrosis. These results strongly suggest that the RhoA/ROCK signaling is crucial for maintaining the structural and functional phenotypes of the vasculature and hence has the potential to be a therapeutic target in cardiovascular diseases.

Methods

Animals and mesenteric artery preparation

All animal procedures for the isolation of arteries from rabbits were approved by the Animal Care and Use Committees of the Boston Biomedical Research Institute. The detailed procedures for isolation of arterial rings and organ culture have been described previously (Huh et al. 2011). Briefly, the rabbits were euthanized with an overdose of halothane. After thoracotomy, the mesenteric arteries were isolated and surrounding fat and fluffy connective tissues were discarded. To choose which segment in a mesenteric arterial tree is apt to conduct the morphofunctional studies using organ culture system, we first analyzed the gross morphology and protein expression of the main artery and first and second branches of the rabbit mesenteric artery. In the hematoxylin and eosin-stained low- and high-magnification images, no morphological difference between the main and branched mesenteric artery was noted except for the number of smooth muscle cell layers in the section (Figure 1A). Moreover, the contractile protein expressions related to the RhoA/ROCK signaling pathway were quite similar (Figure 1B,C). Based on these data, we chose the first branches of the mesenteric artery for organ culture because the inner and outer diameters were more uniform than the main artery and second branches of the mesenteric artery regardless of the region.

Organ culture procedure

To set up the organ culture condition, the first branches of the mesenteric artery were cut into rings (0.5~0.6 mm in outer diameter and 0.75 mm in length). The arterial rings were then placed in a silicone elastomer dish containing 4 ml of Dulbecco's modified Eagle's medium (DMEM) supplemented with 50 U penicillin, 50 μg/ml streptomycin, and 10% FBS (Sigma-Aldrich Co., St. Louis, MO, USA; cat# F4135) with or without 10 μM Y-27632. The arterial rings were maintained at 37°C with 5% CO_2 for 7 days. The culture medium was changed every 2 days.

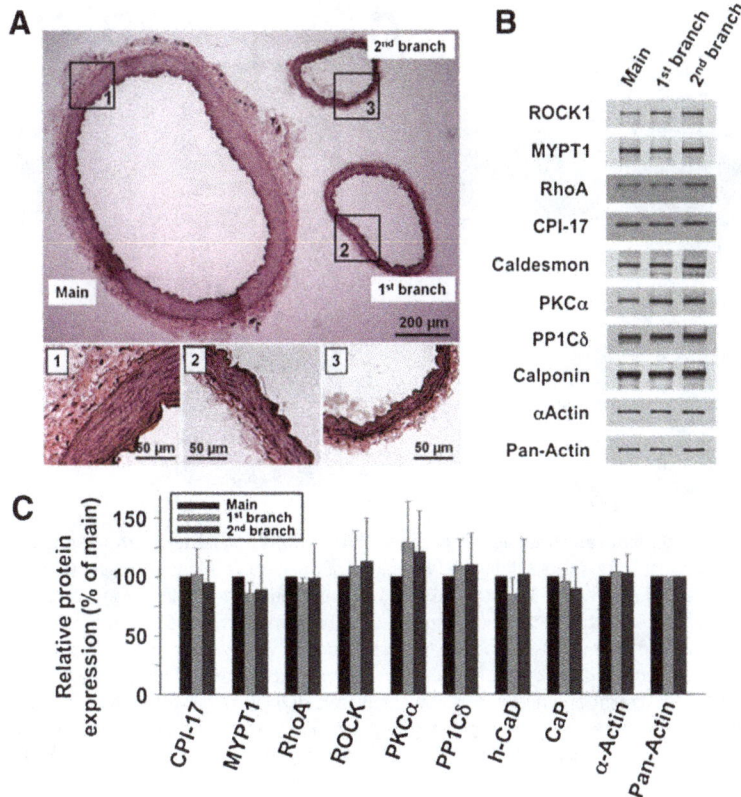

Figure 1 Gross morphology and contractile protein expression in the rabbit mesenteric artery. (A) To compare the gross morphology, the isolated main artery and first and second branches of the rabbit mesenteric artery were embedded in the same OCT block, and the collected cryosections were stained with hematoxylin and eosin. The high-magnification images of the boxed regions (1, 2, 3) are displayed below. **(B, C)** Representative immunoblot images and a quantitative summary for relative contractile protein expression ($n = 4$). The protein expressions of the main artery and first and second branches of the mesenteric artery were compared after matching the pan-actin expression. For actins, protein extracts that were diluted tenfold were loaded.

During the culture, each ring was loosely laid and placed on the silicone elastomer and secured by a pin at the center of the lumen without stretching (see Figure 2A). Once or twice a day during culture, the medium was gently shaken and the rings were moved around the supporting pin to prevent it from attachment to the silicone dish and subsequent cell migration.

Artery immunohistochemistry

For immunohistochemistry analysis, the stretched arterial rings on the force transducer were fixed with 4% para-formaldehyde in phosphate-buffered saline (PBS) solution for 30 min at 30°C. Then, the pre-fixed rings were removed from the apparatus and further fixed with 4% paraformaldehyde for 2 h at 4°C. After three washes in PBS, the rings were embedded in the Tissue-Tek optimal cutting temperature (OCT) compound (Fischer Scientific, Waltham, MA, USA), frozen in liquid nitrogen-cooled isopentane, and kept at −80°C for future use. The cryostat sections (10 μm) were collected on glass slides, washed with PBS to remove OCT compound, treated with 0.2%

Triton X-100, and washed with PBS. The sections were then incubated in the blocking solution (PBS containing 10% goat serum, 5% milk, and 4% bovine serum albumin (BSA)) for 1 h at room temperature and treated with the primary antibody solution (PBS containing 0.1% Tween-20 and 2% BSA) for overnight at 4°C in humidified chambers. After rinsing with PBS containing 0.5% BSA, the sections were treated with fluorescence-conjugated secondary antibody (Invitrogen, Inc., Carlsbad, CA, USA). After washing, the sections were counterstained with 4′,6-diamidino-2-phenylindole (DAPI) and mounted with FluorSave mounting medium (Millipore Corp., Billerica, MA, USA). Epifluorescence images were captured with a Leica DMR fluorescence microscope and a Leica DC300F digital camera system (Leica Microsystems Ltd., Milton Keynes, UK).

Immunoblotting

Immunoblotting experiments were performed as previously described (Kitazawa et al. 2003). Briefly, the arterial rings were isolated from the main artery and first and second

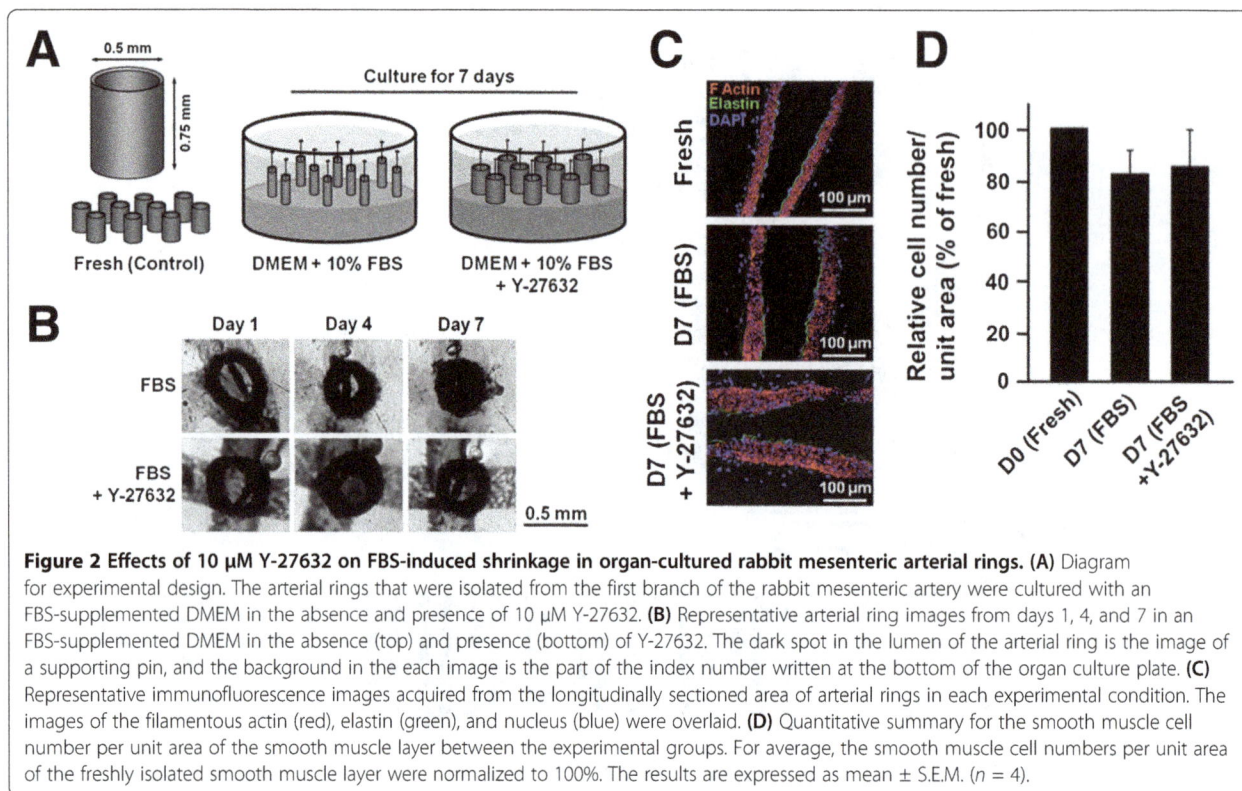

Figure 2 Effects of 10 μM Y-27632 on FBS-induced shrinkage in organ-cultured rabbit mesenteric arterial rings. (A) Diagram for experimental design. The arterial rings that were isolated from the first branch of the rabbit mesenteric artery were cultured with an FBS-supplemented DMEM in the absence and presence of 10 μM Y-27632. **(B)** Representative arterial ring images from days 1, 4, and 7 in an FBS-supplemented DMEM in the absence (top) and presence (bottom) of Y-27632. The dark spot in the lumen of the arterial ring is the image of a supporting pin, and the background in the each image is the part of the index number written at the bottom of the organ culture plate. **(C)** Representative immunofluorescence images acquired from the longitudinally sectioned area of arterial rings in each experimental condition. The images of the filamentous actin (red), elastin (green), and nucleus (blue) were overlaid. **(D)** Quantitative summary for the smooth muscle cell number per unit area of the smooth muscle layer between the experimental groups. For average, the smooth muscle cell numbers per unit area of the freshly isolated smooth muscle layer were normalized to 100%. The results are expressed as mean ± S.E.M. ($n = 4$).

branches of the mesenteric artery, and they were quickly frozen in liquid nitrogen and kept in 10% trichloroacetic acid (TCA)-acetone at –80°C overnight. After the TCA treatment, the tissues were gradually warmed and washed in acetone at room temperature, and allowed to dry. The small dried rings were homogenized in a Laemmli sample buffer (LSB; with final concentrations of 62.5 mM Tris, 1% SDS, 15% glycerol, 30 mM dithiothreitol, and 0.005% bromophenol blue) using a glass-glass mini homogenizer. The homogenates were then centrifuged and the supernatants were collected. The total protein concentration was measured using a Coomassie Plus Protein Assay Reagent Kit (Pierce, Rockford, IL, USA) and adjusted to 2 mg/ml with LSB. For the actin, the samples were diluted tenfold with LSB. The proteins were separated on a 4% to 20% polyacrylamide gradient gel and then transferred to nitrocellulose membranes using a wet transfer method. The membranes were blocked in Tris-buffered saline (TBS) solution containing 0.05% Tween-20 and 5% non-fat milk for 1 h at room temperature. After the treatment with the primary antibody solution, the membranes were incubated with alkaline phosphatase-conjugated secondary antibodies, and the bands were developed with alkaline phosphatase substrate (Sigma-Aldrich Co., St. Louis, MO, USA). The bands were scanned and analyzed using the IPLabGel image analyzing system (Signal Analytics, Vienna, VA, USA).

Apoptosis detection by the TUNEL assay

For the *in situ* apoptosis detection at the cellular level, we used the ApopTag red *in situ* apoptosis detection kit (Chemicon International, Temecula, CA, USA) for the terminal deoxynucleotide transferase dUTP nick end labeling (TUNEL) assay. Briefly, the cryostat sections (10 μm) from the 7 days organ-cultured arterial rings with or without 10 μM Y-27632 were obtained. The sections from the fresh mesenteric artery and the artery cultured in 1 μM angiotensin II-containing medium for 7 days were used as the negative and positive control, respectively. The tissue sections were fixed in 1% paraformaldehyde for 2 h at 4°C. After three washes in PBS (5 min each), the tissues were post-fixed with a mixture of ethanol and acetic acid (at a ratio of 2:1) at –20°C for 5 min and washed three times (5 min each). The tissues were then incubated with the terminal deoxynucleotidyl transferase (TdT) enzyme in a humid atmosphere for 60 min at 37°C. After incubation for 30 min with a rhodamine-labeled anti-digoxigenin conjugate and following counterstaining with DAPI, the sections were mounted with a FluorSave mounting medium (Millipore Corp., Billerica, MA, USA). Slides were viewed and imaged with a fluorescence microscope (Leica DMR) and a digital camera (Leica DC300F) system.

Transmission electron microscopy

For transmission electron microscope (TEM) analysis, mesenteric arterial rings in different organ culture conditions were mounted on the force transducer setup and stretched to adjust the muscle length as described previously (Kitazawa et al. 2009). The fresh arterial rings were stretched to 1.2 times their original slack length and the organ-cultured rings were stretched to the same tension level as that of the fresh one. The stretched arterial rings were fixed for 30 min at 30°C in pre-warmed 0.1 M sodium cacodylate buffer containing 2% glutaraldehyde, 2% paraformaldehyde, and 3.5% sucrose while they were stretched on the force transducer. Then, the pre-fixed rings were removed from the apparatus and further fixed in the same fixing solution for 2 h at 4°C with agitation. After three washes in 0.1 M sodium cacodylate buffer (10 min each), the rings were post-fixed with 1% osmium tetroxide on ice for 2 h and washed three times (10 min each) with a 0.1 M sodium cacodylate buffer. After gradual dehydration in an ethanol and propylene oxide series, the rings were then embedded in an Epon 812 mixture and polymerized in an oven at 60°C for 24 h. The embedded blocks were then sectioned on an ultramicrotome with a diamond knife, and ultrathin sections were collected on collodion-coated copper grids. The sections were stained with 2.5% uranyl acetate (7 min) and Reynolds lead citrate (2 min) and were examined using a TEM (Technai G^2 Spirit Twin, FEI, Hillsboro, OR, USA) at 120 kV.

Antibodies

The following primary antibodies were used in this study: polyclonal anti-MYPT1 (BabCO, Richmond, CA, USA, 1:5,000), anti-CPI-17 IgY (1:5,000), polyclonal anti-caldesmon (from Dr. A. Wang of BBRI, 1:10,000), polyclonal anti-h-calponin (from Dr. E. Mabuchi of BBRI, 1:10,000), polyclonal anti-PKCα (Sigma, St. Louis, MO, USA, 1:1000), polyclonal anti-PP1Cδ (from Dr. Eto of Thomas Jefferson University, 1:5,000), polyclonal anti-ROCK1 (Sigma, St. Louis, MO, USA, 1:2,000), monoclonal anti-RhoA (Santa Cruz Biotechnology, Dallas, Texas, USA, 1:1,000), polyclonal anti pan-actin (Sigma, St. Louis, MO, USA, 1:1,000), monoclonal anti α-actin (Sigma, St. Louis, MO, USA, 1:5,000), and polyclonal anti-platelet endothelial cell adhesion molecule (PECAM)-1 (Santa Cruz Biotechnology, Dallas, Texas, USA, 1:1,000). The secondary antibody against chicken IgY was from Promega (Madison, WI, USA, 1:5,000). The anti-mouse and anti-rabbit IgG secondary antibodies (1:5,000) were from Chemicon (Billerica, MA, USA).

Statistics

The results are expressed as mean ± S.E.M. of n experiments. Statistical significance was evaluated with one-way ANOVA; $P < 0.05$ was considered statistically significant.

Results

Effect of Y-27632 on the FBS-induced contraction of the organ-cultured artery

To examine the morphological changes in the arterial rings during the FBS-supplemented organ culture, the isolated arterial rings were cultured for 7 days with 10% FBS-supplemented DMEM with or without 10 µM Y-27632 (Figure 2A). Then, we obtained an image of each arterial ring every 24 h to analyze the changes in outer and inner diameters (*i.e.*, thickness) of the arterial wall (Figure 2B). We first confirmed the previous findings (Huh et al. 2011) that the arterial rings organ-cultured in the 10% FBS medium in the absence of Y-27632, the inner diameter shrank to about half at day 4, and the lumen was completely obstructed on day 7 (Figure 2B, top panel). On the other hand, 10 µM Y-27632 markedly reduced the gradual shrinkage of the organ-cultured arterial rings and the changes in the arterial wall thickness (Figure 2B, bottom panel).

Then, we check precisely whether the shrinkage of arterial rings cultured with FBS in the absence of Y-27632 was mediated by smooth muscle cell migration to the culture plate from the tunica media possibly by the FBS-mediated phenotypic change. All arterial rings used for the organ culture were fixed and double-stained with anti α-actin antibody and DAPI, and the total cell number in the whole cross-sectional area of the tunica media was counted (Figure 2C). The total number of DAPI-stained nuclei in the tunica media of the arterial rings cultured for 7 days with FBS in the absence of Y-27632 (81 ± 9% of fresh, $n = 4$) and in the presence of Y-27632 (85 ± 14% of fresh, $n = 4$) was not significantly different from that of the freshly isolated mesenteric artery (Figure 2D).

Y-27632 effect on the maintenance of the endothelial cell and smooth muscle cell integrity in the organ-cultured artery

We showed previously that Y-27632 prevented FBS-induced decrease in the eNOS mRNA expression and the acetylcholine (Ach)-induced relaxation (Huh et al. 2011). To examine whether Y-27632 also prevented FBS-induced structural alteration of the endothelial cell layer, the freshly isolated and organ-cultured mesenteric arteries that had been tested for Ach-induced relaxation were fixed and longitudinally (parallel to the longitudinal axis of the smooth muscle) sectioned for immunohistochemical analysis. Figure 3A shows the fluorescence images of the arterial section representing the anti-PECAM-stained endothelial cells, auto-fluorescencing elastin layer, DAPI-stained nucleus, and merged images, respectively. In the freshly isolated mesenteric artery, one continuous PECAM layer (red, left image of top panel) was coincident with one side of the auto-fluorescencing elastin lines (green, left image of second panel), which suggests that the

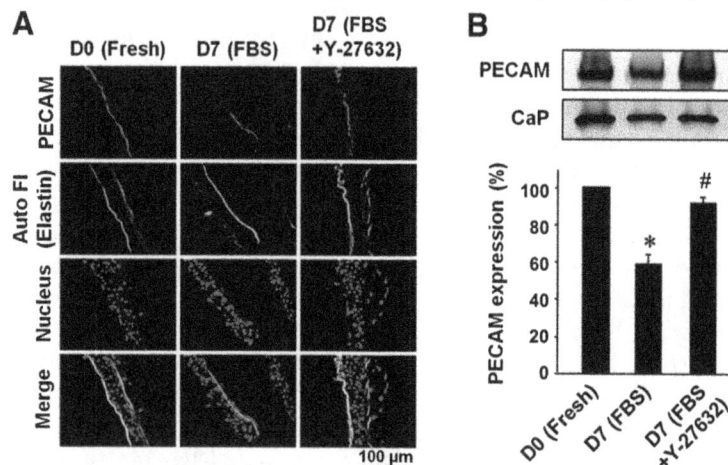

Figure 3 Effect of 10 µM Y-27632 on the maintenance of endothelium integrity in organ-cultured mesenteric arteries. (A) Representative images of the endothelial layer that were immunostained with anti-PECAM (CD31) antibody in the longitudinal sections of freshly isolated and organ-cultured mesenteric arteries. To verify the morphological integrity of the endothelial cell layer, elastin autofluorescence (Auto FI; green) and cell nucleus (DAPI; blue) were also captured with PECAM (red). **(B)** Representative immunoblot images of PECAM and calponin (as a loading marker) and a quantitative summary (lower) for the relative PECAM expression ($n = 3$). The asterisk represents a significant ($P < 0.05$) difference from that of the fresh mesenteric artery. The number sign represents a significant ($P < 0.05$) difference from that of the FBS-supplemented organ-cultured mesenteric artery in the absence of Y-27632.

endothelial cell layer fully covers the tunica intima and is tightly associated with the elastin layer (left image of bottom panel). By contrast, in the organ-cultured artery with FBS in the absence of Y-27632, approximately half of the endothelial cell layers were missing (middle image of top panel) and the remaining PECAM-positive layers were only loosely associated with the auto-fluorescencing elastin layer (middle image of bottom panel). This result suggests that considerable numbers of endothelial cells were already absent and the remaining layers were not tightly associated with the internal elastic layer. In the presence of Y-27632, however, majority of the endothelial cell layers except the small areas unstained to the PECAM (right image of top panel) were tightly adhered to the elastin layer along the luminal side of the vascular wall (right image of bottom panel).

To confirm the immunohistochemical results, we performed immunoblotting and determined the changes in the PECAM expression during organ culture. The PECAM expression in the organ-cultured artery without Y-27632 was significantly decreased to about half, whereas the PECAM expression was maintained at a level similar to that of the fresh mesenteric artery in the presence of 10 µM Y-27632 (Figure 3B). These results are consistent with the previous functional finding on ACh-induced relaxation (Huh et al. 2011).

Y-27632 effect on the apoptotic and ultrastructural degeneration in the organ-cultured mesenteric artery

We showed previously that the artery organ-cultured with FBS for 7 days reduced both depolarization- and agonist-induced contractions and that Y-27632 significantly prevented the FBS-induced reduction of contraction (Huh et al. 2011). To test the hypothesis whether FBS treatment prompted and Y-27632 prevented cell death and/or phenotypic change during organ culture, we performed the apoptosis assay using the TUNEL method (Figure 4). To delineate the smooth muscle-containing tunica media, nuclei were counterstained with DAPI and the elastin autofluorescence was captured. Then, the apoptotic cells in the tunica media of each experimental condition were counted, respectively. The FBS-containing organ culture increased the number of TUNEL-positive apoptotic cells to $17 \pm 2\%$ ($n = 7$) of the total cells counted in the tunica media (second panel of Figure 4A), whereas Y-27632 limited apoptotic cells in the tunica media to $5 \pm 2\%$ (third panel of Figure 4A) as compared to a negligible background level ($1 \pm 0\%$) in the freshly isolated mesenteric artery. We used the arterial rings that were cultured with 1 µM angiotensin II (AT II) as a positive control sample (Best et al. 1999) to confirm the assay efficiency ($n = 7$; bottom panel of Figure 4A). Angiotensin II drastically increased the apoptotic cell fraction to $71 \pm 3\%$ in the tunica media (Figure 4B).

Although RhoA and ROCK are upregulated and MYPT1 is phosphorylated in the chronic FBS organ culture conditions, the extent of contraction induced by agonists such as the α_1-agonist is significantly lower after FBS treatment. This may be partly due to about 50% reduction in expression of the myosin light chain (MLC) and CPI-17 (Huh et al. 2011). We further tested whether the ultrastructural integrity of smooth muscle cells is

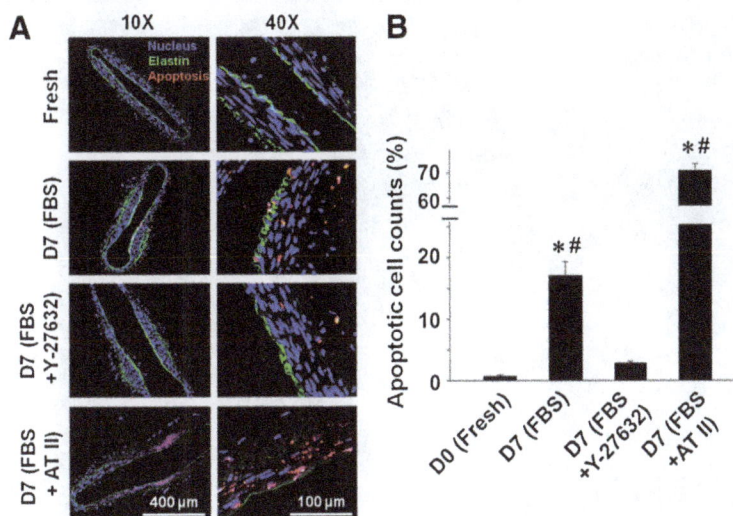

Figure 4 Effect of 10 μM Y-27632 on the apoptotic degeneration of organ-cultured rabbit mesenteric arteries. (A) Cross-sections of the freshly isolated and organ-cultured mesenteric arteries were stained using the TUNEL method. The cell nuclei were counterstained with DAPI, and the elastin autofluorescence was captured to define the tunica media. To validate the TUNEL assay results, we used the arterial rings cultured with 1 μM angiotensin II (AT II) as the positive control sample. The representative TUNEL (red) images merged with the DAPI (blue) and autofluorescence (green) images. **(B)** Statistical analysis of TUNEL data in the tunica media under each condition ($n = 7$). The asterisk represents a significant ($P < 0.05$) difference from that of the fresh mesenteric artery on D0. The number sign represents a significant ($P < 0.05$) difference from that of the FBS-supplemented organ-cultured mesenteric artery in the presence of Y-27632.

impaired with FBS treatment. The mesenteric arterial rings were fixed under optimally stretched conditions (see '*Methods*' section) for TEM analysis (Figure 5). At low-magnification images (Figure 5A,B,C), $33 \pm 2\%$ ($n = 4$) of the smooth muscle cells in the organ-cultured arteries with FBS in the absence of Y-27632 were markedly swollen and had reduced cytoplasmic density with fewer organelles including contractile filaments (Figure 5B). No such cells were found in the fresh mesenteric artery (Figure 5A), which suggests that the necrotic smooth muscle cells were induced during the FBS organ culture. On the other hand, most smooth muscle cells in the organ-cultured arteries with 10 μM Y-27632 maintained structural integrity of smooth muscle cells except with increased extracellular space (Figure 5C), which corresponds to a mild arterial wall thickening (Figure 2B). At higher magnification images (Figure 5D,E,F,G), the bulk of the cytoplasm was filled with contractile thick and thin filaments and dense bodies in the smooth muscle cells of the mesenteric arteries cultured with 10 μM Y-27632 (Figure 5F) as well as those of the fresh mesenteric arteries (Figure 5D). By contrast, in the cultured mesenteric artery without Y-27632, both the contractile filaments and dense bodies were not observed in the swollen and lightly stained cells (Figure 5G, cells marked with an asterisk in Figure 5B), whereas the other darkly stained cells in Figure 5B still retained normal thick and thin filaments and dense bodies in the cytoplasm (Figure 5E), as seen in the fresh mesenteric arteries.

Discussion

In this study, we used an organ culture system of differentiated mesenteric arterial vessels as a physiological model that is reasonably closer to the *in vivo* environment than the cell culture system. The chronic treatment with the growth stimulant FBS impairs contractility, morphological alteration, and DNA synthesis of arteries during organ culture (Huh et al. 2011; Lindqvist et al. 1999; Murata et al. 2005). To evaluate whether the inhibition of RhoA/ROCK signaling pathway prevents the structural alterations of the artery, we organ-cultured arteries in the presence of ROCK inhibitor Y-27632, together with FBS. Y-27632 prevents the detachment and loss of the endothelial cell layer, the disappearance of contractile thick and thin filaments of smooth muscle cells, and the apoptosis and necrosis during chronic FBS-supplemented organ culture.

According to our previous organ culture study, chronic treatment with Y-27632 protects almost all functional integrities of both endothelial and smooth muscle cells, including excitatory agonist-induced contraction and ACh-induced relaxation (Huh et al. 2011). Y-27632 also successfully inhibited the FBS-induced gradual arterial constriction during organ culture for 7 days. In addition, H-1152, another ROCK inhibitor, and simvastatin, which inhibit RhoA geranylgeranylation (Noma et al. 2006), also showed a significant inhibitory effect on the FBS-induced arterial constriction (Huh et al. 2011). Consistent with those previous works, the present study demonstrated that

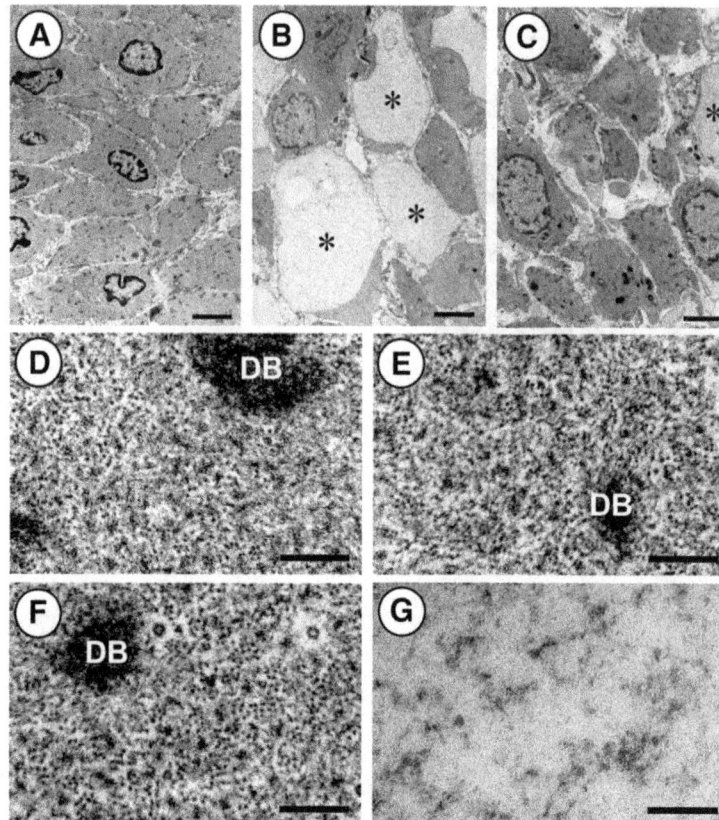

Figure 5 Representative TEM micrographs of cross-sectioned smooth muscle cells from freshly isolated and organ-cultured rabbit mesenteric arteries. Representative low-magnification images of **(A)** freshly isolated, **(B)** FBS-supplemented, and **(C)** FBS with Y-27632-supplemented organ-cultured arterial smooth muscle cells, respectively. **(D-F)** Representative high-magnification images of the darkly stained cells in **(A-C)**, respectively. **(G)** A high-magnification image of the lightly stained swollen cells (cells marked with an asterisk in **(B)**). The asterisk represents the swollen cells in **(B)** and **(C)**. DB, dense body. Scale bars = 5 μm **(A-C)** and 200 nm **(D-G)**.

Y-27632 markedly reduced the morphological alterations in the organ-cultured arterial rings. Interestingly, chronic treatment with Y-27632 did not prevent the FBS-induced overexpression of RhoA and ROCK but rather augmented the expressions to higher levels than those of FBS alone (Huh et al. 2011). Considering that the Y-27632 effectively protects the organ-cultured artery against FBS-induced gradual arterial constriction (Figure 2B), inhibition of the downstream signaling of upregulated RhoA/ROCK may be a major cause for preventing arterial constriction even in higher expression of RhoA and ROCK.

Endothelial cells regulate vascular tone and permeability by producing nitric oxide (NO) (Félétou and Vanhoutte 2006), but the upregulated RhoA/ROCK signaling pathway leads to impaired NO production by suppressing eNOS expression or inhibiting eNOS activity (Noma et al. 2006; Murata et al. 2005; Bolz et al. 2003). Thus, the ROCK inhibitors have been proposed as a therapeutic agent for cardiovascular diseases (Noma et al. 2006; Shimokawa and Takeshita 2005; Loirand et al. 2006). In the case of type 1 diabetes, dysfunction of endothelial cells plays a

key role in the pathogenic disorder of vasculature. The upregulation of ROCK and the reduction of the endothelial NO production are closely associated by the involvement of RhoA/ROCK signaling pathway, while suppressing ROCK activity restores vascular function (Cicek et al. 2013; El-Remessy et al. 2010; Kizub et al. 2010; Yao et al. 2013). In the present study, majority of endothelial cell layers were tightly adhered to the elastin layer in the presence of Y-27632, and PECAM expression was also maintained at a level similar to that of the fresh mesenteric artery (Figure 3). These results are consistent with our previous report on the protective role of Y-27632, such as maintenance of eNOS mRNA expression and endothelial cell-mediated relaxation in the organ-cultured mesenteric artery (Huh et al. 2011). Together with our previous report, the present study shows that chronic treatment with Y-27632 protects not only functional but also structural integrities of endothelial cells. Whether Y-27632-induced protection of eNOS mRNA (Huh et al. 2011) and PECAM expression in endothelial cells is caused by or independent from maintaining adherence of endothelial cell layers to the smooth muscle cell layers is unclear but warrants examination.

According to recent studies, the inhibition of RhoA/ROCK signaling pathway promotes the apoptosis of gastric cancer cells (Xiao-Tao et al. 2012). By contrast, inhibition of RhoA/ROCK signaling decreased the neuronal apoptosis in the ischemic penumbra of the rat brain and the penile apoptosis in the cavernous nerve injury model (Wu et al. 2012; Hannan et al. 2013). In the present study, Y-27632 drastically decreased the FBS-induced apoptosis of smooth muscle cells and endothelial cells in the organ-cultured mesenteric artery (Figure 4). This finding may provide another beneficial effect that the inhibition of RhoA/ROCK signaling pathway by Y-27632 protects the artery by preventing the apoptosis of both smooth muscle cell and endothelial cell even though the mechanism for ROCK-induced apoptosis needed to be further addressed.

In vascular diseases, the phenotype of smooth muscle cells is easily changed from contractile to synthetic/proliferative with the loss of contractile filaments in response to various stimuli including growth factors. These phenotypic changes are associated with the downregulation of α-actin and h-caldesmon and the upregulation of RhoA and ROCK (Gerthoffer 2007; Owens et al. 2004; Woodsome et al. 2006). In our pervious organ culture study, RhoA and ROCK expression and MYPT1 phosphorylation increased, but the expression of α-actin decreased upon chronic FBS treatment. Moreover, the expression of MLC and CPI-17 also decreased to about 50%, both of which have an essential function in the development of vascular smooth muscle contraction (Dimopoulos et al. 2007), but the decrease of MLC and CPI-17 expression was prevented in presence of Y-27632 during organ culture (Huh et al. 2011). To address the effect of the changes in contractile protein expressions on the structural integrity of contractile filaments, we performed the TEM analysis of organ-cultured arterial smooth muscle cells (Figure 5). The TEM analysis showed that smooth muscle cells in the FBS-supplemented organ culture were markedly necrotized with contractile filament loss (Figure 5B,G), consistent with the previous report by Lindqvist et al. (1999). However, 10 μM Y-27632 maintained the structural integrity of the smooth muscle cells with intact actin and myosin filaments (Figure 5C,F). These findings suggest that the overexpression of RhoA and ROCK and the down-expression of α-actin, MLC, and CPI-17 are closely related with the loss of contractile actin and myosin filament in organ-cultured smooth muscle cells, as shown in the TEM results of the present study (Figure 5G). In addition, the inhibition of the upregulated RhoA/ROCK signaling pathway and the maintenance of MLC and

CPI-17 expression by the chronic treatment with Y-27632 may play a significant role in the prevention of FBS-induced contractile filament deterioration.

Conclusions

The findings of the present study provide structural evidence that inactivation of a pathophysiologically important RhoA/ROCK signaling pathway by Y-27632 prevents FBS-supplemented organ culture-induced arterial damages. Y-27632 effectively maintains the arterial diameter and the structural integrity of adhesion of endothelial cells to smooth muscle cells. Also, Y-27632 prevents the apoptosis of both endothelia and smooth muscle cells and the ultrastructural degeneration of contractile actin and myosin filaments in smooth muscle cells. Together, these results suggest that the chronic inhibition of the RhoA/ROCK signaling pathway by inhibitors such as Y-27632 produces preventive effects on FBS-induced vascular complications and could offer a therapeutic means for RhoA/ROCK-mediated vascular diseases, such as hypertension and vasospasm.

Abbreviations
AT II: Angiotensin II; MLC: Myosin light chain; MYPT1: Myosin phosphatase targeting subunit 1; PECAM: Platelet endothelial cell adhesion molecule; RhoA: Ras homolog gene family member A; ROCK: Rho-associated kinase (Rho-kinase).

Competing interests
The authors declare that they have no competing interests.

Authors' contributions
YHH and TK designed research. YHH and TK performed research. YHH, HSK, and TK analyzed data. YHH, HSK, and TK wrote the paper. All authors read and approved the final manuscript.

Acknowledgments
This research was supported by the National Institute of Health grant R01 HL070881 to TK and the Korea Basic Science Institute grant K32601 to HSK.

Author details
[1]Division of Electron Microscopic Research, Korea Basic Science Institute, 169-148 Gwahangno, Yuseong-gu, 305-806, Daejeon, Republic of Korea. [2]Boston Biomedical Research Institute, 64 Grove St, 02472, Watertown, MA, USA. [3]Department of Microbiology and Physiological Systems, University of Massachusetts Medical School, 55 Lake Avenue North, 01655, Worcester, MA, USA.

References
Best PJ, Hasdai D, Sangiorgi G, Schwartz RS, Holmes DR, Simari RD, Lerman A (1999) Apoptosis. Basic concepts and implications in coronary artery disease. Arterioscler Thromb Vasc Biol 19:14–22
Bolz SS, Vogel L, Sollinger D, Derwand R, De Wit C, Loirand G, Pohl U (2003) Nitric oxide-induced decrease in calcium sensitivity of resistance arteries is attributable to activation of the myosin light chain phosphatase and antagonized by the RhoA/Rho kinase pathway. Circulation 107:3081–3087
Brown JH, Del Re DP, Sussman MA (2006) The Rac and Rho hall of fame: a decade of hypertrophic signaling hits. Circ Res 98:730–742
Chen NX, Chen X, O'Neill KD, Atkinson SJ, Moe SM (2010) RhoA/Rho kinase (ROCK) alters fetuin-A uptake and regulates calcification in bovine vascular smooth muscle cells (BVSMC). Am J Physiol Renal Physiol 299:F674–F680

Cicek FA, Kandilci HB, Turan B (2013) Role of ROCK upregulation in endothelilal and smooth muscle vascular functions in diabetic rat aorta. Cardiovasc Diabetol 12:51, 10.1186/1475-2840-12-51

Dimopoulos G, Semba S, Kitazawa K, Eto M, Kitazawa T (2007) Ca^{2+}-dependant rapid Ca^{2+} sensitization of contraction in arterial smooth muscle. Circ Res 100:121–129

El-Remessy AB, Tawfik FE, Matragoon S, Pillai B, Caldwell RB, Caldwell RW (2010) Peroxynitrite mediates diabetes-induced endothelial dysfunction: possible role of Rho kinase activation. Exp Diabetes Res 2010:247861. doi:10.1155/2010/247861

Félétou M, Vanhoutte PM (2006) Endothelial dysfunction: a multifaceted disorder. Am J Physiol Heart Circ Physiol 291:H985–H1002

Gerthoffer WT (2007) Mechanisms of vascular smooth muscle cell migration. Circ Res 100:607–621

Hannan JL, Albersen M, Kutlu O, Gratzke C, Stief CG, Burnett AL, Lysiak JJ, Hedlund P, Bivalacqua TJ (2013) Inhibition of Rho-kinase improves erectile function, increases nitric oxide signaling and decreases penile apoptosis in a rat model of cavernous nerve injury. J Urol 189:1155–1161

Huh YH, Zhou Q, Liao JK, Kitazawa T (2011) ROCK inhibition prevents fetal serum-induced alteration in structure and function of organ-cultured mesenteric artery. J Muscle Res Cell Motil 32:65–76

Kitazawa T, Eto M, Woodsome TP, Khalequzzaman M (2003) Phosphorylation of the myosin phosphatase targeting subunit and CPI-17 during Ca^{2+} sensitization in rabbit smooth muscle. J Physiol 546:879–889

Kitazawa T, Semba S, Huh YH, Kitazawa K, Eto M (2009) Nitric oxide-induced biphasic mechanism of vascular relaxation via dephosphorylation of CPI-17 and MYPT1. J Physiol 587:3587–3603

Kizub IV, Pavlova OO, Johnson CD, Soloviev AI, Zholos AV (2010) Rho kinase and protein kinase C involvement in vascular smooth muscle myofilament calcium sensitization in arteries from diabetic rats. Br J Pharmacol 159:1724–1731

Lindqvist A, Nordström I, Malmqvist U, Nordenfelt P, Hellstrand P (1999) Long-term effects of Ca^{2+} on structure and contractility of vascular smooth muscle. Am J Physiol 277:C64–C73

Loirand G, Guérin P, Pacaud P (2006) Rho kinases in cardiovascular physiology and pathophysiology. Circ Res 98:322–334

Murata T, Suzuki N, Yamawaki H, Sato K, Hori M, Karaki H, Ozaki H (2005) Dexamethasone prevents impairment of endothelium-dependent relaxation in arteries cultured with fetal bovine serum. Eur J Pharmacol 515:134–141

Noma K, Oyama N, Liao JK (2006) Physiological role of ROCKs in the cardiovascular system. Am J Physiol Cell Physiol 290:C661–C668

Owens GK, Kumar MS, Wamhoff BR (2004) Molecular regulation of vascular smooth muscle cell differentiation in development and disease. Physiol Rev 84:767–801

Ozaki H, Karaki H (2002) Organ culture as a useful method for studying the biology of blood vessels and other smooth muscle tissues. Jpn J Pharmacol 89:93–100

Pagiatakis C, Gordon JW, Ehyai S, Mcdermott JC (2012) A novel RhoA/ROCK-CPI-17-MEF2C signaling pathway regulates vascular smooth muscle cell gene expression. J Biol Chem 287:8361–8370

Schwartz SM (1997) Smooth muscle migration in atherosclerosis and restenosis. J Clin Invest 99:2814–2816

Shimokawa H, Rashid M (2007) Development of Rho-kinase inhibitors for cardiovascular medicine. Trends in Pharmacol Sci 28:296–302

Shimokawa H, Takeshita A (2005) Rho-kinase is an important therapeutic target in cardiovascular medicine. Arterioscler Thromb Vasc Biol 25:1767–1775

Singh J, Maxwell PJ 4th, Rattan S (2011) Immunocytochemical evidence for PDBu-induced activation of RhoA/ROCK in human internal anal sphincter smooth muscle cells. Am J Physiol Gastrointest Liver Physiol 301:317–325

Somlyo AP, Somlyo AV (2003) Ca^{2+} sensitivity of smooth muscle and nonmuscle myosin II: modulated by G proteins, kinases, and myosin phosphatase. Physiol Rev 83:1325–1358

Thorne GD, Paul RJ (2003) Effects of organ culture on arterial gene expression and hypoxic relaxation: role of the ryanodine receptor. Am J Physiol Cell Physiol 284:C999–C1005

Uehata M, Ishizaki T, Satoh H, Ono T, Kawahara T, Morishita T, Tamakawa H, Yamagami K, Inui J, Maekawa M, Narumiya S (1997) Calcium sensitization of smooth muscle mediated by a Rho-associated protein kinase in hypertension. Nature 389:990–994

Woodsome TP, Polzin A, Kitazawa K, Eto M, Kitazawa T (2006) Agonist- and depolarization-induced signals for myosin light chain phosphorylation and force generation of cultured vascular smooth muscle cells. J Cell Sci 119:1769–1780

World Health Organization (2011) Global status report on noncommunicable diseases 2010. World Health Organization, Geneva

Wu J, LI J, Hu H, Liu P, Fang Y, Wu D (2012) Rho-kinase inhibitor, fasudil, prevents neuronal apoptosis via the Akt activation and PTEN inactivation in the ischemic penumbra of rat brain. Cell Mol Neurobiol 32:1187–1197

Xiao-Tao X, Qi-Bin S, Yi Y, Bin X, Peng R, Ze-Zhang T (2012) Inhibition of RhoA/ROCK signaling pathway promotes the apoptosis of gastric cancer cells. Hepatogastroenterology 59:2523–2526

Yao L, Chandra S, Toque HA, Bhatta A, Rojas M, Caldwell RB, Caldwell RW (2013) Prevention of diabetes-induced arginase activation and vascular dysfunction by Rho kinase (ROCK) knockout. Cardiovascular Res 97:509–519

13

Co-deposition and distribution of arsenic and oxidizable organic carbon in the sedimentary basin of West Bengal, India

Sayan Bhattacharya[1][*], Gunjan Guha[2], Dhrubajyoti Chattopadhyay[4], Aniruddha Mukhopadhyay[1], Purnendu K Dasgupta[5], Mrinal K Sengupta[5] and Uday C Ghosh[3]

Abstract

Background: The study investigated the extent of soil arsenic (As) contamination in agricultural plots in Bengal Delta through contaminated groundwater irrigation. Edaphic levels of As and oxidizable organic carbon (OOC) were tested along a depth gradient (0 to 160 ft) in agricultural plots.

Methods: Soil samples were collected from surface up to 160 feet depth at every 5 feet. By boreholes drilling and soil arsenic was estimated in ICP-MS. The analysis for estimation of soil OOC was performed. Statistical analyses were performed by one-way ANOVA to determine significant differences between groups at $P<0.05$.

Results: Concentration of As in soil was observed to be highest in surface soil, then decreased with increasing depth till about 40 ft, after which the concentration remained constant. Similar trends were noted for OOC. OOC showed significant ($P < 0.05$) positive correlations with the As levels.

Conclusion: Natural organic matter may enhance the release of As from soils and sediments into the soil solution, and thus can help in As leaching into the groundwater. Detailed investigation of the soil profile and the extent of bioaccumulation in the edible crops are urgently needed in those arsenic-contaminated areas.

Keywords: Arsenic; Oxidizable organic carbon; Bengal delta

Background

Arsenic (As) is a metalloid that executes severe environmental threats due to its extravagant toxicity and colossal abundance. It naturally occurs in over 200 different mineral forms, of which around 60% are arsenates, 20% are sulfides and sulfosalts, and the remaining are arsenides, arsenites, oxides, silicates, and elemental arsenic (Onishi 1969). Its source is mainly geological, but anthropological activities like mining, burning of fossil fuels, and uses of pesticides also lead to arsenic contamination (Bissen and Frimmel 2003). It is a potent endocrine disruptor and can alter hormone-mediated cell signaling even at extremely low concentrations (Kaltreider et al. 2001).

The permissible limit for As in water is 10 ppb, as recommended by the World Health Organization (WHO World Health Organization 2001). Its concentration in most rivers and lakes are below 10 ppb and that of the groundwater is about 1 to 2 ppb, except in areas with volcanic rock and sulfide mineral deposits (World Health Organization 2001). Contamination of As in groundwater has been widely reported in Bangladesh, India, China, Taiwan, Vietnam, USA, Argentina, Chile, and Mexico (WHO World Health Organization 2001). Among these, the Bengal Basin (of eastern India and Bangladesh), which holds more than a hundred million inhabitants, is regarded to be the most acutely arsenic-affected geological province in the world (Mukherjee et al. 2008). The use of As-contaminated groundwater for irrigation of crops in this region elevates arsenic concentration both in the surface soil and the plants (Mandal et al. 1998; Meharg and Rahman 2003; Roychowdhury et al. 2005). Soil arsenic levels are very much related with local well water arsenic concentration, which suggests that the source of soil contamination is the irrigation water (Bhattacharya et al.

* Correspondence: sayan_evs@yahoo.co.in
[1]Department of Environmental Science, University of Calcutta, Kolkata, West Bengal, India
Full list of author information is available at the end of the article

2012). The absorption of arsenic by plants is influenced by the concentration of arsenic in the soil. In Bangladesh, where irrigation is carried out with arsenic-contaminated groundwater, soil arsenic level can reach up to 83 mg/kg (Roychowdhury et al. 2005).

Arsenic is associated with the primary sulfides: (hydro) oxides of Al, Fe, and Mn; clays; sulfates; phosphates; and carbonates in the Bengal Delta Plain (Foster 2003). Arsenic is released in the soil by weathering of arsenopyrite (FeAsS) and sulfide minerals. The grain sizes of soil particles play an important role in controlling the distribution and mobility of arsenic. The surface area of the fine-grained particles is large, and hence, they can adsorb more arsenic (Bhattacharya et al. 2007). For this reason, clay minerals and Fe, Al, and Mn (hydro) oxides are important sinks for arsenic in the aquifers and sediment layers in Bengal Delta. Arsenic in soil normally occurs in pentavalent state under oxidizing conditions while under reducing conditions trivalent As(III) species prevail that is more mobile and bioavailable (Bhattacharya et al. 2012). The enrichment of arsenic in soil is primarily due to its co-precipitation not only with or sorption on poorly crystalline Fe oxyhydroxides which precipitate from the irrigation water but also to sorption on other components (including crystalline Fe oxides and hydroxides, clay minerals, and organic matter) of the soil matrix (Bhattacharya et al. 2007). However, the relation between arsenic and soil organic matter does not necessarily imply that the fixation of arsenic is controlled by organic matter but is rather due to the fact that the vertical distribution of both arsenic and organic matter depends on their penetration depth into the soil (Norra et al. 2005).

Methods

Site of the study

The study was performed at Kalinarayanpur (23°22'N, 88°56'E), Nadia, West Bengal, India. The region is considered to be among the severely arsenic-affected zones of Bengal Delta (Chakraborti et al. 2002). The agricultural system in this region is mostly dependent on irrigation with As-contaminated groundwater obtained from a depth ranging from 70 to 600 ft through tube wells. Huge amount of groundwater is used for agricultural irrigation. Much of this groundwater is contaminated with arsenic, which is deposited in the soil in contact with the irrigation water throughout the year.

Analysis of soil

Analyses of As and OOC in soil were performed in soil samples obtained from agricultural plots irrigated with groundwater. Boreholes were drilled using a conventional household technique by hand percussion and reversed circulation. Though the method allows a continuous recovery of the drilled material, it allows only the collection of disturbed bulk samples (Horneman et al. 2004). Soil samples were collected from surface up to 160 ft depth at every 5 ft. All samples were packed in individual air-tight polyethylene bags and stored at 4°C for further analysis.

Estimation of soil As was performed by inductively coupled plasma-mass spectrometry (ICP-MS). Collected soil samples were dried in open air under diffused sunlight followed by drying in hot-air oven at 50°C for 24 h. Each sample was then ground manually with a mortar and a pestle to form a fine powder, which was passed through a sieve to get homogenized sample particles. Five milliliters

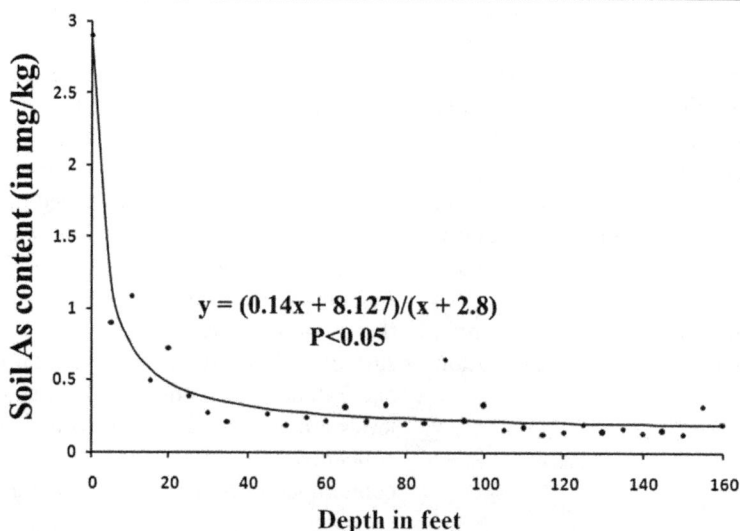

$$y = (0.14x + 8.127)/(x + 2.8)$$
$$P < 0.05$$

Figure 1 Trend of As imbibition along the depth gradient of soil from the surface to 160 ft. As levels were observed to decrease up to a depth of 45 ft (approximately), beyond which the level remained stable. Data expressed at $P < 0.05$.

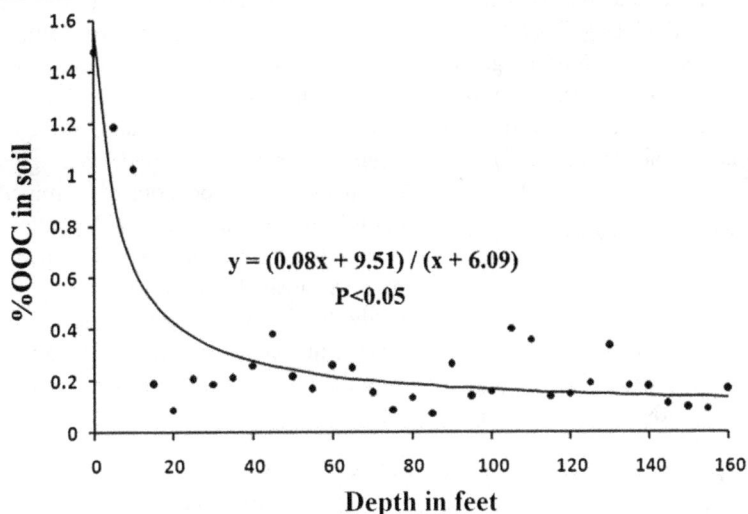

Figure 2 Changes in the percentage of OOC with increasing soil depth. Data are presented at $P < 0.05$.

of 2 M H_2SO_4 was added to about 1 g of prepared soil sample powder, vortexed for 1 min, and centrifuged for 5 min at 1,500 rpm. The supernatant was then decanted and analyzed for total arsenic by ICP-MS using an X Series II ICP-MS (Thermo Scientific Inc., Waltham, MA, USA). The operating conditions for the ICP-MS analysis were as follows: Rf power: 1,400 W, cool gas: 13.0 L/min, auxiliary gas: 0.9 L/min, nebulizer gas: 0.95 L/min, spray chamber temperature: 3°C, sampling and skinner cone: Ni, expansion chamber pressure: 1.9 bar, analyzer chamber pressure: 3.6×10^{-7} mbar, nebulizer back pressure: 2.1 bar, sampling depth: 150 mm, detector mode: pulse, element

monitored: [75]As. The Thermo PlasmaLab ver. 2.5.5.290 (Thermo Scientific Inc.) software was used to evaluate the concentrations of As in the different soil samples obtained from different depths (0 to 160 ft).

The analysis for estimation of soil OOC was performed following the protocol of Walkley and Black (1934). Fresh soil samples were passed through 0.2-mm sieve and were stored in 4°C. Ten milliliters of 1 N $K_2Cr_2O_7$ and 20 ml of concentrated H_2SO_4 were added to 0.5 g of each soil sample and swirled for 1 min. The samples were then kept undisturbed for 30 min; following which, 200 ml of de-ionized water, 10 ml of H_3PO_4, and 1 ml of diphenylamine

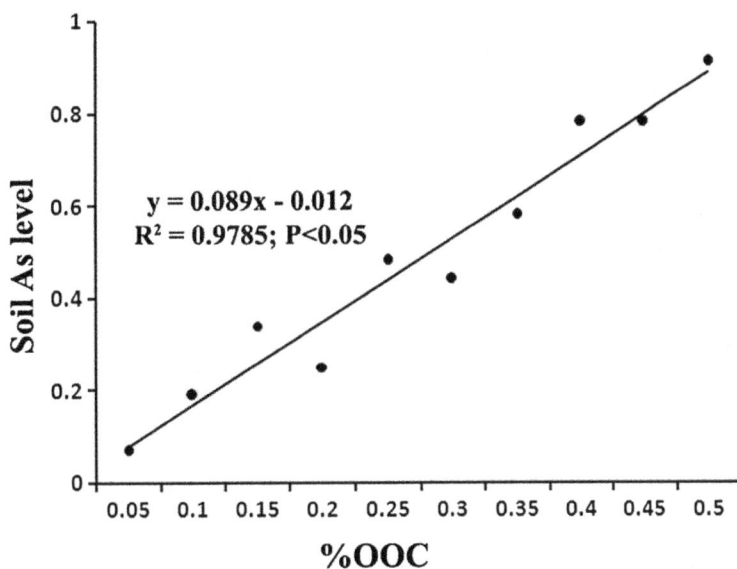

Figure 3 Relationship between the As level in soil with percentage of OOC. OOC shows strong and significant positive correlation with soil As level ($P < 0.05$).

indicator were added to each sample and titrated against 0.5 N ferrous ammonium sulfate $(Fe(NH_4)_2(SO_4)_2.6H_2O)$ solution till the color changed from blue violet to green. The volume of $Fe(NH_4)_2(SO_4)_2.6H_2O$ solution consumed was recorded. Percentages of OOC were calculated for each sample by the following formula (Schollenberger 1927):

$$\%OOC = \left[\left(V_{blank} - V_{sample} \right) \times 0.3 \times M \right] / \left[\text{weight of soil(g)} \right]$$

Statistical analyses

All analyses were carried out in triplicates. Data were presented as mean ± standard deviation. Statistical analyses were performed by one-way ANOVA to determine significant differences between the groups at $P < 0.05$. Estimated correlations were tested for significance by Student's t test at the same confidence limit. MATLAB ver. 7.0 (Natick, MA, USA), SPSS ver. 9.05 (Chicago, IL, USA), and Microsoft Excel 2007 (Roselle, IL, USA) were used for the statistical and graphical evaluations.

Results and discussion

The ICP-MS data revealed that the soil samples had varying levels of As from the surface to a depth of 160 ft. Figure 1 illustrates that As levels significantly ($P < 0.05$) decreased from the surface (2.89 mg/kg of soil) till a depth of about 45 ft (0.3 mg/kg); following which, there was a constant concentration of As along the gradient of increasing depth. Hence, the surface level of As was approximately ten times greater than that observed at a depth of 45 ft or greater. Interestingly, OOC in soil showed a gradual decrease with depth till approximately 45 to 50 ft (Figure 2) and then showed constant levels beyond 45 to 50 ft depth. These results were identical to the trend that was observed for the soil As content.

Along the depth gradient, strong significant correlations ($P < 0.05$) were found between soil As level and %OOC ($R^2 = 0.9785$). This was in synchrony with the fact that As was co-deposited with OOC in soil, and the correlation between them was due to As retention and high OOC inputs in the vegetated zones of the Bengal Basin (Meharg et al. 2006). Furthermore, a previous study (Wang and Mulligan 2006) reported that organic matter might serve as binding agents, thereby reducing As mobility in soil. This study further extended the strong relationship of OCC and As observed in the current study. Figure 3 demonstrates the As-OOC correlation along with their respective regression data.

A number of previous studies proved the relation between natural organic matter and arsenic distribution in soil and sediments. An increase in dissolved organic carbon content can promote both As(V) and As(III) solubilization in soils (Dobran and Zagury 2006). Natural organic matter may enhance the release of As from soils and sediments into the soil solution and thus can help in As leaching

into the groundwater (Wang and Mulligan 2006). The main influencing factors are competition for available adsorption sites, formation of aqueous complexes, and/or changes in the redox potential of site surfaces and As redox speciation (Wang and Mulligan 2006). It has also been observed that hydrogeological features of the sediments, the proportions of Fe minerals and sedimentary organic matter, and the concentration of dissolved humic materials, all influence the accumulation and mobilization of As (Varsányi and Kovács 2006). However, the present study first reported the vertical distribution pattern of arsenic in soil in the Bengal Delta and its strong correlation with oxidizable organic carbon content of soil.

Conclusions

The absorption of arsenic by plants is influenced by the concentration of arsenic in the soil. In Bangladesh, where irrigation is carried out with arsenic-contaminated groundwater, soil arsenic level can reach up to 83 mg/kg (Roychowdhury et al. 2005). Except in the rainy season, the agricultural land in the study area has been exposed to irrigated groundwater round the year. It can be the main reason for which there was maximum amount of arsenic found in the surface soil in the present study (2.89 mg/kg). Sometimes, the farmers used to run the shallow tube wells in the rainy season due to insufficient rain. Most of the vegetables and other crops used by the villagers were cultivated in this area and entered the local market. The bioavailability of arsenic in edible plants cultivated in Bengal Delta must be investigated in detail to understand the importance of arsenic exposure from these food sources. Intensive investigation on a complete food chain is urgently needed in the arsenic contaminated zones, which should be our priority in future researches. Additionally, in-depth study of the chemical interactions between arsenic and organic matter and organic carbon could be beneficial in this regard.

Competing interests
The authors declare that they have no competing interests.

Authors' contributions
SB performed the field sample collection, analysis of arsenic and oxidizable organic carbon in soil, and helped in statistical analysis. GG performed the statistical analysis. DC supervised the work. AM supervised the work. PKD helped in arsenic analysis of soil samples. MKS helped in arsenic analysis of soil samples. UCG supervised the work. All authors read and approved the final manuscript.

Author details
[1]Department of Environmental Science, University of Calcutta, Kolkata, West Bengal, India. [2]Department of Pharmaceutical Sciences, College of Pharmacy, Oregon State University, Corvallis, OR, USA. [3]Department of Chemistry, Presidency University, Kolkata, West Bengal, India. [4]B. C. Guha Centre for Genetic Engineering and Biotechnology, University of Calcutta, Kolkata, West Bengal, India. [5]Department of Chemistry and Biochemistry, University of Texas at Arlington, Arlington, TX, USA.

References

Bhattacharya P, Mukherjee AB, Bundschuh J, Zevenhoven R, Loeppert R (eds) (2007) Arsenic in soil and groundwater environment: biogeochemical interactions, health effects and remediation. Elsevier, Amsterdam

Bhattacharya S, Gupta K, Debnath S, Ghosh UC, Chattopadhyay DJ, Mukhopadhyay A (2012) Arsenic bioaccumulation in rice and edible plants and subsequent transmission through food chain in Bengal basin: a review of the perspectives for environmental health. Toxicol Environ Chem 94(3):429–441

Bissen M, Frimmel FH (2003) Arsenic – a review. Part I: occurrence, toxicity, speciation and mobility. Acta Hydrochim Hydrobiol 31:9–18

Chakraborti D, Rahman MM, Paul K, Chowdhury UK, Sengupta MK, Lodh D, Chanda CR, Saha KC, Mukherjee SC (2002) Arsenic calamity in the Indian subcontinent: what lessons have been learned? Talanta 58:3–22

Dobran S, Zagury GL (2006) Arsenic speciation and mobilization in CCA-contaminated soils: influence of organic matter content. Sci Total Environ 364:239–250

Foster AL (2003) Spectroscopic investigations of arsenic species in solid phases. In: Welch AH, Stollenwerk KG (eds) Arsenic in groundwater: geochemistry and occurrence. Kluwer Academic Publishers, Boston

Horneman A, van Geen A, Kent DV, Mathe PE, Zheng Y, Dhar RK, O'Connel S, Hoque MA, Aziz Z, Shamsudduha M, Seddique AA, Ahmed KM (2004) Decoupling of As and Fe release to Bangladesh groundwater under reducing conditions. Part I: evidence from sediment profiles. Geochim Cosmochim Acta 18:3459–3473

Kaltreider RC, David MA, Lariviere JP, Hamilton JW (2001) Arsenic alters the function of glucorticoid receptor as a transcription factor. Environ Health Persp 109:245–251

Mandal BK, Chowdhury TR, Samanta G, Mukherjee DP, Chanda CR, Saha KC, Chakraborti D (1998) Impact of safe water for drinking and cooking on five arsenic-affected families for 2 years in West Bengal, India. Sci Total Environ 218:185–201

Meharg AA, Rahman MM (2003) Arsenic contamination in Bangladesh paddy field soils: implication for rice contribution to arsenic consumption. Environ Sci Technol 37:229–234

Meharg AA, Scrimgeour C, Hossain SA, Fuller K, Cruickshank K, Williams PN, Kinniburgh DG (2006) Codeposition of organic carbon and arsenic in Bengal delta aquifers. Environ Sci Technol 40:4928–4935

Mukherjee A, von Brömssen M, Scanlon BR, Bhattacharya P, Fryar AE, Hasan MA, Ahmed KM, Chatterjee D, Jacks G, Sracek O (2008) Hydrogeochemical comparison and effects of overlapping redox zones on groundwater arsenic near the Western (Bhagirathi sub-basin, India) and Eastern (Meghna sub-basin, Bangladesh) margins of the Bengal Basin. J Contam Hydrol 99:31–48

Norra S, Berner ZA, Agarwala P, Wagner F, Chandrasekharam D, Stuben D (2005) Impact of irrigation with As rich groundwater on soil and crops: A geochemical case study in West Bengal Delta Plain, India. Appl Geochem 20:1890–1906

Onishi H (1969) Arsenic. In: Wedepohl KH (ed) Handbook of geochemistry. Springer, New York

Roychowdhury T, Tokunaga H, Uchino T, Ando M (2005) Effect of arsenic-contaminated irrigation water on agricultural land soil and plants in West Bengal, India. Chemosphere 55:799–810

Schollenberger CJ (1927) A rapid approximate method for determining soil organic matter. Soil Sci 24:65–68

Varsányi I, Kovács L (2006) Arsenic, iron and organic matter in sediments and groundwater in the Pannonian Basin, Hungary. Appl Geochem 21:949–963

Walkley A, Black IA (1934) An examination of Degtjareff method for determining soil organic matter and a proposed modification of the chromic acid titration method. Soil Sci 37:29–37

Wang S, Mulligan CN (2006) Effect of natural organic matter on arsenic release from soils and sediments into groundwater. Environ Geochem Health 28:197–214

World Health Organization (2001) United Nations synthesis report on arsenic drinking water. World Health Organization, Geneva

14

Occurrence, risk assessment, and source apportionment of heavy metals in surface sediments from Khanpur Lake, Pakistan

Javed Iqbal and Munir H Shah[*]

Abstract

Background: The present study was carried out to assess the seasonal variations, source apportionment, and risk assessment of heavy metals (Cd, Cr, Cu, Fe, Mn, Pb, and Zn) in the surface sediments from the Khanpur Lake, Pakistan.

Methods: Composite samples are collected and processed to measure the concentrations of heavy metals in $Ca(NO_3)_2$ extract and acid extract of the sediments using flame atomic absorption spectrophotometry.

Results: The highest concentrations in acid extracts of the sediments are found for Fe, followed by Mn, while the least concentrations are noted for Cd. Relatively higher extraction efficiencies in $Ca(NO_3)_2$ extract are observed for Pb and Cd, which also reveal extremely severe enrichment in the sediments as shown by the enrichment factor. Geoaccumulation index shows moderate and strong to extreme pollution of Pb and Cd, respectively, whereas potential ecological risk factor exhibits low to very high risk by Cd; the cumulative ecological risk index reveals low to very high risk of contamination in the sediments as a whole. Principal component analysis and cluster analysis reveal dominant anthropogenic contributions of Cd, Pb, Cr, and Zn.

Conclusion: Measured concentrations of Cd, Cr, Cu, Mn, and Pb in the sediments exceed the sediment quality guideline for the lowest effect levels (LEL), while the concentrations of Cd and Pb are also higher than the effects range low (ERL) values, manifesting occasional adverse biological effects to the surrounding flora and fauna. Moreover, the mean effects range medium (ERM) quotient reveals 21% probability of toxicity in the sediments.

Keywords: Sediment; Metal; Risk assessment; Multivariate analysis; AAS; Pakistan

Background

Contamination of aquatic ecosystems with heavy metals has received much attention due to their toxicity, abundance, and persistence in the environment and subsequent accumulation in aquatic habitats (Arnason and Fletcher 2003). Elevated levels of heavy metals in environmental compartments, such as aquatic sediments, may pose a risk to human health due to their transfer in aquatic media and uptake by living organisms, thereby entering the food chain (Sin et al. 2001; Varol and Sen 2012). Heavy metals may enter a freshwater reservoir from a variety of sources, either natural or anthropogenic (Adaikpoh et al. 2005; Akoto et al. 2008). Generally, in natural ecosystems, most of the metals are present in very low concentrations and are mostly derived from rock and soil weathering (Reza and Singh 2010; Varol and Sen 2012). Major anthropogenic sources of heavy metal pollution are mining and smelting activities, atmospheric deposition, disposal of untreated/partially treated urban and industrial effluents, metal chelates from different industries, and haphazard use of heavy metal-containing fertilizers and pesticides during agricultural activities (Martin 2000; Nouri et al. 2008; Reza and Singh 2010).

Sediments are ecologically sensitive components of the aquatic ecosystems and are also a reservoir of the contaminants, which take part considerably in maintaining the trophic status for any water reservoir (Singh et al. 2005). Depending upon the physicochemical conditions, sediments can act both as source and sink for nutrients and heavy metals. Hence, sediments are not only considered as carriers of contaminants but also potential secondary sources of contaminants in an aquatic ecosystem.

* Correspondence: munir_qau@yahoo.com
Department of Chemistry, Quaid-i-Azam University, Islamabad 45320, Pakistan

Consequently, the analysis of sediments is a useful method to study the heavy metal pollution in any area (Gielar et al. 2012; Varol and Sen 2012). The toxicity and mobility of the metals in sediments vary among different chemical forms (Cuong and Obbard 2006; Yu et al. 2010). Therefore, the evaluation of distribution and mobility/potential bioavailability of heavy metals in surface sediments is an important step to evaluate the degree of contamination of an aquatic ecosystem (Martin et al. 2009; Sprovieri et al. 2007). Assessment of biologically available fractions of heavy metals helps to evaluate their potential for mobilization and availability to benthic organisms (Rodrigues et al. 2010). Various chemical extraction methods have been suggested to determine the bioavailable fractions of the metals in sediments. Generally, weak acids/electrolytes are used to extract the bioavailable fractions of the metals in sediments (An and Kampbell 2003).

Major objectives of the present study are (i) to measure the concentrations of heavy metals (Cd, Cr, Cu, Fe, Mn, Pb, and Zn) in sediments during summer and winter; (ii) to determine potential ecological risk using enrichment factor (EF), geoaccumulation index (I_{geo}), potential ecological risk factor (E_i), and potential ecological risk index (RI); (iii) to identify risks of potential toxicity by comparison with sediment quality guidelines (SQGs); (iv) to determine potential bioavailability and mobility of the metals; and (v) to define their natural/anthropogenic contributions using multivariate statistical methods. It is anticipated that the study would provide a baseline data regarding the distribution and accumulation of heavy metals in the sediments and would help reduce the contamination by identifying the major pollution sources.

Methods

Study area

Khanpur Lake (longitude 72°56′E and latitude 33°48′N) is situated on the Haro river near the town of Khanpur, about 40 km northwest of Islamabad, Pakistan (Figure 1). It supplies drinking water to the inhabitants of twin cities of Islamabad and Rawalpindi, Pakistan, and irrigation water to

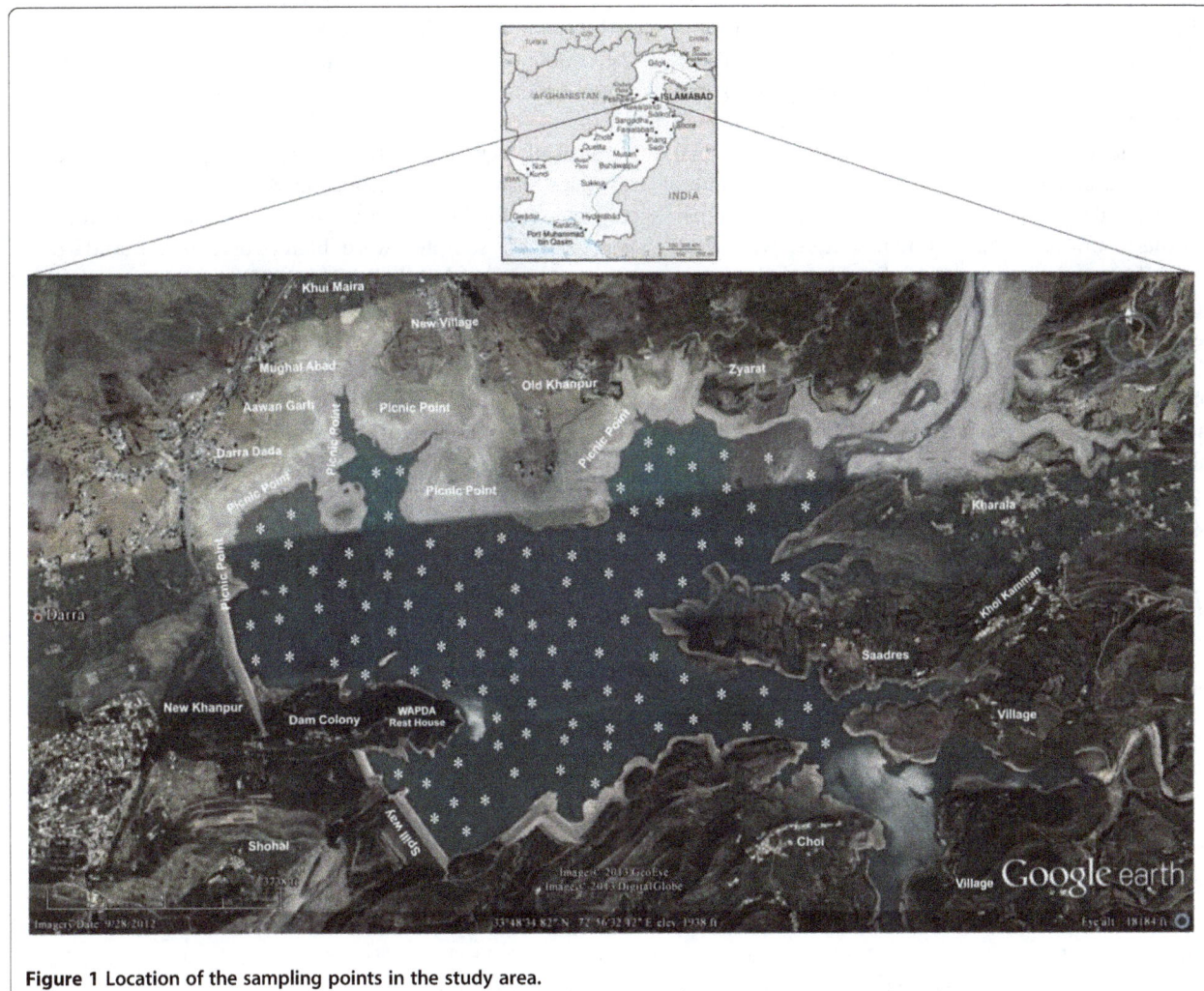

Figure 1 Location of the sampling points in the study area.

the agricultural areas surrounding the cities. It was built in 1983 with the storage capacity of 140 million m^3 of water. It is 51 m high with an average depth of 15 m. The gross storage capacity of the reservoir is 0.132 km^3 with a total catchment area of 798 km^2. The surface area of the reservoir varies from maximum of 1,806 ha to minimum of 215 ha. In past, the lake was leased for commercial exploitation. The area around the lake has been planted with flowering trees and laid out with gardens, picnic spots, and secluded paths. The lake is used for picnics, fishing, boating, sailing, water skating, and diving. Untreated and/or partially treated urban and industrial effluents, road and agricultural run offs, poultry farms wastes, and contaminants released during the recreational use of motorboats are among the suspected sources of pollution in the lake.

Sampling and preservation

A total of 100 composite surface sediment samples from Khanpur Lake, Pakistan, were collected in the summer and winter of 2008. Each sediment sample was a composite of three to five sub-samples from an area of 1 to 2 m^2 and collected using a snapper (Ø 5 cm) in top layer (0 to 10 cm). The sediment samples were taken from the central portion of the snapper with a plastic spatula to avoid any contamination from the metallic parts of the sampler. Before transferring the samples in pre-cleaned Ziploc polythene bags (S. C. Johnson & Son, Inc., Racine, WI, USA), the above water was decanted. The samples were kept in airtight large plastic containers for transport to the laboratory. The sediment samples were then oven-dried, grounded, homogenized, and sealed in precleaned polythene bags and stored in a refrigerator until further processing (Radojevic and Bashkin 1999).

Sample processing and analysis

The samples were processed to assess the $Ca(NO_3)_2$-extractable and acid-extractable fractions of heavy metals. A single-step extraction procedure using 0.1 M $Ca(NO_3)_2$ was applied to the sediment samples at room temperature in order to evaluate the bioavailable metal fractions (An and Kampbell 2003). An aliquot of 5 g of the sample was added to 50-mL solution of 0.1 M $Ca(NO_3)_2$, and the

extraction was performed in pre-cleaned glass vessel by shaking on an auto-shaker at 240 vibrations/min for 16 h. A blank sample was also processed with the same amount of reagents without sediment sample. Three replicate extractions were performed for each sample. The final extracts were separated from the solid residues through filtration using a fine (0.45-μm pore) filter paper (An and Kampbell 2003; Radojevic and Bashkin 1999; Rodrigues et al. 2010). To measure the acid-extractable fractions, 1- to 2-g dried sediment sample was digested in a microwave system using an acid mixture of 9 mL HNO_3 and 3 mL HCl (USEPA 2007). Three replicate extractions were performed for each sample. The digests were then filtered through the fine filter paper and made up to 50 mL with double distilled water and stored at 4°C. A blank sample was also processed with the same amount of chemical reagents without sediment sample. Heavy metals (Cd, Cr, Cu, Fe, Mn, Pb, and Zn) in the sediment samples were analyzed using a flame atomic absorption spectrophotometer (Shimadzu AA-670, Kyoto, Japan). The calibration line method was used for quantification of the metals, and the samples were appropriately diluted whenever required (Radojevic and Bashkin 1999; Shah et al. 2012). The optimum analytical conditions used for the quantification of the selected metals on the spectrophotometer are given in Table 1. During sample collection and analysis, strict QA/QC measures were taken including method blanks, analysis of standard reference material, and analysis of duplicate samples. The reagents for the blanks were prepared during each extraction, and all samples were blank-corrected. Standard reference material (NIST SRM-2709) was also used to ensure the reliability of the metal data as shown in Table 1. The measured metal levels closely matched with the certified values. Moreover, reliability of the finished data was also ensured using known spikes and by conducting interlaboratory comparison, and the results were within ±1.5%. Working standards of the metals were prepared from a stock solution of 1,000 mg/L (E-Merck, Darmstadt, Germany) by successive dilutions. The moisture content of each sediment sample was determined by drying separate 5-g sample in an oven (105°C ± 2°C) to constant weight.

Table 1 Description of optimum analytical conditions and analysis of selected metals in SRM

	Cd	Cr	Cu	Fe	Mn	Pb	Zn
Wavelength (nm)	228.8	357.9	324.8	248.3	279.5	217.0	213.9
HC lamp current (mA)	4.0	5.0	3.0	8.0	5.0	7.0	4.0
Slit width (nm)	0.3	0.5	0.5	0.2	0.4	0.3	0.5
Fuel gas flow rate (L/min)	1.8	2.6	1.8	2	1.9	1.8	2
Detection limit (μg/L)	4.0	6.0	4.0	6.0	3.0	10.0	2.0
SRM-certified level (mg/kg)	0.38	130	34.6	35,000	538	18.9	106
SRM-measured level ± SD (mg/kg)	0.36 ± 0.03	138 ± 8	35.2 ± 1.2	34,300 ± 385	547 ± 11	19.3 ± 1.4	109 ± 3.2

The analytical conditions were maintained on AAS using air-acetylene flame, and the standard reference material is SRM-2709.

From this, a correction to dry mass was obtained, which was applied to all reported metal concentrations. All the measurements were made in triplicate.

Statistical analysis

Statistical analysis can be used to evaluate the complex eco-toxicological processes by showing the relationship and interdependency among the variables and their relative weights. Basic statistical parameters, such as minimum, maximum, mean, median, standard error (SE), and skewness, were computed along with correlation study. Multivariate techniques have been used for evaluation and characterization of analytical data (Fadigas et al. 2010). Principal component analysis (PCA) and cluster analysis (CA) are among the most popular methods. The PCA finds out the diagonalization of the covariance or correlation matrix transforming the original chemical measurements into linear combinations of these measurements, which are the principal components (PCs). It rotates the coordinate space axes so that the explained variance of each PC is maximized. This technique allows for data reduction from higher to lower dimensional spaces to simplify their representation. Nonetheless, CA demonstrates the similarities between variables by examining the interpoint distances representing all possible variables in the higher dimensional space. The PCA was performed using varimax normalized rotation on the dataset, and the CA was applied to the standardized matrix of the samples using Ward's method, and the results are reported in the form of dendrograms. PCA and CA complement each other and have been widely used in environmental studies (Gielar et al. 2012; Iqbal and Shah 2011; Shah et al. 2012; Singh et al. 2005).

Pollutant indicators and risk assessment

To gauge the degree of contamination and to distinguish natural and anthropogenic inputs, EFs, I_{geo}, E_i, and RI are computed (Cukrov et al. 2011; Hakanson 1980; Muller 1969). EFs are calculated (Cukrov et al. 2011; Iqbal and Shah 2011; Luoma and Rainbow 2008; Tessier et al. 2011) by comparing the measured metal levels to the preindustrial levels (Lide 2005). In order to avoid the overestimation or underestimation of the enrichment; geochemical normalization based on the concentration of a conservative element is commonly employed. The purpose of normalization is to correct changes in the nature of sediments, which may influence the contaminant distribution. Various conservative elements may be used: Al, Fe, Th, Ti, Zr, etc. (Larrose et al. 2010; Reimann and de Caritat 2005). Iron is chosen as the conservative element for normalization in this work. The interest of using Fe content is its relationship to the abundance of clay and other aluminum silicates in the sediments. Its contents are influenced by natural sedimentation and the effects of enhanced erosion, but not by pollution (Iqbal and Shah 2011). The normalized

EF is usually computed as double ratios of the target element and Fe as a reference element in the examined sediments and Earth's crust using the following relationship:

$$EF = \frac{[X/Fe]_{sample}}{[X/Fe]_{crust}},$$

where $[X/Fe]_{sample}$ and $[X/Fe]_{crust}$ refer, respectively, to the ratios of mean concentrations (mg/kg, dry weight) of the target element and Fe in the sediments and continental crust (Lide 2005).

The I_{geo} enables the assessment of contamination by comparing the measured and pre-industrial concentrations of the metals in the Earth's crust (Loska et al. 2004; Muller 1969). It is computed using the following relationship:

$$I_{geo} = \log_2\left(C_n \middle/ 1.5B_n\right),$$

where C_n is the measured concentration of the element in the sediment samples, and B_n is the geochemical background value in the Earth's crust (Lide 2005). Factor 1.5 is introduced to minimize the effect of possible variations in the background values which may be attributed to lithogenic variations.

RI is introduced to assess the degree of heavy metal pollution in sediments, which was originally introduced by Hakanson (1980), according to the toxicity of heavy metals and the response of the environment:

$$RI = \sum E_i$$
$$E_i = T_i f_i$$
$$f_i = C_i/C_b,$$

where RI is computed as the sum of all risk factors in sediments, E_i is the monomial potential ecological risk factor for individual factors, and T_i is the metal toxic factor. Based on the standardized heavy metal toxic factor developed by Hakanson (1980), the order of the level of heavy metal toxicity is Cd > Pb = Cu > Cr > Zn. The toxic factors for the metals are 30, 5, 5, 2, and 1, respectively. f_i is the metal pollution factor, C_i is the concentration of metal in the sediments, and C_b is the reference value of a given metal in the Earth's crust (Lide 2005).

Multiple contamination which is often encountered in natural environments affected by human activities is also calculated in terms of mean-effects range medium-quotient (m-ERM-Q) by the following relationship (de Vallejuelo et al. 2010; Long and MacDonald 1998; Tessier et al. 2011):

$$m\text{--}ERM\text{--}Q = \frac{\sum_{i=1}^{n} C_i/ERM_i}{n},$$

where C_i is the concentration of a metal in a sediment, ERM_i is the ERM value for metal i, and n is the number of metals.

Results and discussion
Distribution of heavy metals in the sediments
Concentrations of heavy metals in acid extracts of the sediments during summer and winter in terms of statistical distribution parameters are shown in Table 2. During summer, the data reveal dominant mean level of Fe (4,630 mg/kg), followed by Mn (447.5 mg/kg), while the average concentration of Cd (1.883 mg/kg) is the lowest. On the average basis, the metals follow a decreasing concentration order: Fe > Mn > Zn > Cu > Cr > Pb > Cd. Among the metals, Fe indicates almost comparable mean and median levels with lower skewness, indicating relatively symmetrical distribution in acid extract of the sediments. The counterpart statistical data during winter show the highest average levels of Fe (3,791 mg/kg), followed by Mn (321.4 mg/kg), whereas Pb (18.24 mg/kg) and Cd (2.457 mg/kg) are found at relatively lower levels. On the mean basis, the metals exhibit a decreasing concentration order: Fe > Mn > Zn > Cr > Cu > Pb > Cd. Relatively normal distribution is revealed by Cd and Pb, which are also associated with lower skewness. Maximum dispersion in terms of SE is exhibited by Fe. Overall, significantly elevated average levels of the metals (except Cd and Cr) are noticed during summer compared with winter (Table 2). It could be due to the leaching of the metals into the reservoir from the roadside and agricultural runoffs during wet summer season.

Correlation study
The correlation coefficient matrix of heavy metals in the acid extract of the sediments during summer and winter is given in Table 3. During summer, strong correlations of Fe with Mn and Cu, Cr with Zn, and Cu with Mn are noted. Some other significant relationships of Pb with Cd and Cr are also observed. However, Pb and Zn show negative associations with Cu, Fe, and Mn, revealing their opposing distribution in the sediments during summer. The counterpart data related to the metal levels in the sediments during winter indicate strong correlations of Zn with Cu and Mn, Cu with Mn, and Cr with Cd and Cu, thus manifesting close association of these metals which might share common sources. Some significant correlations for Pb with Cr, Cu, Mn, and Zn are also observed. Fe does not show any significant relationship with other heavy metals in the sediments during winter, suggesting its independent variations in the sediments.

Pollution indices
The range and mean EF values of heavy metals in acid extract of the sediments during summer and winter are shown in Figure 2a. Seven degrees of contamination are commonly defined (Birch et al. 2003): EF < 1 indicates no enrichment, EF < 3 minor enrichment, EF = 3 to 5 moderate enrichment, EF = 5 to 10 moderately severe enrichment, EF = 10 to 25 severe enrichment, EF = 25 to 50 very severe enrichment, and EF > 50 extremely severe enrichment. During summer, on the average basis, Cr reveals moderate enrichment, Cu and Mn indicate moderately severe enrichment, Zn manifests severe enrichment, Pb shows very severe enrichment, and Cd illustrates extremely severe enrichment in the sediments. The geochemical normalization study during winter reveals that Cr, Cu, and Mn indicate moderately severe

Table 2 Statistical summary of heavy metal distribution in acid extract and Ca(NO$_3$)$_2$ extract of the sediments

		Summer (n = 50)					Winter (n = 50)					p value
		Min	Max	Mean	SE	Skew	Min	Max	Mean	SE	Skew	
Acid extract	Cd	0.196	4.500	1.883	0.234	0.584	0.149	5.183	2.457	0.235	−0.084	<0.05
	Cr	11.35	63.45	34.66	2.293	−0.262	23.82	68.97	37.65	1.543	1.790	Non-significant
	Cu	25.15	49.39	36.84	1.285	−0.072	18.22	51.53	28.05	1.314	1.166	<0.05
	Fe	3,835	5,186	4,630	57.83	−0.350	3,523	4,182	3,791	30.95	0.426	<0.05
	Mn	236.2	836.7	447.5	32.97	0.713	167.7	886.0	321.4	26.00	2.234	<0.05
	Pb	9.739	78.48	33.71	3.419	0.771	0.412	39.03	18.24	1.966	−0.003	<0.01
	Zn	70.71	114.4	86.09	2.032	0.650	42.24	115.2	61.90	2.459	2.190	<0.05
Ca(NO$_3$)$_2$ extract	Cd	0.004	0.122	0.058	0.006	−0.081	0.016	0.146	0.071	0.006	0.175	Non-significant
	Cr	0.042	0.546	0.217	0.027	0.794	0.008	0.478	0.230	0.023	0.061	Non-significant
	Cu	0.008	0.220	0.098	0.008	0.212	0.012	0.134	0.073	0.006	−0.201	<0.05
	Fe	0.020	28.50	2.069	0.985	4.415	0.248	1.218	0.658	0.053	0.314	<0.01
	Mn	0.004	0.274	0.078	0.012	1.149	0.010	0.072	0.042	0.003	0.178	<0.01
	Pb	0.206	2.032	1.192	0.085	−0.343	0.140	2.082	1.205	0.087	−0.254	Non-significant
	Zn	0.010	0.656	0.179	0.023	1.972	0.056	0.186	0.118	0.007	0.175	<0.05

The heavy metal distribution is expressed in milligrams per kilogram.

Table 3 Correlation coefficients (r)* matrix for heavy metals in acid extract of sediments during summer and winter

	Cd	Cr	Cu	Fe	Mn	Pb	Zn
Cd	1	0.580	0.389	−0.029	0.473	0.181	0.223
Cr	0.320	1	0.657	0.087	0.319	0.412	0.345
Cu	0.171	0.153	1	0.143	0.685	0.480	0.803
Fe	0.071	−0.056	0.615	1	0.126	0.151	−0.016
Mn	0.086	−0.031	0.863	0.551	1	0.425	0.619
Pb	0.362	0.432	−0.068	−0.187	−0.087	1	0.427
Zn	0.103	0.540	−0.028	−0.208	−0.245	0.062	1

Values for summer are below the diagonal, and those for winter are above the diagonal. *r values >0.330 or <−0.330 are significant at $p < 0.01$.

enrichment; Pb and Zn explicate severe enrichment, and Cd illuminates extremely severe enrichment. Overall, Cd emerge as the major pollutant during both seasons; Pb poses severe to extremely severe enrichment during summer and minor to very severe enrichment during winter. Zn causes severe enrichment during both seasons. Mostly, elevated degree of pollution by the metals is noted during summer than during winter.

The lowest, mean, and highest values of I_{geo} in acid extract of the sediments during summer and winter are illustrated in Figure 2b. The following categorizations are given by Muller (1969) for geoaccumulation index: I_{geo} < 0 indicates unpolluted, I_{geo} = 0 to 1 unpolluted to moderately polluted, I_{geo} = 1 to 2 moderately polluted, I_{geo} = 2 to 3 moderately to strongly polluted, I_{geo} = 3 to 4 strongly polluted, I_{geo} = 4 to 5 strongly to extremely polluted, and I_{geo} > 5 demonstrates extremely polluted. The highest category reflects at least a 100-time enrichment above the background values. As shown in the figure, during summer, Cd and Pb pose strong to extreme contamination and moderate contamination, respectively. However, the remaining metals exhibit practically uncontamination in the sediments. During winter, Cd indicates strong to extreme pollution; Zn causes unpolluted to moderate pollution, whereas Pb shows least to moderate contamination.

Ecological risk assessment

The range and mean E_i values of the heavy metals in acid extract of the sediments during summer and winter are shown in Figure 2c. The following categorization is given by Hakanson (1980) for E_i: E_i < 40 demonstrates low risk, E_i = 40 to 80 moderate risk, E_i = 80 to 160 considerable risk, E_i = 160 to 320 great risk, and E_i > 320 demonstrates very great risk. The categorization related to RI is also suggested by Hakanson (1980): RI < 65 explicates low risk, RI = 65 to 130 moderate risk; RI = 130 to 260 considerable risk, and RI > 260 explicates very high risk. The results elucidate that Cd causes low to very high risk, while the rest of the metals explicate low

risk in the sediments during both seasons. Overall, the cumulative potential risk index (RI = 45.91 to 935 during summer and RI = 31.87 to 1,058 during winter) reveals low to very high risk of the sediments during both seasons. However, relatively higher potential ecological risk is observed during winter compared to summer.

Source apportionment

One of the important aspect of the present study is the source apportionment of the metals in sediments using PCA and CA. The principal component loadings of the heavy metals in acid extract of the sediments during summer and winter are given in Table 4, whereas the corresponding CA is shown in Figure 3. During summer, two PCs are extracted with eigenvalues more than 1, explaining about 60% of the total variance. The first PC (36.14% variance) reveals elevated loadings of Fe, Mn, and Cu, supported by their mutual cluster in CA. These metals are likely to be contributed by lithogenic processes such as soil erosion and rock weathering. The second PC (23.77% variance) shows significant loadings of Pb, Cd, Cr, and Zn supported by their shared cluster and are mainly contributed by automobile emissions, agricultural runoff, and untreated urban wastes. The counterpart data during winter also yield two PCs with eigenvalues greater than 1, explaining more than 66% of the total variance. PC1 (51.09% variance) exhibits higher loadings for Zn, Cu, Cr, Mn, Pb, and Cd, which are predominantly contributed by transportation activities, untreated urban wastes, and agricultural runoff. The cluster analysis also shows a joint cluster for these metals. PC2 (15.22% variance) reveals the natural/lithogenic contribution as manifested by the elevated loadings of Fe only which shows almost independent pattern in CA.

Sediment quality guidelines

The assessment of acid-extractable metal levels in the sediments is the first step to gauge the pollution of the water reservoir. However, it does not provide information on the potential toxicity to the benthic flora and fauna in the reservoir. For this purpose, numerous sediment quality guidelines are used to protect aquatic biota from the harmful and toxic effects related with sediment-bound contaminants (Caeiro et al. 2005; McCready et al. 2006; Spencer and Macleod 2002). These guidelines evaluate the degree to which the sediment-associated chemical status might adversely affect the aquatic organisms and therefore are designed for the interpretation of sediment quality. SQGs have been developed for both freshwater and marine ecosystems to represent threshold chemical concentrations associated with the presence or absence of biological effects on communities (Caeiro et al. 2005; Long and

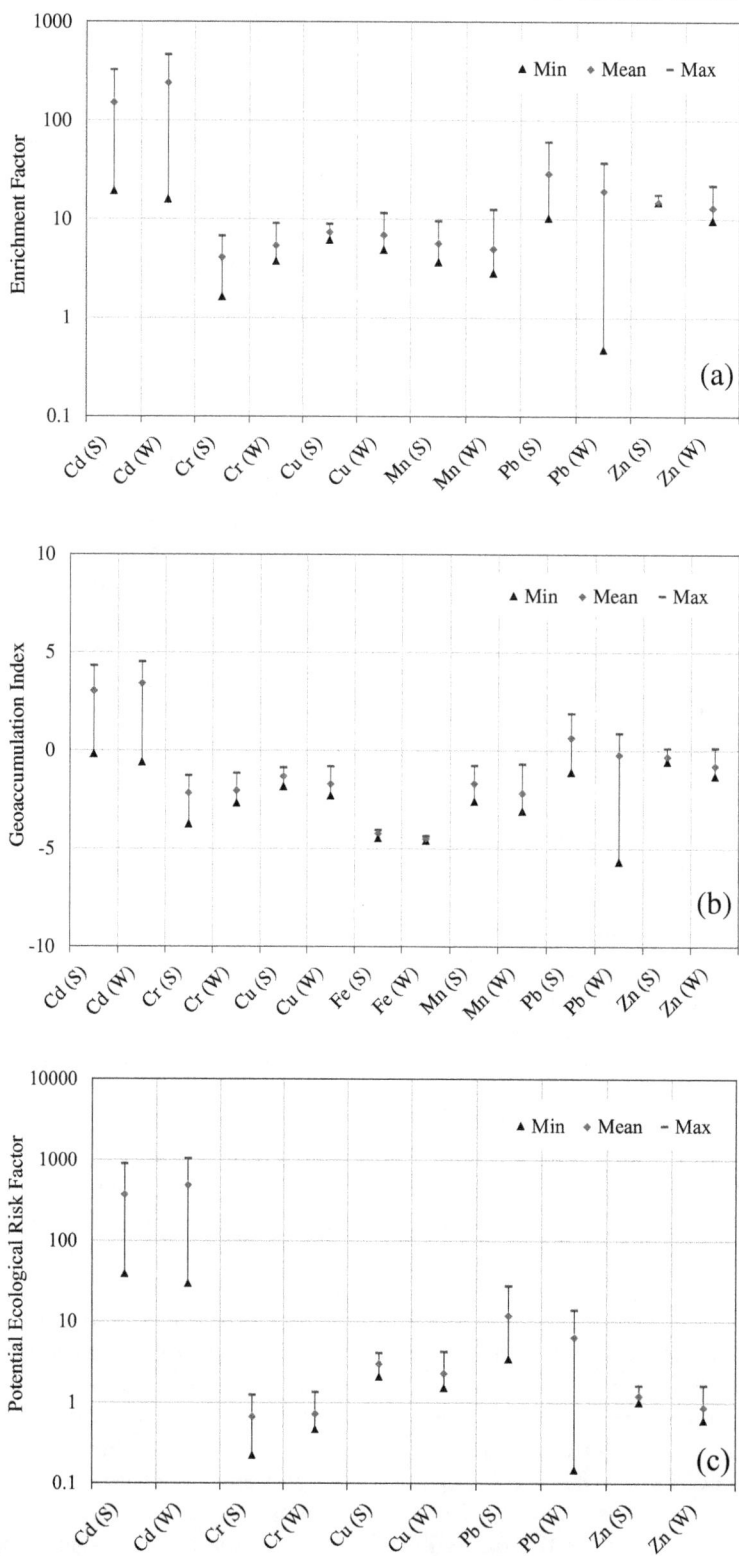

Figure 2 Description of the different parameters. Description of **(a)** enrichment factor (EF), **(b)** geoaccumulation index (I_{geo}) and **(c)** potential ecological risk factor (E_i) of heavy metals in acid extract of sediments during summer (S) and winter (W).

Table 4 Principal component loadings of heavy metals in acid extract of sediments during summer and winter

	Summer		Winter	
	PC1	PC2	PC1	PC2
Eigenvalue	2.530	1.664	3.576	1.065
Percentage of total variance	36.14	23.77	51.09	15.22
Percentage of cumulative variance	36.14	59.91	51.09	66.30
Cd	0.131	0.733	0.506	−0.139
Cr	0.284	0.666	0.838	−0.015
Cu	0.914	0.144	0.884	0.179
Fe	0.800	−0.108	−0.035	0.951
Mn	0.913	0.020	0.835	0.140
Pb	−0.179	0.714	0.557	0.388
Zn	−0.273	0.393	0.876	0.011

MacDonald 1998; MacDonald et al. 2000; Thompson et al. 2005; Wenning et al. 2005). These guidelines have been widely used to screen sediment contamination by comparing the concentrations in sediments with the corresponding quality guidelines in aquatic ecosystems (Caeiro et al. 2005; MacDonald et al. 2000). It is important to determine whether the estimated concentrations of heavy metals in sediments pose a threat to aquatic life, and they are assessed by two sets of sediment quality guidelines: (i) lowest effect level (LEL) and severe effect level (SEL) and (ii) effects range low (ERL) and effects range medium (ERM) (MacDonald et al. 2000). These two sets of numerical SQGs are directly applied to assess the possible risk associated with heavy metal contamination in the sediments. It is interpreted that LEL and ERL as the concentrations below which adverse biological effects rarely occur. Hence, these are considered to

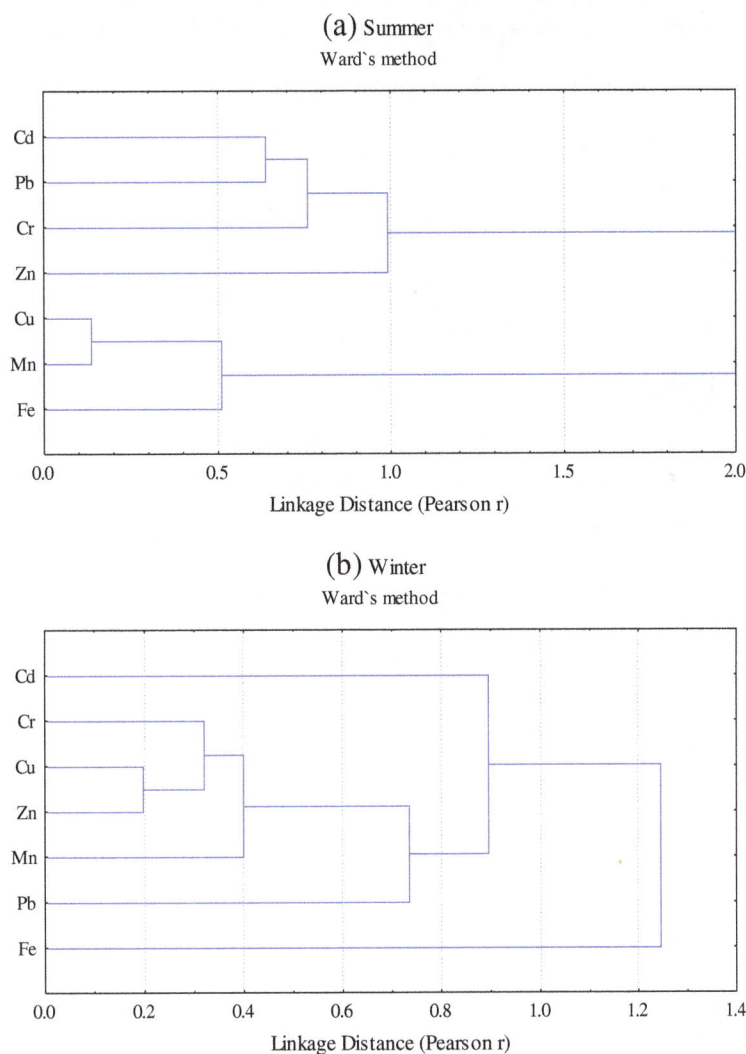

Figure 3 Cluster analyses of heavy metals in acid extract of sediments during (a) summer and (b) winter.

provide a high level of protection for aquatic organisms. Similarly, SEL and ERM refer to the concentrations above which adverse biological effects frequently occur. Hence, these are considered to provide a lower level of protection for aquatic organisms (Long and MacDonald 1998; MacDonald et al. 2000).

The description of SQGs and sediment classification along with the results related to the sediments from Khanpur Lake during summer and winter is presented in Table 5, while the percent contribution of heavy metals towards potential acute toxicity in the sediments during summer and winter is depicted in Figure 4. During summer, the measured levels of Cd, Cr, Cu, Mn, and Pb are found to be higher than the LEL values in 87%, 100%, 100%, 37%, and 37% sediment samples, respectively. It depicts that these metals could pose moderate impact on the biota (Graney and Eriksen 2004). On the other hand, the concentrations of Fe and Zn are found to be lower than the LEL levels in 100% sediment samples, demonstrating that these metals cause little or no

impact on biota in the lake. Similarly, the measured levels of Cd, Cr, Cu, and Zn are found to be lower than the ERL values in 100% sediment samples, revealing that these metals are not associated with adverse health effects to the dwelling biota (MacDonald et al. 2000). However, Pb levels are found to be higher in 37% sediment samples, manifesting that Pb is associated with frequent adverse biological effects to the underlying organisms (MacDonald et al. 2000). Furthermore, potential acute toxicity (ΣTUs) study shows that the mean levels of toxic units (TUs) for heavy metals follow a decreasing order: Cd > Cr > Pb > Zn > Cu. It indicates relatively higher contributions of Cd, Cr, and Pb to ΣTUs (i.e., 31%, 22%, and 21%, respectively; Figure 4) (Pedersen et al. 1998). Nevertheless, Cu (11%) is the minor contributor to ΣTUs compared with the other heavy metals. The levels of ΣTUs range from 0.64 to 3.45 with a mean value of 1.75 in the sediments. Based on the USEPA sediments classification (Giesy and Hoke 1990), Cr, Cu, and Zn show moderate contamination, Mn and Pb exhibit heavy pollution, and Cd

Table 5 Description of sediment classification and sediment quality guidelines in acid extract of sediments in two seasons

		Cd	Cr	Cu	Fe	Mn	Pb	Zn
Sediment classification	Non-polluted	-	<25	<25	<17,000	<300	<40	<90
	Moderately polluted	-	25 to 75	25 to 50	17,000 to 25,000	300 to 500	40 to 60	90 to 200
	Heavily polluted	>6	>75	>50	>25,000	>500	>60	>200
Sediment quality guidelines (SQGs)	LEL	0.6	26	16	20,000	460	31	120
	SEL	10	110	110	40,000	1,100	250	820
	ERL	5	80	70	-	-	35	120
	ERM	9	145	390	-	-	110	270
Percentage of samples (summer)	Non-polluted	100	-	-	100	23	63	67
	Moderately polluted	-	100	100	-	43	23	33
	Heavily polluted	-	-	-	-	34	14	-
	<LEL	13	-	-	100	63	63	100
	≥LEL and <SEL	87	100	100	-	37	37	-
	>SEL	-	-	-	-	-	-	-
	<ERL	100	100	100	-	-	63	100
	≥ERL and <ERM	-	-	-	-	-	37	-
	>ERM	-	-	-	-	-	-	-
Percentage of samples (winter)	Non-polluted	100	3.0	40	100	53	100	97
	Moderately polluted	-	97	57	-	44	-	3.0
	Heavily polluted	-	-	3.0	-	3.0	-	-
	<LEL	10	3.0	-	100	90	93	100
	≥LEL and <SEL	90	97	100	-	10	7.0	-
	>SEL	-	-	-	-	-	-	-
	<ERL	97	100	100	-	-	93	100
	≥ERL and <ERM	3.0	-	-	-	-	7.0	-
	>ERM	-	-	-	-	-	-	-

The units of metals are expressed in milligrams per kilogram. LEL, lowest effect level; SEL, severe effect level; ERL, effect range low; ERM, effect range median.

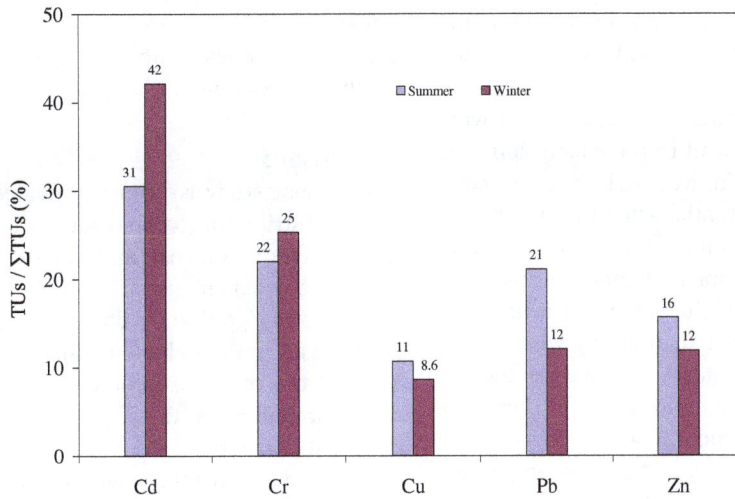

Figure 4 Percent contribution of heavy metals to ΣTUs in acid extract of sediments during summer and winter.

and Fe reveal little or no contamination in the sediments during summer. It demonstrates that Cr, Cu, Zn, Pb, and Mn are the major contributors toward the gross pollution of the water reservoir.

During winter, the measured levels of Cd, Cr, Cu, Mn, and Pb in the sediments are found to be higher than the LEL values in 90%, 97%, 100%, 10%, and 7.0% samples, respectively. It reveals moderate impact on the biota health. The observed values of Fe and Zn are found to be lower than the LEL values in 100% sediment samples, indicating that these metals are not associated with adverse impact on the biota (Graney and Eriksen 2004). The ERL and ERM SQGs manifest that Cd and Pb levels exceed the ERL values in 3.0% and 7.0% sediment

samples, respectively, demonstrating that these metals are associated with occasional adverse health hazards to the surrounding biota (MacDonald et al. 2000). The concentrations of Cr, Cu, and Zn are lower than the ERL values in 100% sediment samples, demonstrating little or no undesirable health hazards. The potential acute toxicity study reveals that the average levels of TUs for heavy metals follow a decreasing order: Cd > Cr > Pb > Zn > Cu. It illustrates that Cd, Cr, and Pb are the major contributors to ΣTUs (i.e., 42%, 25%, and 12%, respectively; Figure 4), while Cu (8.6%) is a minor contributor (Pedersen et al. 1998). The values of ΣTUs range from 0.54 to 3.29 with an average value of 3.29 in the sediments. Based on the USEPA sediments classification

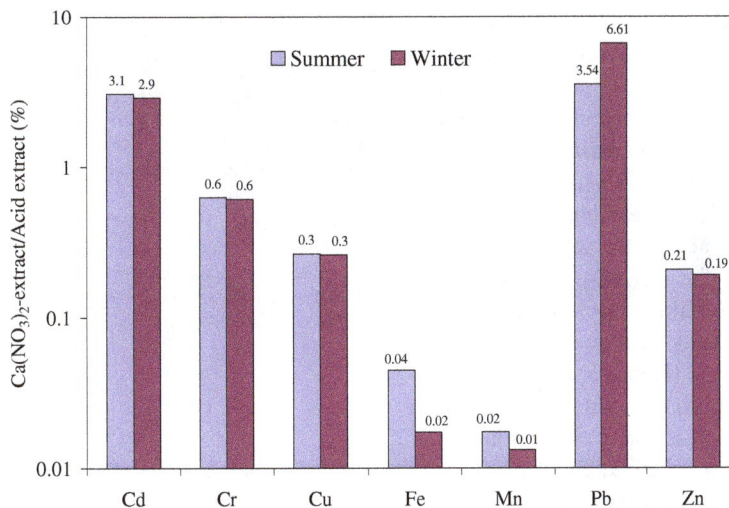

Figure 5 Percent extraction of heavy metals in Ca(NO₃)₂ extract of sediments during summer and winter.

(Giesy and Hoke 1990), Cd, Fe, and Pb may pose little or no pollution. Cr and Zn cause moderate contamination, and Cu and Mn exhibit heavy pollution in the sediments. Consequently, Cr, Cu, Mn, and Zn emerge as the major pollutants in the water reservoir during winter. Overall, the SQG results lead to the conclusion that the metals, such as Cd, Cr, Cu, Mn, and Pb are of concern during both seasons. Potential acute toxicity results demonstrate that Cd, Cr, and Pb are the major toxicants, while Zn and Cu are the minor pollutants during both seasons. However, relatively higher potential acute toxicity is observed during summer than during winter.

From the ecotoxicological dataset obtained for the US Coasts, Long et al. (1998) have defined several classes of toxicity probability for benthic biota: m-ERM-Q < 0.1 has a 9% probability of being toxic (based on amphipod survival test), m-ERM-Q between 0.11 and 0.5 has 21% probability of toxicity, m-ERM-Q between 0.51 and 1.5 has a probability of 49% to be toxic, and m-ERM-Q > 1.50 has 76% probability of toxicity. In the present study, the m-ERM-Q values range from 0.159 to 0.408 and 0.126 to 0.337 with the average values of 0.247 and 0.200 during summer and winter, respectively. Consequently, the metals pose approximately 21% probability of toxicity to the benthic organisms in the lake during both seasons.

Bioavailability of heavy metals in the sediments

Potential toxicity of heavy metals in the sediments is also assessed by the measurement of mobile metal concentrations. The statistical distribution parameters related to the concentrations of heavy metals in $Ca(NO_3)_2$ extract of the sediments during summer and winter are given in Table 2, whereas their percent extraction in $Ca(NO_3)_2$ extract is shown in Figure 5. The summer results reveal Fe as having the highest contributions (2.069 mg/kg), while the measured levels of Mn and Cd are the least. Nonetheless, the winter results demonstrate an elevated concentration of Pb (1.205 mg/kg), and the mean level of Mn is the lowest. On the percent extraction basis, the metals follow identical decreasing sequence during summer and winter: Pb > Cd > Cr > Cu > Zn > Fe > Mn. Moreover, there are no direct relationships among the $Ca(NO_3)_2$-extractable and acid-extractable fractions of the metals in sediments. The $Ca(NO_3)_2$-extractable recoveries are found to be within approximately 11% during summer and 15% during winter of the acid-extractable metal concentrations (Figure 5). Since element bioavailability is related to its solubility, extractable metal concentrations may correspond to the bioavailable concentrations (An and Kampbell 2003). The results demonstrate that Pb and Cd show the maximum extraction efficiencies, mobilities, and bioavailabilities, followed by Cr, while Fe and Mn manifest the least during both seasons.

Accordingly, Pb, Cd, and Cr exhibit higher mobility and higher potential toxicity to the surrounding biota, while Fe and Mn show least mobility and bioavailability to the benthic biota in the water reservoir.

Conclusions

The present study is primarily related to the evaluation of the distribution, correlation, source apportionment, contamination, and risk assessment of the heavy metals in surface sediments from Khanpur Lake, Pakistan. The study shows significantly divergent metal levels for most of the cases in the sediments during summer and winter. Most of the metals exhibit random distribution and diverse correlations in the sediments. Extremely severe enrichment is noted for Cd and Pb, while Zn shows severe enrichment. Moderate pollution is associated with Pb levels; strong to extreme pollution is shown by Cd, which is also associated with very high risk. On the whole, RI shows low to very high risk of contamination in the sediments. Multivariate PCA and CA manifest dominantly anthropogenic contributions of Pb, Cd, Cr, and Zn in the sediments. Comparison of heavy metal contents in the sediments with quality guidelines indicates adverse biological effects to the surrounding flora and fauna due to elevated levels of the metals. The m-ERM-Q study reveals 21% probability of toxicity due to the metals in the sediments. The potential toxicity, mobility, and bioavailability manifest that Cd and Pb are more mobile and available to the benthic flora and fauna. The present investigation clearly indicates that the sediments from freshwater reservoir are contaminated with some toxic heavy metals. Consequently, there is a dire need to reduce/regulate the anthropogenic sources of pollution in the study area.

Competing interests
The authors declare that they have no competing interests.

Authors' contributions
JI performed the field sample collection, extraction/analysis of the metals, and prepared the main draft of the manuscript. MHS designed/supervised the work, performed the statistical analysis, and helped in writing the manuscript. All authors read and approved the final manuscript.

Acknowledgements
The research fellowship awarded by Quaid-i-Azam University, Islamabad, to carry out this project is appreciatively accredited. We are also grateful to the administration of Khanpur Lake, Islamabad, for their assistance and help during the sampling campaign.

References
Adaikpoh EO, Nwajei GE, Ogala JE (2005) Heavy metals concentrations in coal and sediments from River Ekulu in Enugu, coal city of Nigeria. J Appl Sci Environ Manage 9:5–8
Akoto O, Bruce TN, Darko G (2008) Heavy metals pollution profiles in streams serving the Owabi reservoir. Afr J Environ Sci Technol 2:354–359

An YJ, Kampbell DH (2003) Total, dissolved, and bioavailable metals at Lake Texoma marinas. Environ Pollut 122:253–259

Arnason JG, Fletcher BA (2003) A 40+ year record of Cd, Hg, Pb, and U deposition in sediments of Patroon Reservoir, Albany County, NY, USA. Environ Pollut 123:383–391

Birch G et al (2003) A scheme for assessing human impacts on coastal aquatic environments using sediments. In: Coastal GIS. Wollongong University Papers in Center for Maritime Policy, Australia, p 14

Caeiro S, Costa MH, Ramos TB, Fernandez F, Silveira N, Coimbra A, Medeiros G, Painho M (2005) Assessing heavy metal contamination in Sado estuary sediment: an index analysis approach. Ecol Indicat 5:151–169

Cukrov N, Bilinski SF, Hlaca B, Barisic D (2011) A recent history of metal accumulation in the sediments of Rijeka harbor, Adriatic Sea, Croatia. Mar Pollut Bull 62:154–167

Cuong DT, Obbard JP (2006) Metal speciation in coastal marine sediments from Singapore using a modified BCR-sequential extraction procedure. Appl Geochem 21:1335–1346

de Vallejuelo SFO, Arana G, de Diego A, Madariaga JM (2010) Risk assessment of trace elements in sediments: the case of the estuary of the Nerbioi–Ibaizabal River (Basque Country). J Hazard Mater 181:565–573

Fadigas JC, dos Santos AMP, de Jesus RM, Lima DC, Fragoso WD, David JM, Ferreira SLC (2010) Use of multivariate analysis techniques for the characterization of analytical results for the determination of the mineral composition of kale. Microchem J 96:352–356

Gielar A, Rybicka EH, Moller S, Einax JW (2012) Multivariate analysis of sediment data from the upper and middle Odra River (Poland). Appl Geochem 27:1540–1545

Giesy JP, Hoke RA (1990) Freshwater sediment quality criteria: toxicity bioassessment. In: Baudo R, Giesy JP, Muntao M (eds) Sediment: chemistry and toxicity of in-place pollutants. Lewis Publishers, Ann Arbor, MI, p 39

Graney JR, Eriksen TM (2004) Metals in pond sediments as archives of anthropogenic activities: a study in response to health concerns. Appl Geochem 19:1177–1188

Hakanson L (1980) An ecological risk index for aquatic pollution control: a sedimentological approach. Water Res 14:975–1001

Iqbal J, Shah MH (2011) Distribution, correlation and risk assessment of selected metals in urban soils from Islamabad, Pakistan. J Hazard Mater 192:887–898

Larrose A, Coynel A, Schafer J, Blanc G, Masse L, Maneux E (2010) Assessing the current state of the Gironde Estuary by mapping priority contaminant distribution and risk potential in surface sediment. Appl Geochem 25:1912–1923

Lide DR (2005) CRC handbook of Chemistry and Physics, Geophysics, Astronomy, and Acoustics. Section 14, Abundance of elements in the Earth's crust and in the sea, 85th edn. CRC Press, Boca Raton, FL

Long ER, MacDonald DD (1998) Recommended uses of empirically derived, sediment quality guidelines for marine and estuarine ecosystems. Hum Ecol Risk Assess 4:1019–1039

Long ER, Field LJ, McDonald DD (1998) Predicting toxicity in marine sediments with numerical sediment quality guidelines. Environ Toxicol Chem 17:714–727

Loska K, Wiechula D, Korus I (2004) Metal contamination of farming soils affected by industry. Environ Int 30:159–165

Luoma SN, Rainbow PS (2008) Metal contamination in aquatic environments: science and lateral management. Cambridge University Press, Cambridge, UK, pp 91–103

MacDonald DD, Ingersoll CG, Berger TA (2000) Development and evaluation of consensus-based sediment quality guidelines for freshwater ecosystems. Arch Environ Contam Toxicol 39:20–31

Martin CW (2000) Heavy metal trends in floodplain sediments and valley fill, River Lahn, Germany. Catena 39:53–68

Martin J, Cabeza JAS, Eriksson M, Levy I, Miquel JC (2009) Recent accumulation of trace metals in sediments at the DYFAMED site (Northwestern Mediterranean Sea). Mar Pollut Bull 59:146–153

McCready S, Birch GF, Long ER (2006) Metallic and organic contaminants in sediments of Sydney Harbour, Australia and vicinity—a chemical dataset for evaluating sediment quality guidelines. Environ Int 32:455–465

Muller G (1969) Index of geoaccumulation in sediments of the Rhine River. J Geol 2:108–118

Nouri J, Mahvi AH, Jahed GR, Babaei AA (2008) Regional distribution pattern of groundwater heavy metals resulting from agricultural activities. Environ Geol 55:1337–1343

Pedersen F, Sjobrnestad E, Andersen HV, Kjolholt J, Poll C (1998) Characterization of sediments from Copenhagen harbour by use of biotests. Water Sci Technol 37:233–240

Radojevic M, Bashkin VN (1999) Practical environmental analysis. The Royal Society of Chemistry, Cambridge, UK

Reimann C, de Caritat P (2005) Distinguishing between natural and anthropogenic sources for elements in the environment: regional geochemical surveys versus enrichment factors. Sci Total Environ 337:91–107

Reza R, Singh G (2010) Heavy metal contamination and its indexing approach for river water. Int J Environ Sci Technol 7:785–792

Rodrigues SM, Henriques B, Coimbra J, da Silva EF, Pereira ME, Duarte AC (2010) Water-soluble fraction of mercury, arsenic and other potentially toxic elements in highly contaminated sediments and soils. Chemosphere 78:1301–1312

Shah MH, Iqbal J, Shaheen N, Khan N, Choudhary MA, Akhter G (2012) Assessment of background levels of trace metals in water and soil from a remote region of Himalaya. Environ Monit Assess 184:1243–1252

Sin SN, Chua H, Lo W, Ng LM (2001) Assessment of heavy metal cations in sediments of Shing Mun River, Hong Kong. Environ Int 26:297–301

Singh KP, Malik A, Sinha S, Singh VK, Murthy RC (2005) Estimation of source of heavy metal contamination in sediments of Gomti River (India) using principal component analysis. Water Air Soil Pollut 166:321–341

Spencer KL, Macleod CL (2002) Distribution and partitioning of heavy metals in estuarine sediment cores and implications for the use of sediment quality standards. Hydrol Earth Syst Sci 6:989–998

Sprovieri M, Feo ML, Prevedello L, Manta DS, Sammartino S, Tamburrino S, Marsella E (2007) Heavy metals, polycyclic aromatic hydrocarbons and polychlorinated biphenyls in surface sediments of the Naples harbor (southern Italy). Chemosphere 67:998–1009

Tessier E, Garnier C, Mullot JU, Lenoble V, Arnaud M, Raynaud M, Mounier S (2011) Study of the spatial and historical distribution of sediment inorganic contamination in the Toulon bay (France). Mar Pollut Bull 62:2075–2086

Thompson PA, Kurias J, Mihok S (2005) Derivation and use of sediment quality guidelines for ecological risk assessment of metals and radionuclides released to the environment from uranium mining and milling activities in Canada. Environ Monit Assess 110:71–85

USEPA (2007) Microwave assisted acid digestion of sediments, sludges, soils, and oils. Method 3051A. United States Environmental Protection Agency, Office of Solid Waste and Emergency Response, US Government Printing Office, Washington, DC

Varol M, Sen B (2012) Assessment of nutrient and heavy metal contamination in surface water and sediments of the upper Tigris River, Turkey. Catena 92:1–10

Wenning R, Batley G, Ingersoll C, Moore D (2005) Use of sediment quality guidelines and related tools for the assessment of contaminated sediments. Society of Environmental Toxicology and Chemistry (SETAC) Press, USA

Yu R, Hu G, Wang L (2010) Speciation and ecological risk of heavy metals in intertidal sediments of Quanzhou Bay, China. Environ Monit Assess 163:241–252

15

Laser-induced thermal lens spectrometry after cloud point extraction for trace analysis of mercury in water and drug samples

Nader Shokoufi[*], Rasoul Jafari Atrabi and Kazem Kargosha[*]

Abstract

Background: We have developed spectrometric determination of mercury in micro volume using laser-induced thermal lens spectrometry (LI-TLS) after cloud point extraction (CPE). TLS as a sensitive method is particularly suited for small volume samples such as that of the remained phase after CPE.

Results: Under optimum conditions, the calibration graph was linear in the range of 1 to 50 µg L^{-1} with a limit of detection (LOD) of 0.2 µg L^{-1}. The preconcentration factor of 200 was achieved for 10-mL samples containing the analyte, and relative standard deviations were lower than 4%.

Conclusion: Combination of LI-TLS with CPE introduced a powerful method for trace analysis of mercury. The method was successfully applied for the determination of Hg^{+2} in water and drug samples.

Keywords: Thermal lens spectrometry; Mercury; Cloud point extraction; Triton X-114

Background

In recent years, the toxicity and effects of trace elements on human health and the environment are receiving increasing attention in pollution and nutritional studies. Mercury is considered as a chemical pollutant and has become widespread in the environment mainly as a result of anthropogenic activities (Eisler 2004; Pacyna and Pacyna 2006). It has been found that mercury accumulates itself in vital organs and tissues such as the kidney and brain. Mercury and its compounds are considered as health hazards, and reports of mercury poisoning as a result of industrial, agricultural, and laboratory exposure as well as its suicidal use are numerous.

Since mercury concentrations in waters are expected to be very low, some powerful techniques are required that a few of them show enough sensitivity. Different analytical techniques have been used for mercury determination at low concentration including cold vapor atomic absorption spectrometry (CV-AAS) (Pourreza and Ghanemi 2009), cold vapor atomic fluorescence spectrometry (CV-AFS) (Geng et al. 2008), flow injection-inductively coupled plasma-optical emission spectrometry (FI-ICP-OES) (Wuilloud et al. 2002), and high-performance liquid chromatography-inductively coupled plasma mass spectrometry (HPLC-ICP-MS) (Chen et al. 2009).

Laser-induced thermal lens spectrometry (LI-TLS) is a subgroup of photothermal spectrometry methods, which is an ultrasensitive means to measure optical absorbance. The first photothermal spectroscopic method applied for sensitive chemical analysis is thermal (photothermal) lens spectrometry (TLS) (Gordon et al. 1964), and the first measurement of the thermal lens effect was performed by Gordon et al. (1965) using a simple single beam apparatus (Gordon et al. 1965). Thermal lens effect has been developed as a powerful analytical technique for measurement of low absorbance. The utility of the techniques has increased substantially because it has been recently demonstrated that, in addition to ultrasensitivity, the techniques are particularly suited for small volume samples (Kitamori et al. 2004). In fact, the combined ultrasensitivity and small volume capability make it possible to successfully use them as powerful techniques for analytical devices.

The effect of thermal lens is the result of the temperature rise, which is subsequent to absorption of optical radiation and nonradiative relaxation of the excited

* Correspondence: Shokoufi@ccerci.ac.ir; K.Kargosha@ccerci.ac.ir
Analytical Instrumentation & Spectroscopy Laboratory, Chemistry & Chemical Engineering Research Center of Iran, Tehran, Iran

molecules. Due to the Gaussian profile of the excitation laser, the temperature gradient produces a refractive index gradient which is maximum at the beam center and behaves like a converging or diverging lens depending on whether dn/dT is positive or negative, respectively (Georges 1994; Snook and Lowe 1995). A steady-state condition is obtained when the rate of laser heating equals the rate of heat loss due to the thermal conductivity of the solvent and the finite temperature rise (Dovichi and Harris 1979; Imasaka et al. 1980).

The thermal lens signal is sensitive to the thermo-optical properties of the medium, namely the temperature-dependent refractive index dn/dT, the thermal conductivity k, and heat capacity Cp. In addition, the signal depends on the sample absorbance and heat yield (Piepmeier 1986; Bialkowski 1996).

Mercury in different samples is present at lower levels than the detection limits of sensitive analytical methods, and its determination is spectroscopically and chemically interfered with other major constituents. Much effort has been put to solve these difficulties by the development of various techniques. Preconcentration and separation can solve these problems and lead to a higher confidence level and easy determination of trace mercury. The classical liquid-liquid extraction and separation methods are usually time consuming and labor extensive and require relatively large volumes of high-purity solvents. The additional concern is disposal of the solvent used, which creates a severe environmental problem. Cloud point extraction (CPE) is an attractive technique that reduces the consumption and exposure to a solvent, disposal cost, and extraction time. Moreover, small volume of the surfactant-rich phase obtained through this method permits the design of extraction schemes that are simple, cheap, highly efficient, speedy, and of lower toxicity to the environment than those extractions using organic solvents (Paleologos et al. 2001, 2002; Manzoori and Bavili-Tabrizi 2002; Chappuy et al. 2010; Shah and Kazi 2011).

Cloud point methodology was used for the extraction and preconcentration of many metal ions after the formation of complexes. These metal ions can subsequently be analyzed using analytical systems such as AAS, ETAAS, ICP, GC, HPLC, CE, spectrophotometry (Sanz-Medel et al. 1999; Paleologos et al. 2005; Martinez and Gonzalo 2000; Giokas et al. 2001; Stalikas 2002; Afkhami et al. 2006), and specially LI-TLS.

The aim of our research is to combine laser-induced thermal lens spectrometry, as a sensitive method, with cloud point extraction for determination of mercury. The connection of CPE with TLS could be a powerful analytical technique for mercury because a low volume of sample remains after CPE and the use of organic solvent enhancement in TLS.

Experimental

Thermal lens spectrometer setup

A single-laser thermal lens spectrometer was designed in our laboratory (Figure 1). The optical direction was designed in a horizontal-vertical mode. In order to obtain thermal lens signal, the diode laser (532 nm, 50 mW, TEM_{00}) was used as the pump/probe source. The laser beam was focused by an 18-cm focal length lens and passed through the micro cell located at the confocal distance. The laser beam was allowed to irradiate the sample or blocked using a chopper. After filtration, the beam intensity change was measured through a 1-mm pinhole by a photo-detector that was located at a 200-cm distance from the sample cell. An operation amplifier amplified the output signal from the photo-detector, and it was digitized by the analog to a digital converter (ADC). The digital signals were processed in a personal computer by laboratory-developed software. For each sample, the signal was derived from the average of three recordings.

Data treatment

The TLS technique is based on the temperature gradient that is produced in an illuminated sample by nonradiative relaxation of the energy absorbed from a TEM_{00} laser beam. Under CW laser excitation and weak absorbing species, the TLS signal depends on the relative changes in the laser beam center intensity. Initial signal $I_{(0)}$ reflects only the Beer's law response of the sample, and after sufficient time, when a steady-state temperature difference is reached, the intensity at the detector will be $I_{(\infty)}$. Then, $\Delta I/I$

$$\frac{\Delta I}{I} = \frac{I_{(0)} - I_{(\infty)}}{I_{(\infty)}} = \frac{2.303 AP(-dn/dT)}{\lambda k} \tag{1}$$

P is the power of the laser, λ is the wavelength, dn/dT is the temperature coefficient of the refractive index, and k is the thermal conductivity of the solvent (Bialkowski 1996). This can also be rewritten as:

$$\frac{\Delta I}{I} = 2.303 EA \tag{2}$$

where

$$E = \frac{P(-dn/dT)}{\lambda k} \tag{3}$$

E represents the so-called enhancement factor compared to conventional transmission measurements. It is clear that, in addition to the sample absorbance and excitation laser power, the thermal lens signal is directly affected by the thermo-optical properties of the solvent such as the dn/dT and k values of the solvent (Dovichi and Bialkowoki 1987; Franko and Tran 1996). Nonpolar organic solvents such as benzene, carbon tetrachloride,

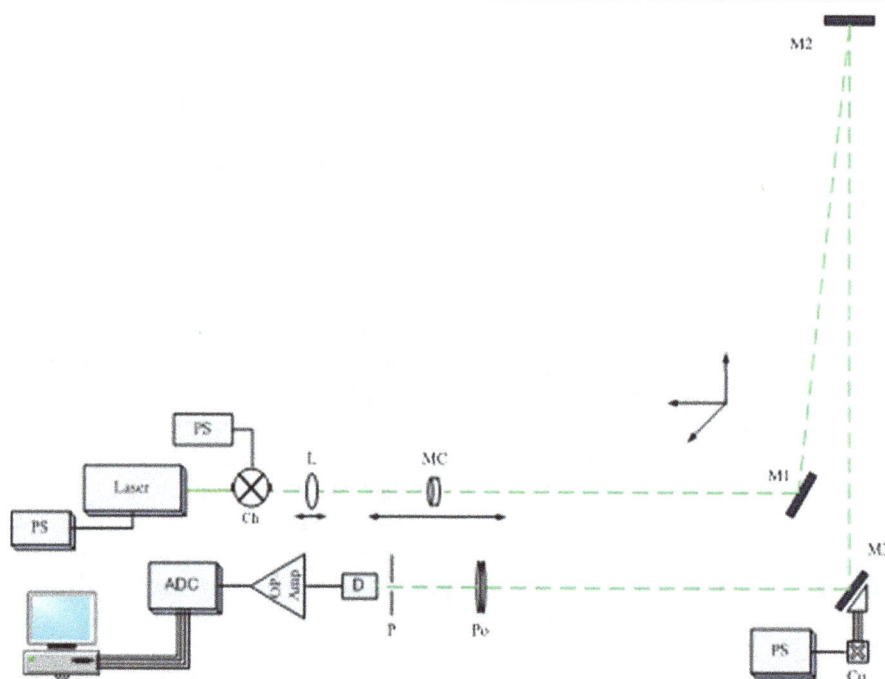

Figure 1 Schematic diagram of single-laser thermal lens spectrometer. PS, power supply; Ch, chopper; L, lens; MC, microcell; M1,M2,M3, mirror; Co, controller; Po, polarizer; P, pinhole; D, detector; Op-Amp, operational amplifier; ADC, analog digital converter; PC, personal computer.

and hexane should provide good media for thermal lens measurements owing to their high dn/dT and low k values. Conversely, water, which is the most powerful and widely used solvent in spectrochemical analysis, is considered to be the worst medium for thermo-optical techniques because it has very low dn/dT and high k values.

Reagents and apparatus

All chemicals used were of analytical reagent grade. The nonionic surfactant, Triton X-114 (Acros, Fair Lawn, NJ, USA), was used without further purification. Stock standard solution of mercury at a concentration of 1,000 mg L^{-1} was prepared from $HgCl_2$ (Merck, Darmstadt, Germany). Working standard solutions were obtained through appropriate dilution of the stock standard solution. A solution of 9.9×10^{-4} mol L^{-1} dithizone (Merck) was prepared in the pure ethanol (Merck). The NaOH solution was prepared by dissolving the appropriate NaOH (Merck) amount in the high-purity water. The materials and vessels used for the trace analysis were kept in 10% nitric acid for at least 24 h and subsequently washed four times with doubly distilled water.

A thermostatic bath (IKA, Königswinter, Germany) maintained at the desired temperature was used for cloud point extraction experiments, and phase separation was performed using a centrifuge (Sigma, Seelze, Germany). A 100-μL quartz micro cell with 1-mm light path was

used as a determination cell (Hellma GmbH, Müllheim, Germany).

Conclusion

The lack of sensitivity associated with the need for detection of micro amounts of samples is probably one of the most frequently encountered problems in trace chemical analysis. As a whole, it was shown that TLS solved these problems, because in addition to the major advantage of thermal lens spectrometry (sensitivity), it gives the possibility to analyze a very low volume of the sample (15 μL); on the other hand, the volume of the remained phase after cloud point extraction is in the micro liter range. Therefore, TLS is suitable for analysis of the remained phase after the cloud point extraction, and this combination method (CPE-LI-TLS) exhibits high enhancement factor because of the preconcentration enhancement factor in CPE and the enhancement factor of TLS. In fact, the combined ultrasensitivity and small volume capability make it possible to be successfully used as powerful techniques for analytical devices. Moreover, in this work, by using an alkaline aqueous solution, the selectivity and sensitivity of the TLS method were improved. In alkaline media, the Hg(II)-dithizone secondary (M_2L_2) complex is formed which has higher absorption than the ML complex and enhances the M_2L_2 complex to background absorption at 539 nm.

In general, this improved method provides good accuracy, recovery (95.3%), and precision. Also, it presents a detection limit of 0.2 ng mL^{-1} and enhancement factor of 179 compared with other determination methods. However, it seems that any other real samples which are available in low volumes, such as biological and biofluid samples, can be analyzed for their mercury contents by the proposed method.

Methods

Extraction and preconcentration procedure
Aliquots of 10.0 mL of the sample containing mercury, Triton X-114 (0.1% (w/v)), dithizone (1.96 × 10^{-6} mol L^{-1}), and NaOH (0.5% (w/v)) were kept in a thermostated bath at 54°C for 6 min. Separation of the aqueous and surfactant-rich phases was accomplished by centrifugation for 5 min sat 3,500 rpm. In this condition (alkaline media), no overlapping was observed between blank and complex spectra. The supernatant aqueous phase was separated completely. Then, the surfactant-rich phase was heated to dry in the oven at 100°C to remove the remaining water. Water is a poor solvent owing to its low dn/dT and high k or Cp values; therefore, for reaching a suitable thermal lens signal, all of the water remaining in the surfactant-rich phase must be eliminated. To decrease the viscosity of the dried surfactant-rich phase and to facilitate sample handling, 50 μL of ethanol was added. The 15-μL homogenized resultant solution was introduced into the micro cell located at the thermal lens spectrometer. The thermal lens spectrometer was set at a wavelength of 532 nm and a chopper frequency of 0.55 Hz. Intensity at the detector over time was obtained during thermal lens effect for each sample, and then the thermal lens signal was calculated by Eq. 1.

Results and discussion

Selection of wavelength
The absorption spectrum of the mercury-dithizone secondary complex in the remaining phase after dissolving using ethanol was studied (Figure 2). This complex shows absorption in λ_{max} = 539 nm. This was well matched to the wavelength of the laser beam used (λ = 532 nm).

Time of steady-state thermal lens
The steady-state thermal lens effect is due to the formation of the temperature-dependent refractive index gradient. In order to optimize the required time for the steady-state thermal lens effect, the laser beam path to the sample was continuously blocked and opened by a chopper at different frequencies. The intensity of detector output over time was evaluated for 25 μg L^{-1} of mercury-dithizone secondary complex after CPE at a 532-nm laser wavelength. It was found out that the frequency of 0.55 Hz (time of 1.82 s) is suitable for the

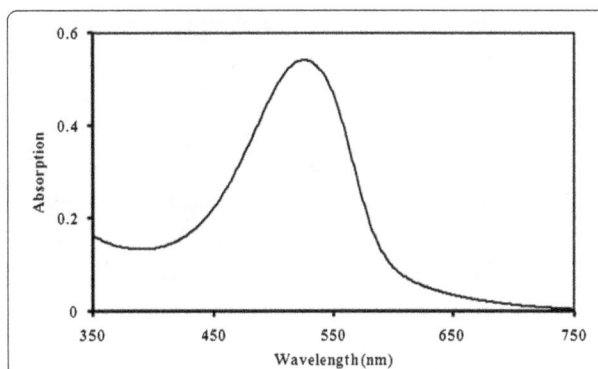

Figure 2 Absorption spectra of Hg(II)-dithizone complex. Conditions: 0.1% (w/v) Triton X-114, 0.5% (w/v) NaOH, 1.96 × 10^{-6} M dithizone, and 100 μg L^{-1} Hg^{+2}.

steady-state condition which is favorable to build up and decay the thermal lens effect.

Optimization of experimental parameters
Because of many difficulties such as high background absorption, high CPE temperature, and unpractical complete separation of the surfactant-rich phase from aqueous solution with the use of dithizone in complexation and CPE of the mercury at neutral or acidic media, we carried out our experiments in alkaline media with the addition of NaOH solution (5% w/v) to the sample solutions prior to complexation and CPE. The addition of NaOH by acting as salting-out agent markedly facilitated the phase separation process and decreased the temperature of the cloud point, since it altered the density of the bulk aqueous phase. Moreover, the alkaline solution resulted in the Hg(II)-dithizone secondary (M$_2$L$_2$) complex that had higher absorption than the Hg(II)-dithizone primary complex formed in neutral solution and enhanced the Hg(II)-dithizone complex to background absorption ratio at 539 nm. This effect was examined with the addition of NaOH, ranging from 0.05 to 0.8% (w/v). As shown in Figure 3, the maximum

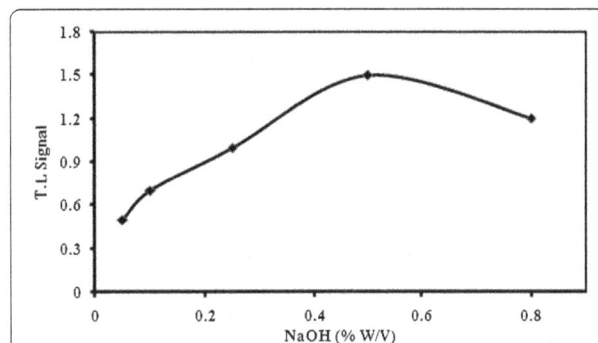

Figure 3 Effect of the concentration of NaOH on thermal lens signal of 25 μg L^{-1} Hg^{+2}. Other conditions: 0.1% (w/v) Triton X-114, 1.97 × 10^{-6} M dithizone, and equilibration temperature 54°C.

Table 1 Comparison of the LI-TLS and other methods of mercury determination

Determination technique	Sample volume (mL)	Solvent volume (mL)	EF[a]	LDR[b] ($\mu g\,L^{-1}$)	LOD[c] ($\mu g\,L^{-1}$)	Reference
HG-AAS	100	-	20	0.05 to 100	0.039	Song et al. (2006)
CV-AAS	3,000	-	150	-	0.005	Shamsipur et al. (2005)
ICP-OES	10	0.5	18.7	0.25 to 100	0.056	Li and Hu (2007)
Spectrophotometry	50	1	11	5.0 to 80.0	0.83	Niazi et al. (2009)
LI-TLS	10	0.05	179	1.0 to 50.0	0.2	Present work

[a]Enhancement factor. [b]Linear dynamic range. [c]Limit of detection.

TLS signal for Hg^{+2} was achieved in the 0.5% (w/v) concentration of NaOH. Moreover, the addition of a salt could markedly facilitate the phase separation process, since it altered the density of the bulk aqueous phase.

Optimal equilibration time and equilibration temperature are necessary to complete reactions and to achieve easy phase separation and preconcentration as efficiently as possible. Therefore, the change of thermal lens signal in equilibration temperature was studied within the range of 35°C to 75°C. It was found that a temperature of 54°C is adequate for extraction of Hg^{+2}. The dependence of thermal lens signal upon equilibration time was studied in the range of 3 to 15 min, and the optimum time of 6 min was chosen as a suitable time.

The concentration of the dithizone as a chelating agent was subsequently studied for its effect on the thermal lens signal of Hg^{+2}. The variation of the thermal lens signal over the dithizone concentration was evaluated in the range of 0.49×10^{-6} to 2.97×10^{-6} mol L^{-1} for 10 mL aliquots of solutions containing 25 $\mu g\,L^{-1}$ Hg. A concentration of 1.96×10^{-6} mol L^{-1} was chosen as the optimum dithizone concentration.

The effect of Triton X-114 concentration was evaluated, and it was found that a quantitative extraction of Hg^{+2} can be obtained with the Triton X-114 concentration in the range of 0.025% to 0.3% (w/v). The maximum of thermal lens signal observed for the Triton X-114 concentration is higher than 0.1% (w/v). In order to achieve a good enrichment factor, 0.1% (w/v) was chosen as the optimal Triton X-114 concentration.

Figures of merit

From measurements made under the optimum conditions described above, the calibration graph was linear in the range of 1 to 50 $\mu g\,L^{-1}$. The calibration equation is $A = 6.42 \times 10^{-2}C + 11.9 \times 10^{-2}$ with a correlation coefficient of 0.9957, where A is the thermal lens signal for Hg^{+2} in the surfactant-rich phase and C is the concentration of Hg^{+2}($\mu g\,L^{-1}$) in the sample solution. The limit of detection, defined as $C_L = 3S_B/m$, where C_L, S_B, and m are the limit of detection, standard deviation of the

blank, and the slope of the calibration graph, respectively, was 0.2 $\mu g\,L^{-1}$.

The reproducibility of the method was studied for five replicate determinations of Hg^{+2} in an aqueous sample spiked with 25 $\mu g\,L^{-1}$ of Hg^{+2} after extractions. The relative standard deviation (RSD) was lower than 4%. The enhancement factor (calculated as the ratio of the slopes of the calibration graph with and without preconcentration) and preconcentration factor (calculated as the ratio of sample volume before and after preconcentration) were 179 and 200, respectively.

Table 1 compares the sample volume, solvent volume, enhancement factor, limit of detection, and linear dynamic range of other determination methods with LI-TLS after cloud point extraction of mercury. It shows that in comparison with other sensitive determination methods such as hydride generation atomic absorption spectrometry (HG-AAS) and cold vapor atomic absorption spectrometry (CV-AAS), the LI-TLS has lower sample volume and needed solvent volume, higher enhancement factor, and good limit of detection.

Interferences

The effect of interfering ions was studied at different concentrations on the thermal lens signal of a solution containing 25 $\mu g\,L^{-1}$ of each analyte. An ion was considered to interfere when its presence produced a variation in the thermal lens signal of the sample of more than 5%. Among the interfering ions tested - CO_3^{-2}, Cl^-, Ca^{2+}, SO_4^{2-}, NH_4^+, Mg^{2+}, K^+, and Na^+ at the concentration of 500 times higher than that of the Hg^{+2} concentration;

Table 2 Determination of mercury in water samples

Sample	Added ($\mu g\,L^{-1}$)	Found ($\mu g\,L^{-1}$)[a]	Recovery (%)
Soft water	-	ND	-
	10	10.14 ± 0.25	101.4
	25	25.12 ± 0.32	100.5
Physiology serum	-	ND	-
	10	9.53 ± 0.27	95.3
	25	23.55 ± 0.29	94.2

[a]Mean ± S.D. ($n = 3$).

Cr^{3+}, Zn^{2+}, Al^{3+}, and Fe^{3+} at the concentration of 200 times higher than that of the Hg^{+2} concentration; and F^-, Cd^{2+}, Ni^{2+}, Fe^{2+}, and Pb^{2+} at the concentration of 80 times higher than that of the Hg^{+2} concentration - no interfering was observed.

Application

To verify the accuracy of the method, the developed method was applied for the determination of Hg^{+2} in the soft water and physiology serum. Along with the samples, several known amounts of Hg^{+2} were spiked to examine the reliability of the method. For this purpose, 10 mL of each sample was preconcentrated, following the proposed procedure. The results are shown in Table 2. As shown in Table 2, the recovery of the spiked amounts was in the range of 94.2% to 101.4%, which demonstrates the reliability of the proposed method.

Abbreviations
CMC: critical micelle concentration; CPE: cloud point extraction; CPT: cloud point temperature; CV-AAS: cold vapor atomic absorption spectrometry; CV-AFS: cold vapor atomic fluorescence spectrometry; FI-ICP-OES: flow injection-inductively coupled plasma-optical emission spectrometry; HPLC-ICP-MS: high-performance liquid chromatography-inductively coupled plasma mass spectrometry; LI-TLS: laser-induced thermal lens spectrometry; LOD: limit of detection.

Acknowledgements
Support of this investigation by the Research Council of Chemistry & Chemical Engineering Research Center of Iran (CCERCI) and Iran National Science Foundation (INSF) gratefully acknowledged.

References

Afkhami A, Madrakian T, Siampour H (2006) Flame atomic absorption spectrometric determination of trace quantities of cadmium in water samples after cloud point extraction in Triton X-114 without added chelating agents. J Hazard Mater B 138:269–272

Bialkowski SE (1996) Photothermal spectroscopy methods for chemical analysis. John Wiley, New York, Chichester, p 584

Chappuy M, Caudron E, Bellanger A, Pradeau D (2010) Determination of platinum traces contamination by graphite furnace atomic absorption spectrometry after preconcentration by cloud point extraction. J Hazard Mater 176:207–212

Chen H, Chen J, Jin X, Wei D (2009) Determination of trace mercury species by high performance liquid chromatography–inductively coupled plasma mass spectrometry after cloud point extraction. J Hazard Mater 172:1282–1287

Dovichi NJ, Bialkowoki SE (1987) Thermo-optical spectrophotometries in analytical chemistry. Crit Rev Anal Chem 17:357–423

Dovichi NJ, Harris JM (1979) Laser induced thermal lens effect for calorimetric trace analysis. Anal Chem 51:728–731

Eisler R (2004) Mercury hazards from gold mining to humans, plants and animals. Rev Environ Contam Toxicol 181:139–198

Franko M, Tran CD (1996) Analytical thermal lens instrumentation. Rev Sci Instrum 67:1–18

Geng W, Nakajima T, Takanashi H, Ohki A (2008) Determination of mercury in ash and soil samples by oxygen flask combustion method-cold vapor atomic fluorescence spectrometry (CVAFS). J Hazard Mater 154:325–333

Georges J (1994) A single and simple mathematical expression of the signal for cw-laser thermal lens spectrometry. Talanta 41:2015–2023

Giokas DL, Paleologos EK, Tzouwara-Karayanni SM, Karayannis MI (2001) Single-sample cloud point determination of iron, cobalt and nickel by flow injection analysis flame atomic absorption spectrometry application to real samples and certified reference materials. J Anal At Spectrom 16:521–526

Gordon JP, Leite RCC, Moore RS, Porto SPS, Whinnery JR (1964) Long transient effects in lasers with inserted liquids samples. Bull Am Phys Soc 9:501

Gordon JP, Leite RCC, Moore RS, Porto SPS, Whinnery JR (1965) Long transient effects in lasers with inserted liquids samples. J Appl Phys 36:3

Imasaka T, Miyaishi K, Ishibashi N (1980) Application of the thermal lens effect for determination of iron(II) with 4,7-diphenyl-1,10-phenanthroline disulfonic acid. Anal Chim Acta 115:407–410

Kitamori T, Takeshi M, Hibara A, Sato K (2004) Thermal lens microscopy and microchip chemistry. Anal Chem 76:52–60 A

Li Y, Hu B (2007) Sequential cloud point extraction for the speciation of mercury in seafood by inductively coupled plasma optical emission spectrometry. Spectrochim Acta Part B 62:1153–1160

Manzoori JL, Bavili-Tabrizi A (2002) The application of cloud point preconcentration for the determination of Cu in real samples by flame atomic absorption spectrometry. Michrochim Acta 72:1–7

Martinez RC, Gonzalo ER (2000) Surfactant cloud point extraction and preconcentration of organic compounds prior to chromatography and capillary electrophoresis. J Chromatogr A 902:251–265

Niazi A, Momeni-Isfahani T, Ahmari Z (2009) Spectrophotometric determination of mercury in water samples after cloud point extraction using nonionic surfactant Triton X-114. J Hazard Mater 165:1200–1203

Pacyna EG, Pacyna JM (2006) Mercury emissions to the atmosphere from anthropogenic sources in Europe in 2000 and their scenarios until 2020. Sci Total Environ 370:147–156

Paleologos EK, Stalikas CD, Karayannis MI (2001) An optimised single-reagent method for the speciation of chromium by flame atomic absorption spectrometry based on surfactant micelle-mediated methodology. Analyst 126:389–393

Paleologos EK, Giokas DL, Tzouwara-Karayanni SM, Karayannis MI (2002) Micelle mediated methodology for the determination of free and bound iron in wines by flame atomic absorption spectrometry. Anal Chim Acta 458:241–248

Paleologos EK, Giokas DL, Karayannis MI (2005) Micelle-mediated separation and cloud-point extraction. Trends Anal Chem 24:426–436

Piepmeier EH (1986) Analytical applications of lasers. Chemical analysis. Wiley, New York, p 703

Pourreza N, Ghanemi K (2009) Determination of mercury in water and fish samples by cold vapor atomic absorption spectrometry after solid phase extraction on agar modified with 2-mercaptobenzimidazole. J Hazard Mater 161:982–987

Sanz-Medel A, Campa MRF, Gonzalez EB, Fernandez-Sanchez ML (1999) Organised surfactant assemblies in analytical atomic spectrometry. Spectrochim Acta Part B 54:251–287

Shah F, Kazi TG (2011) Cloud point extraction for determination of lead in blood samples of children, using different ligands prior to analysis by flame atomic absorption spectrometry: a multivariate study. J Hazard Mater 192:1132–1139

Shamsipur M, Shokrollahi A, Sharghi H, Eskandari MM (2005) Solid phase extraction and determination of sub-ppb levels of hazardous Hg^{+2} ions. J Hazard Mater B 117:129–133

Snook DD, Lowe RD (1995) Thermal lens spectrometry. Analyst 120:2051–2068

Song JY, Hou M, Zhang LX (2006) Determination of mercury at trace level in natural water samples by hydride generation atomic absorption spectrophotometry after cloud point extraction preconcentration. Chin Chem Lett 17:1217–1220

Stalikas CD (2002) Micelle-mediated extraction as a tool for separation and preconcentration in metal analysis. Trends Anal Chem 21:343–355

Wuilloud JCAD, Wuilloud RG, Silva MF, Olsina RA, Martinez LD (2002) Sensitive determination of mercury in tap water by cloud point extraction pre-concentration and flow injection-cold vapor-inductively coupled plasma optical emission spectrometry. Spectrochim Acta B 57:365–374

16

^{40}Ar/^{39}Ar age determination using ARGUS VI multiple-collector noble gas mass spectrometer: performance and its application to geosciences

Jeongmin Kim[*] and Su-in Jeon

Abstract

Background: Ar/^{39}Ar dating technique has been used to determine the age for low-temperature geological event. The introduction of the multiple collectors and the improvement in sensitivity in the noble gas mass spectrometry enable the single-grain ^{40}Ar/^{39}Ar age determination.

Findings: The protocol for ^{40}Ar/^{39}Ar age determination and performance of the new high-sensitivity noble gas mass spectrometer (ARGUS VI) combined with the infra-red laser heating device are shown.

Conclusions: ARGUS VI can produce the precise single-grain ^{40}Ar/^{39}Ar age of 28.4 ± 0.5 Ma (1σ, MSWD = 138, $n = 24$) within the recommended values of Fisher Canyon sanidine (FCS) (28.294 ± 0.036 Ma). For the young samples, we can also get the age of 1.185 ± 0.004 Ma (1σ, MSWD = 6.05, $n = 26$) of Alder Creek sanidine (ACS), equivalent to the recommended age (1.193 ± 0.001 Ma).

Keywords: ^{40}Ar/^{39}Ar age; Noble gas mass spectrometer; Quaternary; Geochronology; CO_2 laser

Findings

Introduction

Recent development in mass spectrometry, such as the increase of sensitivity and the adoption of multiple collector system, enables to analyze the isotopic ratio for the single grain and spot analysis. Since the introduction of in-situ age dating for the small geological samples such as SHRIMP (sensitive high-resolution ion micro probe), the complex geological events can be revealed more precisely. The U-Pb spot age dating for zircon grains suggests the new concept of early Earth (e.g., Wilde et al. 2001) and provides the important keys to solve the juxtaposed geological events in a certain region (e.g., Cheong et al. 2014). However, to reveal the full history of a certain orogeny, it is indispensable to get the age information for the low- to mid-temperature geological activity. The Ar isotope system, whose closure temperatures range from ca. 130 to 690°C (e.g., Faure and Mensing 2005), can provide good clue to solve it. In addition, for the basaltic rocks with rare minerals adequate for U-Pb dating, the Ar

ages have become the good proxy for the eruption age. For these purpose, K-Ar age dating method was introduced in early 50s and had provided age information related to various studies, such as the geomagnetic polarity timescale, one of the essential foundations of plate tectonic concepts (McDougall, 2014). However, the possible inhomogeneity due to the separate analyses of K and Ar and no information about the Ar-loss and/or Ar-gain after the closure in Ar isotope system should be considered for the proper interpretation of K-Ar ages. The ^{40}Ar/^{39}Ar technique, a variant of K-Ar dating system, has now commonly replaced the conventional K-Ar method to overcome these disadvantages. In addition, the introduction of laser heating/ablation technique enables to measure Ar age of single grains and monitor the variation in Ar isotope within minerals. In this technical note, we describe the ^{40}Ar/^{39}Ar age dating protocol using the newly installed multi-collector noble gas mass spectrometer in Korea Basic Science Institute (KBSI) and also report its performance for some geological samples.

Principle of ^{40}Ar/^{39}Ar age determination

^{40}Ar/^{39}Ar age dating is based on the same principle of K-Ar method, which uses the decay of ^{40}K to ^{40}Ar through

* Correspondence: j-mkim@kbsi.re.kr

Division of Earth and Environmental Sciences, Korea Basic Science Institute, 162 Yeongudanji-ro, Ochang-eup, Cheongwon-gun, Chungcheongbuk-do 363-886, Korea

the electron capture and positron emission. The concentration of parent ^{40}K is measured indirectly the neutron irradiation process which converts a part of ^{39}K to ^{39}Ar through the ^{39}K(n, p)^{39}Ar reaction. The constant ^{40}K/^{39}K ratio in samples enables to estimate the content of ^{40}K in samples through the measurement of ^{39}Ar, if we exactly know how many portions of ^{39}K in samples are converted to ^{39}Ar during the neutron irradiation. The ^{40}Ar/^{39}Ar age is calculated using the following formula (McDougall and Harrison 1999):

$$t = \frac{1}{\lambda} \cdot \ln\left(1 + J \cdot \frac{^{40}Ar^*}{^{39}Ar_K}\right)$$

where

$$J = \frac{^{39}K}{^{40}K} \cdot \frac{\lambda}{\lambda_e + \lambda_e'} \cdot \Delta T \cdot \int \phi(E)\sigma(E)dE$$

t age
λ total decay constant of ^{40}K (5.534×10^{-10}/year)

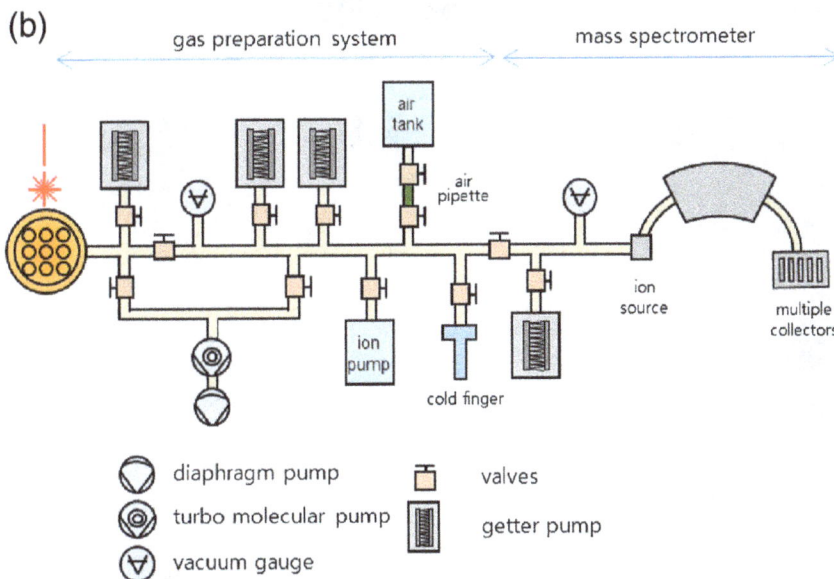

Figure 1 ARGUS VI system in KBSI. (a) Photo of whole ^{40}Ar/^{39}Ar age dating system and (b) schematic illustrations for the gas preparation system.

λ_e decay constant of ^{40}K to ^{40}Ar by electron capture (0.572×10^{-10}/year)

λ_e' decay constant of ^{40}K to ^{40}Ar by emission of a positron (0.0088×10^{-10}/year)

^{40}Ar*   radiogenic ^{40}Ar

^{39}Ar$_K$ ^{39}Ar derived from ^{39}K by the reaction with fast neutron

ΔT duration of irradiation

$\phi(E)$ neutron flux density at energy E

$\sigma(E)$ neutron capture cross section at energy E

As the precise values of $\phi(E)$ and $\sigma(E)$ are difficult to obtain, standard minerals of age-known are used to calculate J values using the following equation:

$$J = \frac{\exp(\lambda t) - 1}{^{40}Ar * /^{39}Ar_K}$$

During neutron irradiations, the isotopes of Ca, K, and Cl in the samples commonly produce interfering Ar isotopes through interactions of neutron. Reactor-derived ^{40}Ar from ^{40}K, and ^{39}Ar and ^{36}Ar from ^{42}Ca and ^{40}Ca, respectively, are corrected by analyzing K- and Ca-salts irradiated together with samples. For the detailed calculation for these interfering isotopes, refer to McDougall and Harrison (1999).

Availability and requirements
Outline of the Ar age dating system
^{40}Ar/^{39}Ar age dating system in KBSI can be divided into the following three parts: (1) laser heating system, (2) gas preparation bench, and (3) high-sensitivity noble gas mass spectrometer (Figure 1).

Laser heating system
Laser heating system (Fusions 10.6, Photon Machines) consists of an integrated CO_2 laser source and beam-delivery system with CCD camera and sample chamber. The laser source generates the continuous CO_2 laser with the wavelength of 10.5 μm. Gantry type beam delivery system consists of three mirrors, one iris and one focusing lens. Whole laser delivery system also navigates 3-dimensionally over sample chamber. The intensity and beam size of laser including the movement of beam delivery system are controlled by the Chromium II software. The sample chamber is made up with a 114 mm diameter Conflat flange and differentially pumped ZnS viewport by turbomolecular pump. A copper sample holder of 43 mm diameter and 5 mm in thickness with 133 holes is loaded into sample chamber for single grain analysis. KBr glass covers the sample holder to prevent the spillover and scattering of sample from the hole during heating. After sample loading, sample chambers are heated by infrared lamp during several hours to achieve good vacuum level. Released gases from sample by laser beam diffuse from sample chamber into gas preparation system.

Gas preparation system
The outline of the gas preparation system is shown in Figure 1b. All pipe line is made from internally polished stainless steel. In order to purify argon from the extracted gases, three SORB-AC getter pumps (NP10) are used. They are constructed from a cartridge of getter material (ST101 alloy of zirconium with 16% aluminum) placed around an axial heater. At room temperature, these getters pump out hydrogen and carbon monoxide which are major background gases in the mass spectrometer.

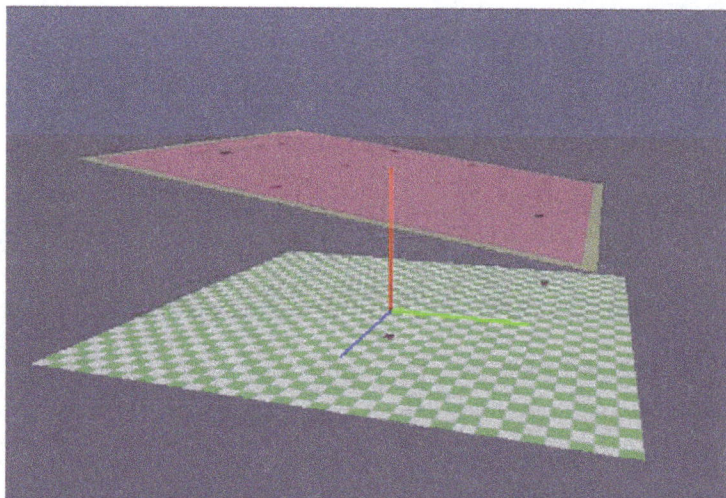

Figure 2 3-Dimensional variation in J values across one irradiation disk.

Table 1 Representative analysis of $^{40}Ar/^{39}Ar$ age for Fisher Canyon sanidine (FCS) and Alder Creek sanidine (ACS)

Lab ID no.	J (×10⁻³)±1σ		^{40}Ar ±1σ		^{39}Ar ±1σ		^{38}Ar ±1σ		^{37}Ar ±1σ		^{36}Ar ±1σ		^{39}Ar Mol ×10⁻¹⁴	Ca/K	%$^{40}Ar^*$	Age (Ma) ±1σ	
Relative isotopic abundances													**Derived results**				
FCS																	
84-02	0.2055	0.0003	56,989	8	748	3	57	3	22	2	1.27	0.09	0.025	0.161	99.4	27.83	0.10
84-03	0.2055	0.0003	96,025	11	1,239	3	87	3	26	2	1.77	0.09	0.041	0.116	99.5	28.32	0.06
84-04	0.2055	0.0003	79,364	11	1,033	7	65	9	9	3	0.56	0.14	0.034	0.050	99.8	28.17	0.20
84-05	0.2055	0.0003	36,927	10	478	7	29	9	7	3	0.14	0.14	0.016	0.087	99.9	28.32	0.42
84-06	0.2055	0.0003	70,660	11	913	7	49	9	10	3	0.87	0.14	0.030	0.065	99.6	28.34	0.22
87-01	0.1959	0.0003	60,413	8	759	3	44	3	14	3	1.99	0.08	0.025	0.100	99.0	27.61	0.10
87-02	0.1959	0.0003	63,977	16	583	3	30	3	0	3	2.47	0.09	0.019	0.000	98.9	33.90	0.18
87-03	0.1959	0.0003	54,528	7	686	3	49	4	11	3	0.83	0.08	0.023	0.091	99.6	27.72	0.14
87-05	0.1959	0.0003	103,304	9	1,264	3	77	2	0	2	3.63	0.09	0.042	0.000	99.0	28.32	0.08
87-06	0.1959	0.0003	46,783	10	581	7	36	8	4	3	0.54	0.13	0.019	0.043	99.7	28.10	0.32
87-07	0.1959	0.0003	35,618	9	445	7	33	8	1	4	0.58	0.13	0.015	0.011	99.5	27.91	0.42
87-09	0.1959	0.0003	78,316	11	969	7	55	8	8	3	1.29	0.14	0.032	0.045	99.5	28.17	0.19
93-01	0.1816	0.0001	63,628	8	724	2	37	3	5	2	1.79	0.09	0.024	0.038	99.2	28.28	0.10
93-02	0.1816	0.0001	127,664	12	1,453	3	70	4	9	3	1.53	0.10	0.048	0.034	99.7	28.41	0.07
93-03	0.1816	0.0001	108,347	10	1,228	3	65	4	7	2	0.59	0.09	0.041	0.034	99.8	28.59	0.08
93-04	0.1816	0.0001	124,168	11	1,424	3	78	4	14	3	2.15	0.09	0.047	0.056	99.5	28.16	0.07
93-05	0.1816	0.0001	51,898	7	595	3	37	3	1	2	1.06	0.08	0.020	0.007	99.4	28.16	0.15
93-06	0.1816	0.0001	100,604	10	1,167	3	77	3	13	3	0.95	0.09	0.039	0.064	99.7	27.90	0.08
93-07	0.1816	0.0001	81,501	9	935	5	52	3	3	3	0.99	0.08	0.031	0.019	99.6	28.18	0.14
93-08	0.1816	0.0001	108,398	9	1,225	5	72	3	51	2	1.04	0.09	0.041	0.233	99.7	28.65	0.11
93-09	0.1816	0.0001	100,317	10	1,154	3	78	3	8	3	3.84	0.09	0.038	0.037	98.9	27.91	0.09
93-10	0.1816	0.0001	120,283	10	1,386	3	96	3	29	2	0.72	0.08	0.046	0.119	99.8	28.13	0.06
93-11	0.1816	0.0001	89,456	9	1,022	3	77	3	25	3	2.80	0.09	0.034	0.139	99.1	28.16	0.08
93-12	0.1816	0.0001	141,018	12	1,631	4	92	3	16	2	0.31	0.08	0.054	0.056	99.9	28.05	0.06
														Average		28.38	0.47
ACS																	
75-01	0.2020	0.0003	11,629	11	3,238	6	184	6	18	5	3.14	0.14	0.108	0.023	92.0	1.176	0.006
75-03	0.2020	0.0003	4,741	5	1,419	4	76	3	14	2	0.27	0.09	0.047	0.047	98.5	1.169	0.008
75-04	0.2020	0.0003	11,887	4	3,330	5	166	3	45	3	2.44	0.10	0.111	0.066	94.0	1.195	0.004
75-05	0.2020	0.0003	11,502	6	3,112	4	172	4	5	3	3.72	0.11	0.103	0.007	90.3	1.189	0.004
75-06	0.2020	0.0003	12,491	5	3,668	3	198	3	15	2	0.90	0.10	0.122	0.020	97.9	1.186	0.003
75-08	0.2020	0.0003	16,880	19	3,793	7	215	6	17	6	14.91	0.30	0.126	0.022	73.5	1.170	0.010
79-01	0.1910	0.0002	13,275	12	3,623	6	245	6	22	5	1.48	0.14	0.120	0.026	96.8	1.194	0.005
79-02	0.1910	0.0002	13,908	12	3,866	6	209	6	25	5	0.39	0.14	0.128	0.027	99.3	1.202	0.004
79-03	0.1910	0.0002	25,059	10	6,147	5	382	3	41	4	11.51	0.15	0.204	0.033	86.3	1.187	0.003
79-04	0.1910	0.0002	17,716	6	4,857	5	275	3	37	4	1.96	0.10	0.161	0.037	96.8	1.189	0.003
79-05	0.1910	0.0002	14,645	5	3,944	4	241	3	17	3	2.89	0.11	0.131	0.021	94.2	1.177	0.003
79-06	0.1910	0.0002	7,592	5	2,118	5	130	5	14	4	0.63	0.10	0.070	0.032	97.6	1.178	0.006
79-07	0.1910	0.0002	11,883	6	3,176	5	175	3	19	3	2.72	0.09	0.105	0.029	93.2	1.175	0.004
79-08	0.1910	0.0002	10,674	5	2,921	3	207	3	18	2	1.68	0.10	0.097	0.031	95.4	1.173	0.004
79-09	0.1910	0.0002	7,120	6	1,973	3	117	4	18	4	0.33	0.10	0.066	0.043	98.8	1.200	0.006

Table 1 Representative analysis of $^{40}Ar/^{39}Ar$ age for Fisher Canyon sanidine (FCS) and Alder Creek sanidine (ACS)
(Continued)

83-01	0.1788	0.0002	21,443	11	5,512	7	415	6	35	4	3.32	0.14	0.183	0.027	95.5	1.172	0.003
83-02	0.1788	0.0002	13,561	12	3,278	6	286	7	23	4	4.76	0.15	0.109	0.030	89.6	1.171	0.005
83-03	0.1788	0.0002	18,842	6	4,812	5	305	4	26	3	1.93	0.11	0.160	0.026	97.0	1.199	0.003
83-04	0.1788	0.0002	25,856	5	6,779	6	406	4	40	3	1.77	0.11	0.225	0.029	98.1	1.180	0.002
83-05	0.1788	0.0002	27,183	9	5,713	7	489	6	24	5	19.39	0.31	0.190	0.021	78.7	1.186	0.005
83-06	0.1788	0.0002	12,743	9	3,243	7	187	5	12	5	2.09	0.27	0.108	0.018	95.1	1.180	0.008
83-07	0.1788	0.0002	12,061	14	3,075	6	174	6	13	5	2.18	0.27	0.102	0.021	94.6	1.172	0.009
83-08	0.1788	0.0002	18,775	18	4,755	7	272	6	21	6	2.12	0.27	0.158	0.022	96.7	1.205	0.006
83-09	0.1788	0.0002	17,799	18	4,521	7	269	6	18	6	1.97	0.27	0.150	0.020	96.8	1.202	0.006
83-10	0.1788	0.0002	6,962	17	1,784	6	111	6	14	6	0.44	0.26	0.059	0.040	98.2	1.210	0.015
													Average			1.185	0.004

The getter can be run at 400°C to enhance the pumping of less reactive gases such as hydrocarbons. The vacuum level of gas preparation system reaches approximately 2×10^{-9} mbar by the turbomolecular pump and ion pump. The air of 0.1 cm^3 from the automatic pipette system consisting of standard volume and two pneumatic valves is routinely measured to derive the discrimination factor.

Mass spectrometer

The purified argon gas through gas preparation system is directly introduced into the mass spectrometer (ARGUS VI). ARGUS VI is an all-metal single focusing, 13 cm radius, 90° extended geometry magnetic sector mass spectrometer designed for operation in static mode (Mark et al. 2009). A Nier-type electron bombardment source is equipped with y- and z-focusing. All source parameters, including the 5 kV acceleration potential, are computer-controlled and monitored by electronic readbacks. This system is equipped with five Faraday detectors and one compact discrete dynode (CDD) detector for the simultaneous measurement of five isotopes of argon (^{36}Ar, ^{37}Ar, ^{38}Ar, ^{39}Ar, and ^{40}Ar). The ^{40}Ar beam is measured in the high-mass (H1) detector fitted with a 10^{11} ohm resistor. The ^{39}Ar to ^{37}Ar beams are collected in other Faraday detectors fitted with 10^{12} ohm resistor. The ^{36}Ar beam is measure by CDD detector. The amplifiers for Faraday detectors are kept in housing evacuated to 1–2 mbar in order to reduce electronic noise. Sensitivity at 200 μA trap current for Ar reaches to 1.15 Amps/Torr and the mass resolution ($M/\Delta M$) is ca. 250. The background value during sample analysis is $<1.0 \times 10^{-16}$ mols ^{40}Ar and $<1.8 \times 10^{-18}$ mols ^{36}Ar in static mode.

System integration and data reduction

Laser heating system, gas preparation system, and mass spectrometer are integrated and controlled by the Mass-Spec software (Alan Deino Software). It communicates with Chromium II software in laser heating system and Qtegra software in ARGUS VI and controls each parameter in source electronics, magnet control, and steering plate in front of multiple collectors as well as the manipulation of gas preparation system. In addition, it enables the scheduled automatic analysis for tens of analysis. Fitting of raw data, age calculation, and data presentation are also performed using MassSpec software.

Experimental procedures for $^{40}Ar/^{39}Ar$ age
Neutron irradiation

As suggested in the above section, neutron irradiation process is indispensable for the $^{40}Ar/^{39}Ar$ dating. For irradiation, several grains of sample were dropped in each hole on aluminum sample disk with 10 holes of 5-mm diameter along with K_2SO_4 and CaF_2 monitors. Age-known standard minerals were placed in the center or margin of sample disk as the neutron flux monitor. Fisher Canyon sanidine (FCS, 28.294 ± 0.036 Ma: Renne et al. 2010) and Alder Creek sanidine (ACS, 1.193 ± 0.001 Ma: Nomade et al. 2005) were used as flux monitor minerals. Two or three sample disks were stacked and wrapped by 0.25-mm-thick cadmium foil in order to reduce the effect of thermal neutrons. Samples were irradiated for 100 h at IP-4(c) position of hydraulic rabbit irradiation facility in HANARO research reactor with the total neutron flux of ca. $3.7 \times 10^{13} \text{n/cm}^2\text{s}$ in Korea Atomic Energy Research Institute (KAERI).

Argon isotope measurement

One or several grains of irradiated samples are placed in a hole on the 133-hole sample holder and then evacuated to ultra high vacuum of $\sim 10^{-9}$ mbar. Multiple run sequences for sample analysis including air and blank measurement are executed on the MassSpec software. The heating of samples and subsequent cleaning of extracted gas and the measurement of isotopes are sequentially performed by previously defined procedure file. After heating samples for 6 seconds and purifying gases for another 4 minutes with 3 SAES getters, the argon gas is introduced to the mass spectrometer. Each signal of argon isotopes is measured by the multiple data collection using five detectors using H1 to CDD detector or by the peak jumping method using CDD for ^{36}Ar to ^{39}Ar and H1 detector for ^{40}Ar, respectively. For the baseline, the signals at mass 35.7 and 40.3 are measured at the start and end of the analysis. Blank analyses are made every three unknown sample analysis with the identical condition to the actual samples except for laser heating. For the mass discrimination, standard air is measured after every 10 analysis. Variation of blank

and mass discrimination factor during analysis are corrected by parabolic regression. Net intensity of each Ar peaks is extrapolated to the zero time when the argon gas was introduced into the mass spectrometer. After correcting the blank and the mass discrimination, J values and ages are calculated for the standard and unknown samples.

Determination of J value

As neutron flux changes considerably across the sample disk, monitor minerals such as FCS and ACS sanidine are placed in three positions to correct the horizontal gradient in neutron flux. For real samples, the corrected

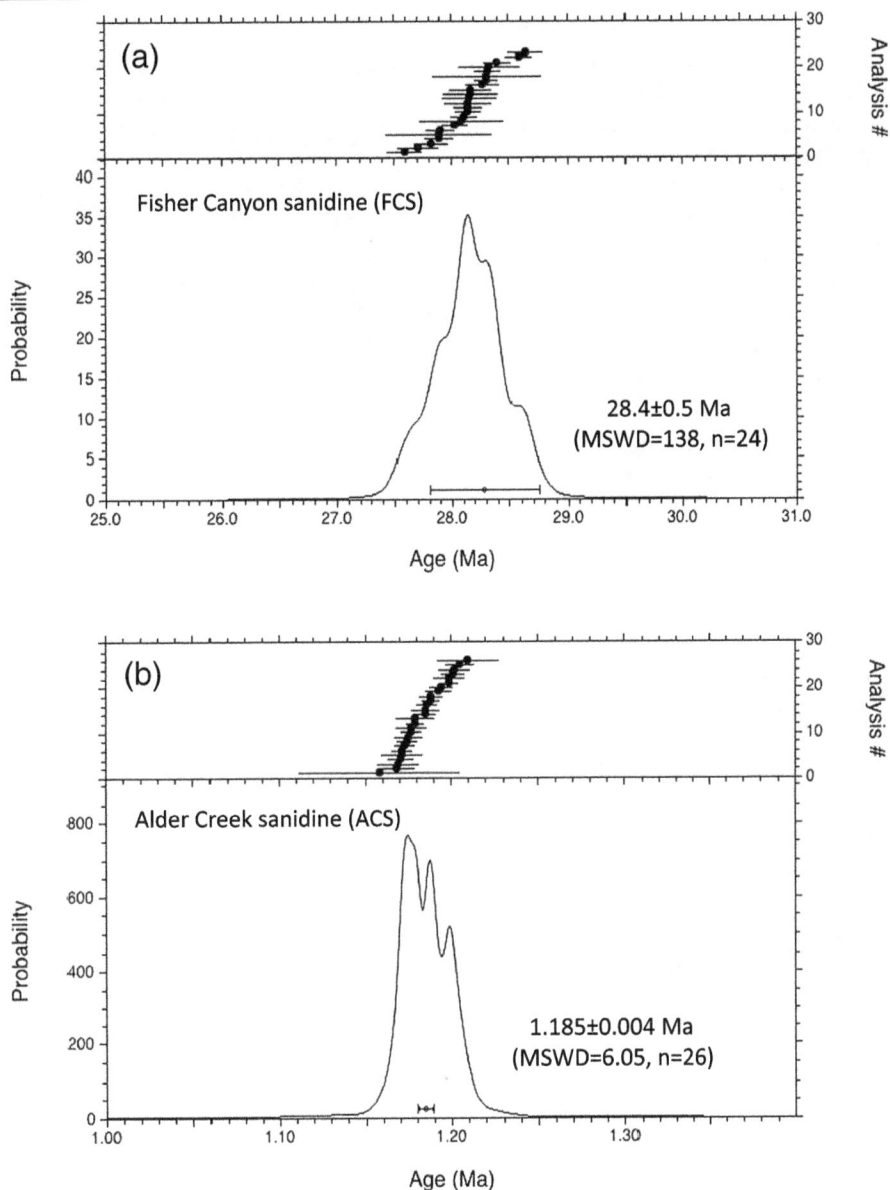

Figure 3 Age probability diagrams for the ^{40}Ar/^{39}Ar **age of standard materials. (a)** Fisher Canyon Sanidine. **(b)** Alder Creek Sanidine.

J value along the different position of samples in the sample disk is applied to the age calculation according to the 3-dimensional modeling for the horizontal flux gradient (Figure 2).

Results and discussion

Single-grain total fusion $^{40}Ar/^{39}Ar$ analyses were repeatedly performed on two standard materials in order to evaluate the performance of ARGUS VI system (Table 1). Figure 3 shows the age probability diagrams for the total fusion ages from single grains of FCS and ACS. The measured value of FCS (Figure 3a) is 28.4 ± 0.5 Ma (1σ, MSWD = 138, $n = 24$) which is consistent with recommended value (28.294 ± 0.036 Ma: Renne et al. 2010) within uncertainty. Figure 3b shows the single-grain total fusion ages of ACS sanidine (1.185 ± 0.004 Ma, 1σ, MSWD = 6.05, $n = 26$), which also coincide with recommended age value (1.193 ± 0.001 Ma, Nomade et al. 2005). Such performance shows that the abovementioned $^{40}Ar/^{39}Ar$ age protocol can be successfully applied to the Quaternary rocks such as ACS.

For age dating of young rocks (<1 Ma), the exact derivation of radiogenic ^{40}Ar ($^{40}Ar*$) is indispensable to get meaningful $^{40}Ar/^{39}Ar$ ages, so that it needs to minimize the reactor-derived ^{40}Ar from ^{40}K. In this study, the $(^{40}Ar/^{39}Ar)_K$ with the 0.25-mm-thick cadmium shielding reaches approximately 0.08, which is higher than the majority of reactor facilities used for $^{40}Ar/^{39}Ar$ age dating (McDougall and Harrison 1999) owing to high slow/fast neutron flux of >780 in Hanaro (Nam S, personal communication). To reduce the incidence of slow neutrons and, thus, the production rate of ^{40}Ar, thicker cadmium shielding should be highly considered.

Conclusions

$^{40}Ar/^{39}Ar$ dating system with multi-collector mass spectrometer and CO_2 laser heating device has been established at KBSI in order to date young and very small amounts of minerals and volcanic rocks. Preliminary experiments on sanidine standards show the reliable age results of total-fusion dating for single grains, i.e., 28.4 ± 0.5 Ma (1σ, MSWD = 138, $n = 24$) for FCS sanidine, and 1.185 ± 0.004 Ma (1σ, MSWD = 6.05, $n = 26$) for ACS sanidine as Quaternary age standard, within recommended values (28.294 ± 0.036 Ma and 1.193 ± 0.001 Ma, respectively).

Competing interests
The authors declare that they have no competing interests.

Authors' contribution
JK conceived of the study and carried out the experiment using ARGUS VI as well as writing manuscript. SJ participated in the preparation of experiment and helped to draw the figures. Both authors read and approved the final manuscript.

Acknowledgements
This study was supported by the KBSI grant (G34200). We thank Dr. Alan Deino in Berkeley Geochronology Center and Mr. Karl Burke in Thermo for their effort to install MassSpec software and ARGUS VI mass spectrometer. We appreciate Mr. Nam Seongsu who organized the neutron irradiation process in HANARO. We are also grateful to reviewers who read the manuscripts and give useful comments.

References
Cheong CS, Kim N, Kim J, Yi K, Jeong YJ, Park CS, Li HK, Cho M (2014) Petrogenesis of Late Permian sodic metagranitoids in southeastern Korea: SHRIMP zircon geochronology and elemental and Nd–Hf isotope geochemistry. J Asian Earth Sci. doi:10.1016/j.jaes.2014.06.005

Faure G, Mensing TM (2005) Isotopes: principles and applications, 3rd edn. John Wiley & Sons, Hoboken

Mark DF, Barford D, Sturat FM, Imlach J (2009) The ARGUS multicollector noble gas mass spectrometer: performance for $^{40}Ar/^{39}Ar$ geochronology. Geochem Geophys Geosys. doi:10.1029/2009GC002643

McDougall I (2014) Perspectives on $^{40}Ar/^{39}Ar$ dating. In: Jourdan F, Mark DF, Verati C (eds) Advances in $^{40}Ar/^{39}Ar$ dating: from archeology to planetary science, vol 378. Geological Society Special Publication, London, pp 9–20

McDougall I, Harrison TM (1999) Geochronology and thermochronology by the $^{40}Ar/^{39}Ar$ method, 2nd edn. Oxford Univ. Press, New York

Nomade S, Renne PR, Vogel N, Deino AL, Sharp WD, Becker TA, Jaouni AR, Mundil R (2005) Alder Creek sanidine (ACs-2): a Quaternary $^{40}Ar/^{39}Ar$ dating standard tied to the Cobb Mountain geomagnetic event. Chem Geol 218:315–338

Renne PR, Mundil R, Balco G, Min K, Ludwig KR (2010) Joint determination of ^{40}K decay constants and $^{40}Ar*-^{40}K$ for the Fish Canyon sanidine standard, and improved accuracy for $^{40}Ar/^{39}Ar$ geochronology. Geochim Cosmochim Acta 74:5349–5367

Wilde SA, Valley JW, Peck WH, Graham CM (2001) Evidence from detrital zircons for the existence of continental crust and oceans on the Earth 4.4 Gyr ago. Nature 409:175–178

17

Validation and measurement uncertainty evaluation of the ICP-OES method for the multi-elemental determination of essential and nonessential elements from medicinal plants and their aqueous extracts

Marin Senila[1*†], Andreja Drolc[2†], Albin Pintar[2], Lacrimioara Senila[1] and Erika Levei[1]

Abstract

Background: The paper presents the development, validation, and evaluation of measurement uncertainty of a method for quantitative determination of essential and nonessential elements in medicinal plants and their aqueous extracts by using inductively coupled plasma optical emission spectrometry.

Methods: The detailed validation of the analytical procedure and calculation of the measurement uncertainty budget allowed the recognition of the methods' critical points.

Results: The obtained limit of quantification, repeatability, and measurement uncertainty were satisfactory. The trueness of the method was verified by recovery estimation using certified reference materials. The recovery rates of all metals were between 95% and 105%.

Conclusions: The paper presents for the first time all the steps needed to evaluate the measurement uncertainty and validate the determination method of selected elements in medicinal plants and their aqueous extracts. In summary, the obtained results demonstrate that the method can be applied effectively for the designed purpose.

Keywords: ICP-OES; Medicinal plants; Multi-elemental analysis; Validation; Measurement uncertainty

Background

The inductively coupled plasma optical emission spectrometry (ICP-OES) is a strong tool for the determination of various elements in liquid and solid samples. Elevated concentrations of essential elements (e.g., Fe, Mn, Zn, Cr, Cu) and low concentrations of nonessential elements (e.g., Cd, Ni, As) may present a potential hazard for human health. When preparing tea by infusion of plants, metals can be leached into the water and consumed by humans. Therefore, the metal contents in the plant infusion should comply with the limit values set by the Drinking Water Directive (Council of the European Union 1998).

Numerous plant species used as remedies in traditional medicine are grown as spontaneous flora (Chuparina and Aisueva 2011). In Romania, among the most used medicinal plants in folk medicine are chamomile (*Matricaria recutita*), milfoil (*Achillea millefolium*), rattle (*Hypericum perforatum*), brotherwort (*Thymus serpyllum*), pot marigold (*Calendula officinalis*), linden (*Tilia platyphyllos*), and peppermint (*Mentha piperita*). Chamomile is used for its anxiolytic, antiseptic, and anti-inflammatory properties, while milfoil is used for its strong astringent effect and to treat a variety of illnesses and disorders from stomach aches to circulatory disorders. Rattle is usually used to treat digestive and neurological disorders; brotherwort has antiseptic properties and is used for acne and allergies treatment; pot marigold has anti-inflammatory and antitumor properties; linden is used for its calming effects and to ease

* Correspondence: marin.senila@icia.ro
†Equal contributors
[1]INCDO-INOE 2000, Research Institute for Analytical Instrumentation, 67 Donath, Cluj-Napoca 400293, Romania
Full list of author information is available at the end of the article

cold and flu symptoms; while peppermint is used to relieve stomach aches, nausea, fever, stress, and to boost the immune system.

Since, all over the world, there are numerous metal-polluted sites (European Environment Agency 2007), metals from soil can be transferred to the plants (Moreno-Jimenez et al. 2009; Malandrino et al. 2011; Senila et al. 2012; Rodrigues et al. 2012) and may have adverse effects on consumers' health, as local residents use the plants in their diet, mainly for tea preparation. Consequently, there is a need to develop reliable methods for the determination of metals in medicinal plants and their aqueous extracts.

The ICP-OES method has become a routine analytical technique for metal determination; however, the information related to method validation are scarce, and research on this field is still needed (Mermet 2005). Several studies present the use of ICP-OES for metals determination in tea or other food samples (Mitic et al. 2012; Froes et al. 2014). For consistent interpretation of the measurement results, it is necessary to evaluate the confidence that can be placed in, therefore, the presentation of an analytical result which must be accompanied by indication of the data quality. This information is essential for the interpretation of the analytical result (Kessel 2002; Drolc and Pintar 2011). Method validation is an essential component of the measures that a laboratory should implement in order to produce reliable analytical data (EURACHEM 1998). Besides common method performance characteristics obtained in the validation process, testing laboratories shall have and apply procedures for estimating the uncertainty of measurements (International Organization for Standardization 2005). This clearly means that the analytical result cannot be viewed only as a separate value. The International Organization for Standardization (ISO) guide (International Organization for Standardization 1995) recommends the calculation of uncertainty using a model equation, based on its uncertainty components, and by using the law of propagation of uncertainty in order to combine them into uncertainty. It has subsequently been interpreted for analytical chemistry (Ellison et al. 2012). There are several possibilities to estimate the uncertainty, as reported in the literature (Ellison et al. 2012; International Organization for Standardization 1995; Magnusson et al. 2012; Baralkiewicz et al. 2013). The measurement uncertainty is estimated mainly by the top-down or bottom-up approaches. In the top-down approach, the major sources of uncertainty are identified and evaluated, while in the bottom-up approach, all the uncertainty sources are systematically evaluated and only those with significant contributions are used to derive the measurement uncertainty. The top-down approach is time-consuming and requires extensive knowledge of the analytical procedure, but it enables identification of major uncertainty sources and consequently reduction of total

measurement uncertainty. Another relatively quick and easy way of uncertainty estimation is the in-house validation that includes the determination of the method performance parameters (Baralkiewicz et al. 2013).

In spite of a several papers published on the topic, there is a lack of fully validated methods for metal determination in medicinal plants and their extracts. The purpose of the present work was to perform a detailed validation of the analytical procedure and estimate the measurement uncertainty budget for determination of some essential (Fe, Mn, Zn, Cr, Cu, Al, Mg) and toxic (Pb, Cd, Ni, As) elements in the medicinal plants and their aqueous extracts. The method was validated according to the international guidelines ISO/IEC 17025:2005 (International Organization for Standardization 2005). The assessment of uncertainty was carried out using modelling approach and a full combined uncertainty calculation, including possible sources of uncertainty.

Methods
Instrumentation
Analyses were carried out using a dual viewing inductively coupled plasma optical emission spectrometer Optima 5300DV (PerkinElmer, Waltham, MA, USA) coupled to an ultrasonic nebulizer CETAC 6000AT+ (CETAC, Omaha, NE, USA). The operating conditions employed for ICP-OES determination were 1,300 W RF power, 15 L min^{-1} plasma flow, 2.0 L min^{-1} auxiliary flow, 0.8 L min^{-1} nebulizer flow, and 1.5 mL min^{-1} sample uptake rate. Axial view was used for metals determination, while 2-point background correction and six replicates were used to measure the analytical signal. In order to eliminate the memory effect caused by the use of ultrasonic nebulization, the delay time for washing between samples and signal measurement was set to 180 s. The measurement of a blank solution after measuring 1 mg L^{-1} calibration standard indicated the lack of memory effect. High-purity argon (99.995%) supplied by Linde Gas SRL (Timis, Timisoara, Romania) was used to sustain plasma and as carrier gas. A closed-vessel MWS-3+ microwave system (Berghof, Germany) with temperature control mode was used for sample digestion. All PTFE digestion vessels were previously cleaned in a bath of 10% (v/v) nitric solution for 48 h to avoid cross-contamination.

Reagents and CRMs
Multi-elemental solutions of 1,000 mg L^{-1} ICP Standard Certipur® (Merck, Darmstadt, Germany) containing the analysed elements (Fe, Mn, Al, Mg, Pb, Zn, Cr, Cu, Cd, Ni, and As) were used for calibration. Analytically graded 65% HNO$_3$ and 30% H$_2$O$_2$ (Merck, Germany) were used for sample digestion. Ultrapure water obtained by a Milli-Q system (Millipore, Molsheim, France) was used for all

Table 1 Wavelengths for selected elements, LoD, and LoQ in aqueous extracts and in dry plants

Element	Wavelength (nm)	Plant aqueous extracts[a]			Plant dry mass[b]		
		LoD (mg L^{-1})	LoQ (mg L^{-1})	Target value[c] (mg L^{-1})	LoD (mg kg^{-1})	LoQ (mg kg^{-1})	Target value[d] (mg kg^{-1})
Cd	228.805	0.13×10^{-3}	0.43×10^{-3}	0.50×10^{-3}	0.019	0.063	0.05×10^{-3}
Cr	267.713	0.60×10^{-3}	2.00×10^{-3}	5.00×10^{-3}	0.075	0.25	-
Cu	327.398	0.75×10^{-3}	2.50×10^{-3}	200×10^{-3}	0.12	0.40	-
Fe	238.205	2.24×10^{-3}	7.46×10^{-3}	20×10^{-3}	0.25	0.83	-
Al	396.153	1.59×10^{-3}	5.30×10^{-3}	20×10^{-3}	0.20	0.67	-
Mg	258.213	0.60×10^{-3}	2.00×10^{-3}	-	0.11	0.37	-
Pb	220.355	0.29×10^{-3}	0.97×10^{-3}	1.00×10^{-3}	0.043	0.14	0.5×10^{-3}
Mn	257.611	0.25×10^{-3}	0.83×10^{-3}	5.00×10^{-3}	0.030	0.10	-
Ni	231.604	0.57×10^{-3}	1.90×10^{-3}	2.00×10^{-3}	0.13	0.43	-
Zn	213.856	1.52×10^{-3}	5.06×10^{-3}	-	0.22	0.73	-
As	193.759	0.30×10^{-3}	1.00×10^{-3}	1.00×10^{-3}	0.045	0.15	-

[a]Calculated for the extraction method (1 g of plant digested with nitric acid and perhydrol in 100 volumetric flask); [b]calculated for the extraction method (0.5 g of plant extracted in water in 100 volumetric flask); [c]10% of the limit values according to Drinking Water Directive; [d]10% of the limit values according to the European Pharmacopeia.

dilutions and infusions. For metals' determination in plants and their aqueous extracts, the calibration standards were prepared by diluting the reference multi-elemental standard solution in 8% (v/v) nitric acid and 0.5% (v/v) nitric acid, respectively, in order to assure the similar concentration of nitric acid in samples and in calibration standards.

A vegetable certified reference material (CRM) IAEA-359 Cabbage (IAEA, Austria) and a water CRM trace metals 1-WP QC11132 (Sigma Aldrich, Steinheim, Germany) were used for the quality control of metals' determination.

Plant samples

Seven medicinal plants (chamomile, milfoil, rattle, brotherwort, pot marigold, linden, peppermint) were randomly collected from spontaneous flora grown in NW Romania. Three specimens of each plant species were intensely rinsed with tap water and distilled water and dried in an oven at 40°C until weight is constant. In order to accelerate the digestion process, samples were grinded to powder with a kitchen mixer grinder and sieved through a 100-μm mesh. Five sub-samples from each plant species were used for analysis.

Microwave digestion procedure

An amount of 0.5 g of plant powder was weighted into dry, clean PTFE vessels then 6 mL of HNO_3 and 2 mL of H_2O_2 were added. Vessels content were mixed and kept at room temperature for 12 h, then the vessels were introduced in the microwave digestion system and digested using a four-step digestion program: (1) 5 min at 280 W, (2) 5 min at 700 W, (3) 10 min at 1,050 W, (4) 1 min at 0 W. The resulting solutions were cooled, diluted to 50 mL with distilled water, filtered, and then

analysed by ICP-OES. In order to evaluate the accuracy of the method, the vegetable CRM was analysed in the same experimental conditions as the samples.

Aqueous extracts

An amount of 1 g of plant powder was prepared for infusion in 200 mL of boiling ultrapure water for approximately 10 min. The obtained infusions were filtered, evaporated to approximately 10 mL on a hot plate, then 1 mL of HNO_3 was added and the samples were digested in the microwave digestion system using the same digestion program as for solid samples. After cooling, the obtained solutions were filtered and diluted with ultrapure water in 100 mL volumetric flasks and analysed by ICP-OES.

Table 2 Confirmation of LoQ in aqueous extracts and in plant dry mass

Element	Plant aqueous extracts		Plant dry mass	
	RSD (%)	Recovery (%)	RSD (%)	Recovery (%)
Cd	9.61	95.6	10.5	98.9
Cr	11.8	91.2	10.2	95.6
Cu	8.86	104	9.95	101
Fe	15.2	112	12.6	115
Al	9.12	104	9.05	110
Mg	8.24	98.6	8.87	96.6
Pb	11.6	94.6	12.2	97.2
Mn	10.1	108	8.96	110
Ni	14.2	114	15.1	104
Zn	8.89	98.8	10.4	96.3
As	14.5	106	12.8	112

Table 3 Calibration curves for working range LoQ to 1.00 mg L^{-1}

Element	a value	b value	r^2 value	PG
Cd	22,300	901,200	0.9999	1.96
Cr	12,090	868,000	0.9999	2.88
Cu	5,345	1,045,000	0.9997	4.33
Fe	23,700	1,503,900	0.9999	4.18
Al	−11,100	1,922,000	0.9999	3.12
Mg	−5,600	3,459,000	1.0000	1.72
Pb	129	184,800	0.9998	4.56
Mn	141,500	8,094,000	1.0000	1.96
Ni	15,400	111,300	0.9999	4.06
Zn	56,500	1,159,000	0.9999	3.36
As	−33	32,400	0.9999	4.67

Results and discussion

Method validation

The validation of the analytical procedure for quantitative determination of elements in medicinal plants and their aqueous extracts was performed by evaluating selectivity, working and linear ranges, limit of detection (LoD), limit of quantification (LoQ), trueness, and precision (repeatability and reproducibility) (EURACHEM 1998).

Selectivity

Selectivity is the ability of a method to accurately quantify the analyte in the presence of interferences, under the stated conditions of the assay for the sample matrix being studied (EURACHEM 1998). The selectivity in the case of ICP-OES method is related to possible interferences of the emission spectrum at specific wavelengths. The emission lines used for quantitation of each element, based on known interferences and baseline signal at selected wavelengths observed empirically during the measurements, are presented in Table 1. Matrix effects were studied by standard addition method, by adding a spike of 1 mg L^{-1}

of each element to the original samples. The recoveries were within 90% and 110% for all the studied elements.

LoD and LoQ in aqueous extracts and in dry plants

The LoD indicates the level at which detection becomes problematic, while LoQ is the lowest concentration of the analyte that can be determined with an acceptable level of repeatability, precision, and trueness. LoD was estimated from the calibration function for a signal equal to the net signal of blank and three times its standard deviation, while LoQ was estimated from the calibration function for a signal equal to the net signal of blank and ten times its standard deviation (EURACHEM 1998; Miller and Miller 2000). Standard deviation of the blank resulted from the analysis of ten independent reagent blank solutions, each measured once on the same day. As the metal content in tea is not legislated, the performance criteria targeted for the LoD for aqueous plant extracts were 10% of the limit values (µg L^{-1}) for drinking water: As - 10; Cd - 5; Cr - 50; Cu - 2,000; Pb - 10; Ni - 20; Fe - 200; Al - 200; and Mn - 50 (Council of the European Union 1998). The European Pharmacopeia (Council of Europe 2011) proposed a limit of 5 mg kg^{-1} for Pb and 0.5 mg kg^{-1} for Cd in herbal drugs. For plant samples, the performance criteria targeted for the LoD were 10% of these values. Data in Table 1 showed that the performance targets were achieved by our methods. In order to experimentally confirm LoQ, six standard solutions with concentrations close to the LoQ were prepared and analysed. The targeted repeatability expressed as relative standard deviation (RSD) and targeted recovery were 20% and 90% to 115%, respectively. The measured RSD and recovery are presented in Table 2.

Working and linear range

Working range is the range of analyte concentrations over which the method is linear. At the lower end of the concentration range, the limiting factor is LoQ, while at the upper end limitations are imposed by various effects

Table 4 Results of analysis of water CRM trace metals 1-WP QC11132 (Sigma Aldrich)

Element	Found value (µg L^{-1})	s (µg L^{-1})	Certified value (µg L^{-1})	U$_{CRM}$ (µg L^{-1})	Recovery (%)
Cd	390	3.10	385	5.07	101
Cr	862	25.7	864	11.6	99.8
Cu	592	8.00	603	7.80	98.2
Fe	1,350	27.0	1,340	24.4	101
Al	452	12.9	463	11.4	97.6
Pb	908	30.2	929	14.4	97.7
Mn	1,200	7.15	1,200	15.5	100
Ni	1,730	21.1	1,710	20.4	101
Zn	918	8.04	915	16.2	100
As	414	7.25	416	8.25	99.5

Table 5 Results of analysis of IAEA-359 cabbage CRM (IAEA)

Element	Certified content, $\mu g\ g^{-1}$	Found content[a], $\mu g\ g^{-1}$
Cd	0.115 to 0.125	0.116 ± 0.014
Cr	1.24 to 1.36	1.25 ± 0.088
Cu	5.49 to 5.85	5.44 ± 0.41
Fe	144.1 to 151.9	147 ± 4.53
Mn	31.3 to 32.5	32.4 ± 2.15
Ni	1.00 to 1.10	1.03 ± 0.094
Zn	37.9 to 39.3	37.8 ± 1.89
As	0.096 to 0.104	0.100 ± 0.012
Mg	2,110 to 2,210	$2,114 \pm 23.6$

[a]Values are expressed in microgram per gram dry weight and reported as average $\pm s$; $n = 5$; 95% confidence level.

depending on the instrument response. Although generally, three or four calibration standards are used to evaluate the linear range of ICP-OES method in order to evaluate the appropriate measurement uncertainty budget; in our study, linearity was evaluated from the regression function of calibration using eight standards, the lowest concentration close to the LoQ, while the others were 0.05, 0.10, 0.20, 0.40, 0.60, 0.80, and 1.00 mg L^{-1} for each element. The fit for purpose working range was selected to be between LoQ of each element and 1.00 mg L^{-1}.

Ten replicates of the lowest and ten at the highest concentration of the working range were measured. To check the homogeneity of variances, the standard deviations (s_1) and (s_2) of the lowest and the highest concentrations from calibration curves and the PG ratios (s_1^2/s_2^2 or s_2^2/s_1^2) were calculated and compared with the critical value $F_{9;9;0.99} = 5.35$. The values for intercept (a), slope (b), correlation coefficient (r^2), and PG ratio are presented in Table 3. The experimental data showed that the variances are homogenous; therefore, linear regression curve can be used (International Organization for Standardization 1990).

Trueness
The most frequent approach to estimate trueness of the method is CRM analysis. Six parallel samples of water and vegetable CRMs were analysed in order to determine the method's trueness (Tables 4 and 5). These results showed that the recoveries for all elements were generally within ±5% of the certified values. The Student's t test confirmed that the obtained recoveries are not significantly different from 100%.

Precision
The most common measures of precision are repeatability and reproducibility (Tables 6 and 7), which were estimated considering within and between days variation, respectively. The results obtained in repeatability were conducted on six parallel samples by a single operator using the same equipment. The set targets for concentrations lower than 100 $\mu g\ L^{-1}$ were standard deviation of repeatability (s_r) below 10% and limit of repeatability (r) below 28%, while for concentrations higher than 100 $\mu g\ L^{-1}$, s_r below 7% and r below 20%.

Measurement uncertainty
Measurement uncertainty was evaluated based on the bottom-up approach (International Organization for Standardization 1995). All the contributions were obtained from calibration certificates and from statistical analysis of repeated measurements. Trueness of the method was calculated from results of CRM analysis, while repeatability was evaluated from precision experiments. The uncertainty of volumetric operations (volumetric flasks, pipettes) was calculated by using manufacturer data on calibration uncertainty (from certificates), the uncertainty associated with the use of glassware at a temperature different from that of calibration, and the repeatability of volumetric deliveries. Uncertainty of balances was calculated from data obtained from

Table 6 Results from the repeatability study for two levels of concentration

Element	Average ($\mu g\ L^{-1}$)	s_r (%)	r (%)	Average ($\mu g\ L^{-1}$)	s_r (%)	r (%)
Cd	23.6	6.8	19	211	3.6	10
Cr	24.6	7.1	20	208	3.7	10
Cu	26.1	6.4	18	198	4.4	12
Fe	25.5	9.5	27	221	6.3	18
Al	25.0	6.2	17	213	3.8	11
Mg	25.3	4.1	11	200	2.9	8.1
Pb	24.6	7.1	20	222	3.2	9.0
Mn	24.8	7.4	21	232	4.1	11
Ni	26.3	7.8	22	215	5.2	15
Zn	25.1	6.9	19	225	4.7	13
As	24.3	8.8	25	208	6.5	18

s_r, standard deviation of repeatability; r, limit of repeatability ($s_r \times 2.8$).

Table 7 Results obtained for the reproducibility by ICP-OES

Element	Average (μg L^{-1})	s_R (%)	R (%)
Cd	107	9.8	27
Cr	98.5	10	28
Cu	101	12	34
Fe	94.8	14	39
Al	96.8	6.9	19
Mg	101	5.8	16
Pb	105	8.6	24
Mn	111	11	31
Ni	105	14	39
Zn	114	8.8	25
As	95.6	16	45

s_R, standard deviation of reproducibility; R, limit of reproducibility ($s_R \times 2.8$).

calibration certificates (declared uncertainty) and the repeatability of weighing. After estimation, all sources of uncertainty were combined according to the law of propagation of uncertainties, obtaining the combined standard uncertainty ($u(C_a)$). The final result was reported as expanded uncertainty ($U(C_a)$), calculated as $U(C_a) = k \times u(C_a)$, where k is the coverage factor, corresponding to a 95% confidence level.

The identified main sources of measurement uncertainty were uncertainty of calibration reference materials (C_i), uncertainty of delivered volumes, uncertainty of measured intensities of the reference solutions (A_i), and recovery of the method (Figure 1).

The contributions of repeatability to the measurement uncertainty were combined into one contribution for the overall experiment and were obtained from the method validation study performed in the laboratory. Recovery accounts for possible interferences in the method when samples of selected matrix are analysed. With these corrections, the concentration of each element (C) in a sample was expressed by the model:

$$C = \frac{A-a}{b} \frac{1}{R} F_{rep} F_{dil} \qquad (1)$$

where R is the method recovery, F_{dil} is the dilution factor, F_{rep} is the repeatability factor, A is the area of the sample, while a and b are the linear regression coefficients. The sources of uncertainty and uncertainty components in determining elements are schematically presented in the cause and effects diagram (Figure 1). Calculations were made by using GUM Workbench software version 1.3 (Metrodata GmbH, Grenzach-Wyhlen, Germany) which is a standard application program with the possibility that user can define any model equation in order to enable various uncertainty calculations. The software was checked and validated before use in order to demonstrate that it is suitable for intended use. The results of the measurement uncertainty are listed in Table 8. Results revealed that for all the metals tested in plant aqueous extracts, extended measurement uncertainty is lower than 10% and therefore fulfills the requirements stated in the Drinking Water Directive (Council of the European Union 1998).

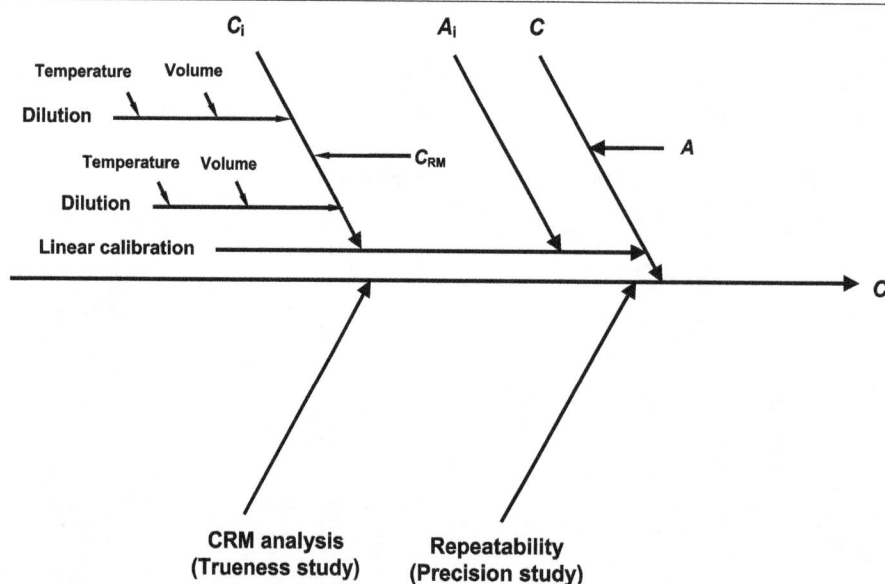

Figure 1 Cause and effects diagram of uncertainties. The uncertainties in the measurement of mass concentration of elements in aqueous herbal extracts are obtained using ICP-OES.

Table 8 Measurement uncertainty of elements determination in plant aqueous extracts and plant dry mass

Element	Measurement uncertainty, % ($k = 2$)	
	Plant aqueous extracts	Plant dry mass
Cd	5.8	8.7
Cr	7.5	8.8
Cu	5.7	8.9
Fe	8.2	8.3
Mn	6.2	6.4
Ni	6.6	8.8
Zn	5.9	6.3
As	7.7	9.8

The uncertainty components of Cr concentration in plant aqueous extracts and in plant dry mass as a case study are presented in Table 9.

The relative uncertainty variance contributions are used to illustrate the relative impact of different uncertainty components. The relative contribution (r_i) of an uncertainty component x_i to the combined standard uncertainty is defined as follows:

$$r_i = \frac{\left(\frac{\delta y}{\delta x_i}\right)^2 \cdot u(x_i)^2}{u(y)^2} \tag{2}$$

where $u(x_i)$ is the standard uncertainties of the input parameters, and $\partial y / \partial x_i$ is the sensitivity coefficient.

The importance of uncertainty sources is determined by their quantitative effect on the measurement result. In case of Cr both for extracts and dry plants, the largest contribution comes from $u(R)$ and from repeatability ($u(F_{rep})$), while uncertainty contributions from other input quantities are of minor importance.

Results on real samples (aqueous plant extract and dry plant)

The concentrations of essential and nonessential elements in the dry mass of the analysed plant samples are presented in Table 10. The As concentrations were, in all cases, below the LoQ, while Pb and Cd concentrations

Table 9 Uncertainty components of Cr in plant aqueous extracts and mass fraction in plant dry mass

Symbol	Unit	Plant aqueous extracts			Plant dry mass		
		Value	Standard uncertainty	r_i	Value	Standard uncertainty	r_i
C_1	mg L^{-1}	0.00	0.004	0.5	0.00	0.004	0.5
C_2	mg L^{-1}	0.05	0.004	0.5	0.05	0.004	0.5
C_3	mg L^{-1}	0.10	0.004	0.4	0.10	0.004	0.5
C_4	mg L^{-1}	0.20	0.004	0.4	0.20	0.004	0.4
C_5	mg L^{-1}	0.40	0.004	0.3	0.40	0.004	0.3
C_6	mg L^{-1}	0.60	0.004	0.3	0.60	0.004	0.3
C_7	mg L^{-1}	0.80	0.004	0.0	0.80	0.004	0
C_8	mg L^{-1}	1.00	0.004	0.1	1.00	0.004	0.1
A_1	-	20,190	3,028	0.4	20,190	3,028	0.4
A_2	-	55,030	1,045	0.0	55,030	1,045	0
A_3	-	95,540	840	0.0	95,540	840	0
A_4	-	181,800	1,181	0.0	181,800	1,181	0
A_5	-	355,180	1,882	0.0	355,180	1,882	0
A_6	-	529,800	3,284	0.2	529,800	3,284	0.2
A_7	-	703,882	3,097	0.1	703,882	3,097	0.1
A_8	-	880,011	5,104	0.1	880,011	5,104	0.1
A	-	255,400	1,127	0.1	31,140	4,670	0.1
R	-	1.00	0.015	20.4	1.00	0.027	35.8
F_{rep}	-	1.00	0.029	76.2	1.00	0.029	60.7
C_m	g L^{-1}	-	-	-	10	0.001	0.0
Result		0.283 mg L^{-1}	0.019 mg L^{-1} (6.7%, $k = 2$)		2.53 mg kg^{-1}	0.202 mg kg^{-1} (9.0%, $k = 2$)	

C_1 to C_8 are the concentrations of calibration standard solutions; A_1 to A_8 are the respective standard solutions; A is the emission intensity of the sample; R is the recovery from CRM; and F_{rep} is repeatability factor; C_m is the concentration of measured sample in digested solution.

Table 10 Contents of metals in dry plant mass

	Mg	Al	Cd	Cr	Cu	Fe	Pb	Mn	Ni	Zn	As
Chamomile	2,390 ± 18.5	152 ± 8.25	0.084 ± 0.011	2.50 0.22	11.2 ± 0.078	144 ± 5.56	0.90 ± 0.11	219 ± 5.95	3.05 ± 0.23	22.4 ± 1.36	<0.15
Milfoil	1,620 ± 12.6	41.1 ± 3.12	0.25 ± 0.022	0.88 ± 0.08	15.7 ± 0.12	42.1 ± 3.27	0.68 ± 0.063	83.2 ± 4.12	5.14 ± 0.48	46.2 ± 3.17	<0.15
Rattle	1,380 ± 11.9	126 ± 4.88	0.14 ± 0.015	1.55 ± 0.16	19.1 ± 0.15	189 ± 8.48	0.69 ± 0.077	120 ± 9.26	2.87 ± 0.22	55.9 ± 3.38	<0.15
Brotherwort	2,230 ± 12.5	186 ± 10.2	0.35 ± 0.029	3.18 ± 0.31	21.9 ± 0.18	224 ± 11.4	2.23 ± 0.17	296 ± 11.3	6.02 ± 0.23	25.8 ± 1.57	<0.15
Pot marigold	3,120 ± 20.8	172 ± 11.4	0.11 ± 0.010	2.88 ± 0.23	29.1 ± 0.19	234 ± 20.7	1.33 ± 0.13	178 ± 9.59	4.08 ± 0.24	51.1 ± 2.96	<0.15
Linden	1,450 ± 11.8	72.6 ± 3.33	0.071 ± 0.008	1.50 ± 0.11	9.22 ± 0.066	64.9 ± 6.23	0.44 ± 0.042	71.9 ± 3.77	0.63 ± 0.071	18.8 ± 1.07	<0.15
Peppermint	2,660 ± 24.1	144 ± 6.03	0.41 ± 0.036	3.61 ± 0.30	19.9 ± 0.18	306 ± 12.2	2.53 ± 0.21	255 ± 14.4	2.11 ± 0.11	64.4 ± 3.22	<0.15

Values are expressed as milligrams per kilogram (mean ± standard deviation of five replicates).

were below the proposed limits by European Pharmacopoeia of 5 and 0.5 mg kg^{-1}, respectively. Ni concentrations varied between 0.63 and 6.02 mg kg^{-1}. These results were in the same order of magnitude with those reported by other authors for herbal drugs collected in Europe (Razic et al. 2006; Basgel and Erdemoglu 2006; Gentscheva et al. 2010).

The contents of essential elements such as Mg (1,450 to 3,120 mg kg^{-1}), Al (41.1 to 186 mg kg^{-1}), Cr (0.88 to 3.61 mg kg^{-1}), Cu (9.22 to 29.1 mg kg^{-1}), Fe (42.1 to 306 mg kg^{-1}), Mn (71.9 to 296 mg kg^{-1}), and Zn (22.4 to 64.4 mg kg^{-1}) were similar with those reported in the literature (Gentscheva et. al. 2010; Maharia et al. 2010; Chuparina and Aisueva 2011; Miranda and Pereira-Filho 2013).

The content of metals in the aqueous plant extracts offers information about the uptake of these elements by drinking of a cup of tea. The concentration of essential and nonessential elements in the aqueous extracts (Table 11) offers information about the uptake of these elements following tea consumption. As and Cd concentrations were lower that the LoQ in all the analysed samples. Also, Pb concentrations were generally below the LoQ, while Ni concentrations ranged between 2.70 and 18.2 μg L^{-1}, below the maximum value of 20 μg L^{-1} established for this element by EU Drinking Water Directive 98/83/EC. Also, the concentrations of essential elements that have established maximum values for drinking water were generally below these limits, except manganese extracted from brotherwort which slightly exceeded 50 μg L^{-1}.

The higher metal concentrations in plant extracts were found for Mg (1,590 to 7,800 μg L^{-1}), but this element have no maximum admitted limit for drinking water. Our results for the concentrations of Al, Cu, Mg, and Fe are in line with those reported by Froes et al. (2014), but Mn concentrations were generally lower, in our case.

By comparing the results presented in Tables 10 and 11, taking into account the mass of dry plants and final volume of infusion, it can be observed that among the analysed elements, Zn, Cu, and Ni were highly extracted in the aqueous extracts (31% to 64%, 23% to 71%, and respectively, 19% to 73%), in function of the plant species, while Pb and Fe had low solubility (below 10%).

Conclusions

A fully validated method for metal analysis in medicinal dry plant mass and its extracts is presented. The fast and accurate ICP-OES method enables the quantification of selected metals in aqueous and dry samples.

The validation results are presented and organized in tables in order to provide an easy overview of the method's performance. The experimentally determined validation parameters for medicinal extracts were then compared to the criteria stated in the Drinking Water Directive. Measurement uncertainty was determined on basis of modelling approach. Detailed uncertainty budget is presented for Cr in aqueous and dry samples. Systematic uncertainty budgets such as these in the design presented facilitate the uncertainty evaluation process and make it easier to compare the contributions of uncertainty

Table 11 Contents of metals in the aqueous extracts

	Mg	Al	Cd	Cr	Cu	Fe	Pb	Mn	Ni	Zn	As
Chamomile	6,010 ± 110	115 ± 5.03	<0.43	2.20 ± 0.21	12.8 ± 0.80	45.5 ± 1.64	<0.33	49.0 ± 3.15	11.1 ± 1.10	70.0 ± 8.58	<1.00
Milfoil	1,060 ± 86.2	91.2 ± 6.22	<0.43	<2.00	27.0 ± 1.56	18.8 ± 1.10	<0.33	39.2 ± 3.24	7.15 ± 0.61	103 ± 7.12	<1.00
Rattle	2,220 ± 103	109 ± 4.10	<0.43	2.21 ± 0.22	64.1 ± 2.90	44.2 ± 1.30	<0.33	35.6 ± 2.27	2.70 ± 0.11	185 ± 10.1	<1.00
Brotherwort	4,600 ± 185	190 ± 8.85	<0.43	3.15 ± 0.30	76.2 ± 3.95	27.0 ± 1.81	0.66 ± 0.08	51.5 ± 4.03	18.2 ± 1.20	68.8 ± 5.22	<1.00
Pot marigold	7,800 ± 230	95.6 ± 5.22	<0.43	3.32 ± 0.34	70.0 ± 3.65	34.5 ± 2.16	0.53 ± 0.08	28.6 ± 2.05	9.58 ± 0.74	163 ± 6.96	<1.00
Linden	1,590 ± 81.6	53.9 ± 2.76	<0.43	<2.00	10.6 ± 0.96	22.2 ± 1.31	<0.33	25.9 ± 1.89	3.55 ± 0.24	29.3 ± 2.11	<1.00
Peppermint	5,650 ± 211	127 ± 3.05	<0.43	4.25 ± 0.39	70.7 ± 3.62	51.0 ± 3.10	0.83 ± 0.09	45.5 ± 2.63	6.21 ± 0.53	196 ± 11.2	<1.00

Values are expressed as micrograms per liter (mean ± standard deviation of five replicates) per 1 g of plant extracted in 200 mL water.

components to the total uncertainty budget and offer a tool for improvement of the method performance. In addition, the use of commercial software can facilitate the calculations in order to make the entire process more user-friendly.

In dry plants, the concentrations of Pb and Cd were below the proposed limits by European Pharmacopoeia of 5 and 0.5 mg kg^{-1}, respectively. The concentrations of essential and nonessential elements in tea infusion of the analysed samples were generally lower than the maximum values established by EU Drinking Water Directive 98/83/EC; thus, these tea can be considered safely for consumption. However, depending on the metal pollution in the sites where the medicinal plants are grown and the uptake of metals in these plants, the concentrations of metals in water extracts can determine the exceeding of the limits for drinking water due to the relatively high extractability of metals like Zn, Cu, and Ni.

Competing interests

The authors declare that they have no competing interests.

Authors' contributions

Samples preparation was done by LS and EL, and instrumental determination by MS. Statistical interpretation was done by AD and AP. All authors contributed in experimental design, writing, proofing, read, and approval of the manuscript.

Acknowledgements

The authors gratefully acknowledge the financial support from the Ministry of Education, Science and Sport, from the Metrological Institute of the Republic of Slovenia (MIRS) and Romanian financing authority CNCS-UEFISCDI, Partnership, project VULMIN, Contract No. 52/2012.

Author details

[1]INCDO-INOE 2000, Research Institute for Analytical Instrumentation, 67 Donath, Cluj-Napoca 400293, Romania. [2]Laboratory for Environmental Sciences and Engineering, National Institute of Chemistry, 19 Hajdrihova, Ljubljana SI-1000, Slovenia.

References

Baralkiewicz D, Pikosz B, Belter M, Marcinkowska M (2013) Speciation analysis of chromium in drinking water samples by ion-pair reversed-phase HPLC–ICP-MS: validation of the analytical method and evaluation of the uncertainty budget. Accred Qual Assur 18:391–401

Basgel S, Erdemoglu SB (2006) Determination of mineral and trace elements in some medicinal herbs and their infusions consumed in Turkey. Sci Total Environ 359:82–89

Chuparina EV, Aisueva TS (2011) Determination of heavy metal levels in medicinal plant Hemerocallis minor Miller by X-ray fluorescence spectrometry. Environ Chem Lett 9:19–23

Drolc A, Pintar A (2011) Measurement uncertainty evaluation and in-house method validation of the herbicide iodosulfuron-methyl-sodium in water samples by using HPLC analysis. Accredit Qual Assur 16:21–29

Ellison SLR, Rosslein M, Williams A (2012) EURACHEM/CITAC, quantifying uncertainty in analytical measurement, 3rd edn. LGC, Teddington

EURACHEM (1998) The fitness for purpose of analytical methods. Eurachem LGC, Teddington

European Environment Agency (2007) http://www.eea.europa.eu/data-and-maps/indicators/#c5=&c7=all&c0=10&b_start=0&c6=progress+in+management+of+contaminated+sites. Accessed 28 May 2013

Council of Europe (2011) European pharmacopoeia, 7th edn. Council of Europe, Strasbourg, France

Froes RES, Neto WB, Beinner MA, Nascentes CC, da Silva JBB (2014) Determination of inorganic elements in teas using inductively coupled plasma optical emission spectrometry and classification with exploratory analysis. Food Anal Methods 7:540–546

Gentscheva GD, Stafilov T, Ivanova EH (2010) Determination of some essential and toxic elements in herbs from Bulgaria and Macedonia using atomic spectrometry. Eurasian J Anal Chem 5:104–111

International Organization for Standardization (1990) Water quality–calibration and evaluation of analytical methods and estimation of performance characteristics - part 1: statistical evaluation of the linear calibration function. ISO 8466–1. ISO, Geneva

International Organization for Standardization (1995) Guide to the expression of uncertainty in measurement (GUM). JCGM 100:2008

International Organization for Standardization (2005) General requirements for the competence of testing and calibration laboratories; ISO/IEC 17025. European Committee for Standardization, Brussels

Kessel W (2002) Measurement uncertainty according to ISO/BIPM-GUM. Thermochim Acta 382:1–16

Maharia RS, Dutta RK, Acharya R, Reddy AVR (2010) Heavy metal bioaccumulation in selected medicinal plants collected from Khetri copper mines and comparison with those collected from fertile soil in Haridwar, India. J Environ Sci Heal B 45:174–181

Malandrino M, Abollino O, Buoso S, Giacomino A, La Gioia C, Mentasti E (2011) Accumulation of heavy metals from contaminated soil to plants and evaluation of soil remediation by vermiculite. Chemosphere 82:169–178

Mermet JM (2005) Is it still possible, necessary and beneficial to perform research in ICP-atomic emission spectrometry? J Anal Atom Spectrom 20:11–16

Miller JN, Miller JC (2000) Statistics and chemometrics for analytical chemistry. Prentice Hall, Harlow, London, New York

Miranda K, Pereira-Filho ER (2013) Sequential determination of Cd, Cu and Pb in tea leaves by slurry introduction to thermospray flame furnace atomic absorption spectrometry. Food Anal Methods 6:1607–1610

Mitic S, Obradovic MV, Mitic MN, Kostic DA, Pavlovic AN, Tosic SB, Stojkovic MD (2012) Elemental composition of various sour cherry and table grape cultivars using inductively coupled plasma atomic emission spectrometry method (ICP-OES). Food Anal Methods 5:279–286

Moreno-Jimenez EM, Penalosa JM, Manzano R, Carpena-Ruiz RO, Gamarra R, Esteban E (2009) Heavy metals distribution in soils surrounding an abandoned mine in NW Madrid (Spain) and their transference to wild flora. J Hazard Mater 162:854–859

Magnusson B, Näykk T, Hovind H, Krysell M (2012) Guide handbook for calculation of measurement uncertainty in environmental laboratories. Nordtest, Oslo

Council of the European Union (1998) Council directive 98/83/EC on the quality of water intended for human consumption. Official Journal of the European Communities L 330:32–54

Razic SS, Dogo SM, Slavkovic LJ (2006) Multivariate characterization of herbal drugs and rhizosphere soil sample according to their metallic content. Microchem J 84:93–101

Rodrigues SM, Pereira ME, Duarte AC, Romkens PFAM (2012) Soil–plant–animal transfer models to improve soil protection guidelines: a case study from Portugal. Environ Int 39:27–37

Senila M, Levei EA, Senila LR (2012) Assessment of metals bioavailability to vegetables under field conditions using DGT, single extractions and multivariate statistics. Chem Cent J 6:119

18

Evaluation of antimicrobial and antioxidant activities from *Toona ciliata* Roemer

Kumara Shanthamma Kavitha and Sreedharamurthy Satish[*]

Abstract

Background: In the present study, the different solvent extracts viz., petroleum ether, chloroform, ethyl acetate and methanol of the medicinal plant *Toona ciliata* (leaf and flower) were evaluated for phytochemical analysis, antimicrobial and antioxidant activities.

Methods: A qualitative phytochemical study was conducted to know the presence or absence of phytoconstituents in the test extracts. Antibacterial and antifungal activities were determined using disc diffusion assay against human and phytopathogens. MIC was carried out using Micro-broth dilution method for pathogenic bacteria and fungi. Radical scavenging activity was also studied using DPPH and ABTS method.

Results: The study revealed the presence of carbohydrates, proteins, phytosterols, flavonoids, glycosides, tannins and phenolic compounds. Ethyl acetate and methanol extracts showed moderate activity against test phytopathogenic bacteria compared to tetracycline. Moderate activity was found against *Proteus mirabilis* and least activity against *Klebsiella pneumoniae*, *Salmonella typhi* and *Staphylococcus aureus* with ethyl acetate and methanol extracts. Ethyl acetate and methanol extracts of *Toona ciliata* exhibited lowest MIC varied from 10-2.5 mgml^{-1} against test human pathogenic and phytopathogenic bacteria Significant antifungal activity against *Microsporum canis* was observed in methanol extract with an MIC of 1.25 mgml^{-1} compared to miconozole. All the test extracts showed significant DPPH and ABTS radical scavenging activity in comparison with BHT.

Conclusions: The present study concludes that the plant Toona ciliata could be exploited for the isolation of bioactive compounds which could be a potential source for antimicrobials and antioxidants.

Keywords: *Toona ciliata*, Antioxidant activity, Antimicrobial activity, Phytochemical analysis

Background

Phytobiology perceives medicinal plants as a source of bioactive compounds which can be traced since evolution. The diverse living systems bear a rich biodiversity in nature. Since ancient era before scientific knowledge would evolve plants performed myriad functions on the biosphere. Among which use of plants in curing illness has been well documented. But it was after the advance technology and improved scientific knowledge transformed plants as source of therapeutic agents as they can serve the purpose with lesser side effects that are often associated with synthetic antimicrobials (Nagumanthri et al. 2012). It was estimated that current global market for plant-derived drugs is worth more than 20 billion and the market

continues growing (Lin et al. 2013). Perusal of literatures reports the medicinal properties of most of the plants bearing biological activity one such species is *Toona ciliata* (Meliaceae) which has been exploited for many traditional uses like construction purpose, dye preparation, furniture, medicines etc., (Negi et al. 2011). *Toona ciliata* along with Siderin, a compound isolated from petroleum ether extract showed significant antibacterial activity and also exhibited significant cytotoxicity (Chowdhury et al. 2003).

The plant extract also showed gastro protective activity (Malairajan et al. 2006). The inhibitive effects on formed protein non-enzymatic glycation an end product was studied from the ethanolic leaf extract (Shao-Hong 2010) Cedrelone, a tetra nortriterpenoid, isolated from *Toona ciliata* (Gopalakrishnan et al. 2000). Compounds such as 12-Deacetoxytoonacilin and 6α-acetoxy-14α,15α-epoxyazadirone were isolated from the seeds (Neto et al.

* Correspondence: satish.micro@gmail.com
Department of Studies in Microbiology, Herbal Drug Technological, Laboratory, University of Mysore, Manasagangotri, Mysore 570 006, Karnataka, India

1995), 12-α-Hydroxystigmast-4-en-3-one was isolated from the petroleum ether extract of *Toona ciliata* together with two steroids and three C-methyl coumarins (Chowdhury et al. 2002), norlimonoids and limonoids from the leaves and stems (Liao et al. 2007), three new norlimonoids (1–3), two new tirucallane-type triterpenoids (4 and 5), and a new pimaradiene-type diterpenoid (6), along with two known limonoids and eight known tirucallane-type triterpenoids, from the leaves and twigs (Chen et al. 2009), toonaciliatone, methyl-3α-acetoxy-1-oxomelic-14(15)-enate, perforin A and cholest-14-ene-3,7,24,25-tetrol-21,23-epoxy-21-methoxy-4,4,8-trimethyl-3-(3-methyl-2-butenoate) from the leaves (Ning et al. 2010) and protolimonoids and norlimonoids from the stem bark of *Toona ciliata* (Wang et al. 2011). Although, several synthetic antioxidants and drugs are commercially available, natural products still substitute most of the chemical agents. In the present study, solvent extracts such as petroleum, chloroform, ethyl acetate and methanol of *Toona ciliata* were evaluated for the qualitative phyto-chemical analysis, *in vitro* antimicrobial and antioxidant activity which may lead to the finding of more effective agent for the management of diseases and effective potential source of natural antioxidant that may help in preventing various oxidative stresses.

Methods

Preparation of the extract

Plant material of *Toona ciliata* leaf and flower were washed with distilled water and shade dried. The dried leaves and flower were ground together to a fine powder using Waring blender. The coarsely powdered sample (50 g) was filled in the thimble and extracted successively with petroleum ether, chloroform, ethyl acetate and methanol using a Soxhlet extractor. The filtrate was evaporated to dryness under reduced pressure using rotary vacuum evaporator. The extracts were stored in ambient bottles until further use (Satish et al. 2007).

Preliminary phytochemical screening

The freshly prepared crude solvent extracts of *Toona ciliata* were qualitatively tested for the presence of phytochemical constituents such as alkaloids, flavones, terpenoids, phenols, tannins etc., by standard methods (Harborne 1973; Sofowara 1993; Ghani 2003).

Bacterial strains

Authenticated cultures of Gram positive bacteria such as *Bacillus subtilis* (MTCC 121), *Listeria monocytogenes* (MTCC 839), *Staphylococcus aureus* (MTCC 7443), *Staphylococcus epidermidis* (MTCC 435), Gram negative - *Escherichia coli* (MTCC 7410), *Enterobacter aerogenes* (MTCC 7325), *Klebsiella pneumoniae* (MTCC 7407), *Proteus mirabilis* (MTCC 425), *Pseudomonas aeruginosa* (MTCC 7903), *Salmonella typhimurium* (MTCC 1254),

Vibrio parahaemolyticus (MTCC 451) and *Erwinia carotovora* (MTCC 1428) were procured from MTCC, Chandigarh, India. Authentic pure cultures of phytopathogens *Xanthomonas axonopodis* pv. *malvacearum*, *Xanthomonas campestris* pv. *vesicatoria* and *Xanthomonas oryzae* pv. *oryzae* were procured from DANIDA research laboratory, University of Mysore, India.

Fungal strains

Four plant fungi *Aspergillus niger*, *Aspergillus flavus*, *Drechslera* and *Fusarium verticillioides* and three human dermatophytic fungi *Candida albicans*, *Microsporum canis* and *Microsporum gypsum* were used.

Preparation of inoculums

Bacterial and fungal inoculums were prepared from 24 h old pure culture grown on nutrient agar for bacteria and a week old culture on potato dextrose agar for fungi. Bacterial colonies were pre-cultured in nutrient broth medium and kept overnight, then centrifuged at 10,000 rpm for 5 min. Pellet was suspended in sterilized distilled water and the cell turbidity was assessed spectroscopically in comparable to that of the 0.5 McFarland standards (approximately 1.5×10^8 CFU/ml) whereas, the fungal spores was scraped from the mother culture and dispensed in sterilized distilled water. Then the spore density was adjusted spectrophotometrically to obtain approximately 10^5 spores/ml final concentration. Then the inoculums were used for the antibacterial and antifungal assays (Mahesh et al. 2008).

Antibacterial and antifungal activity

Antibacterial and antifungal activity of *Toona ciliata* solvent extracts was determined using a modified Kirby Bauer disc diffusion method. Briefly, 100 µl of the test bacteria/fungi was spread onto the nutrient agar and potato-dextrose agar plates respectively. The different test solvent extracts (petroleum ether, chloroform, ethyl acetate and methanol) were loaded to the sterilized sterile 6 mm discs, allowed to dry and then the impregnated discs with 50 µl (100 mgml⁻¹ concentration) onto the inoculated plates. The plates were allowed to stand at 4°C for 2 h before incubation with the test microbial agents. Bacterial plates were incubated at 37°C for 24 h and at room temperature for 3–4 days for fungi. The diameters of the inhibition zones were measured in mm. All the assays were done in triplicate and the results were given in mean ± SD. Standard antibiotics such as gentamicin and tetracycline for pathogenic bacteria, bavistin and miconozole for pathogenic fungi served as positive controls (Bauer et al. 1966).

Table 1 Qualitative chemical analysis of test solvent extracts of *Toona ciliata* leaf and flower

SI no.	Tests	Solvent extracts of *Toona ciliata*				
		Petroleum ether	Chloroform	Ethyl acetate	Methanol	Aqueous
I	**Carbohydrates test**					
	a) Molisch's test	+	+	+	+	+
	b) Fehling's test	+	+	+	+	+
II	**Proteins & Aminoacids**					
	a) Ninhydrin test	-	-	-	-	-
	b) Biuret test	-	+	+	-	-
	c) Sodium bicarbonate test	-	-	-	-	-
	d) Tannic acid test	-	-	-	-	-
	e) Xanthoprotein test	+	+	+	+	+
III	**Alkaloids**					
	a) Wagner's test	+	+	+	+	+
	b) Dragendorff's test	+	+	+	+	+
	c) Mayer's test	+	+	+	+	+
IV	**Saponins**					
	a) Foam test	-	-	-	-	+
V	**Flavonoids**					
	a) Ferric chloride test	-	-	+	+	+
	b) Shinoda test	-	+	+	+	+
	c) Alkali & acid test	+	-	+	-	-
VI	**Tannins & phenolic compounds**					
	a) Ferric chloride test	-	-	+	+	+
	b) Heavy metals test					
	i) Copper sulphate test	-	+	+	+	+
	ii) Potassium ferricyanide test	-	+	+	-	+
	iii) Nitric acid test	+	+	+	+	+
VII	**Glycosides**					
	a) Modified Borntrager's test	+	-	-	-	+
VIII	**Phytosterols**					
	a) Libermann- Burchard's test	+	+	+	+	+
	b) Terpenoids					
	i) 2,4-DNPH test	+	+	+	+	+
	c) Anthraquinones					
	i) Borntrager's test	+	+	+	+	+
	d) Steroids					
	i) Salkowski's test	+	+	+	+	+

Note: +: Present; -: Absent.

Microbroth dilution method

Minimal inhibition concentration was determined by 2,3,5- triphenyl tetrazolium chloride (TTC) and (3-(4,5-dimethylthiazol-2-yl)-2,5-diphenyltetrazolium bromide (MTT) assay using microtitre ELISA plate for bacteria and fungi respectively (Sette et al. 2006; Buatong et al. 2011). The 96 wells were filled with Muller–Hinton broth and Sabouraud's broth medium containing different concentration of solvent extracts, standard reference antibiotics such as gentamycin and miconozole against bacteria and dermatophytic fungi respectively. Antibacterial activity was detected by adding 0.5% TTC (Merck) aqueous solution and antifungal activity by adding 10 µl of a (3-(4,5-dimethylthiazol-2-yl)-2,5-diphenyltetrazolium bromide (MTT) solution [5 mgml^{-1} MTT in phosphate buffered saline (PBS), pH 7.4]. MIC was

Table 2 Antibacterial activity measured as zone of inhibition at 50 µl (100 mg/ml) of solvent extracts of *Toona ciliata* (leaf and flower) and standard antibiotics

SL no.	Pathogenic bacteria	Solvent extracts (50 µl) zone of inhibition in mm (MIC in mg/ml)				Antibiotics
	Human pathogens	Petroleum ether	Chloroform	Ethyl acetate	Methanol	Gentamicin
1	*Bacillus subtilis*	0.00 ± 0.00	0.00 ± 0.00	0.00 ± 0.00	0.00 ± 0.00	34.66 ± 0.57
		(ND)	(ND)	(ND)	(ND)	(0.156)
2	*Escherichia coli*	0.00 ± 0.00	0.00 ± 0.00	0.00 ± 0.00	0.00 ± 0.00	25.33 ± 0.57
		(ND)	(ND)	(ND)	(ND)	(0.625)
3	*Klebsiella pneumoniae*	0.00 ± 0.00	0.00 ± 0.00	0.00 ± 0.00	9.00 ± 0.00	21.33 ± 1.15
		(ND)	(ND)	(ND)	(5)	(1.25)
4	*Listeria monocytogenes*	0.00 ± 0.00	0.00 ± 0.00	0.00 ± 0.00	0.00 ± 0.00	25.00 ± 1.00
		(ND)	(ND)	(ND)	(ND)	(0.3125)
5	*Pseudomonas aeruginosa*	0.00 ± 0.00	0.00 ± 0.00	0.00 ± 0.00	0.00 ± 0.00	25.00 ± 1.00
		(ND)	(ND)	(ND)	(ND)	(0.3125)
6	*Proteus mirabilis*	0.00 ± 0.00	0.00 ± 0.00	20.66 ± 0.57	15.33 ± 0.57	30.66 ± 0.57
		(ND)	(ND)	(2.5)	(2.5)	(0.156)
7	*Salmonella typhi*	0.00 ± 0.00	0.00 ± 0.00	10.00 ± 0.00	10.00 ± 0.00	30.00 ± 0.00
		(ND)	(ND)	(5)	(5)	(0.156)
8	*Staphylococcus aureus*	0.00 ± 0.00	0.00 ± 0.00	0.00 ± 0.00	12.66 ± 0.57	24.00 ± 1.00
		(ND)	(ND)	(ND)	(5)	(0.625)
9	*Staphylococcus epidermidis*	0.00 ± 0.00	0.00 ± 0.00	13.33 ± 0.57	14.00 ± 0.00	28.66 ± 1.15
		(ND)	(ND)	(5)	(5)	(0.156)
10	*Vibrio parahaemolyticus*	0.00 ± 0.00	0.00 ± 0.00	0.00 ± 0.00	0.00 ± 0.00	25.33 ± 0.57
		(ND)	(ND)	(ND)	(ND)	(0.3125)
	Plant pathogens					Tetracycline
11	*Erwinia carotovora*	0.00 ± 0.00	0.00 ± 0.00	10.66 ± 0.57	11.66 ± 0.57	34.00 ± 1.00
		(ND)	(ND)	(5)	(5)	(0.156)
12	*Xanthomonas axonopodis* pv. *malvacearum*	0.00 ± 0.00	0.00 ± 0.00	15.33 ± 0.57	13.66 ± 0.57	30.33 ± 0.57
		(ND)	(ND)	(5)	(5)	(0.156)
13	*Xanthomonas oryzae* pv. *oryzae*	0.00 ± 0.00	0.00 ± 0.00	12.33 ± 0.57	8.33 ± 0.57	30.33 ± 0.57
		(ND)	(ND)	(5)	(5)	(0.156)
14	*Xanthomonas campestris* pv. *vesicatoria*	0.00 ± 0.00	0.00 ± 0.00	10.66 ± 0.57	12.00 ± 0.00	31.00 ± 0.00
		(ND)	(ND)	(5)	(5)	(0.156)

Values are the mean of triplicates ± SD p < 0.05.
ND Not Done.

defined as the lowest concentration of extract that inhibited visible growth, as indicated by the TTC and MTT staining (dead cells will not be stained).

DPPH radical scavenging activity

The antioxidant activity of *Toona ciliata* test solvent extracts was determined in terms of radical scavenging ability by DPPH method. Stock solution of 0.1 mM DPPH in methanol was diluted using methanol. 1.0 ml of solvent extracts solution of differing concentrations (50–250 µgml^{-1}) was added to 1.0 ml of DPPH and made volume up to 3 ml. A negative control (reaction mixture without test extract) was also used in this test. The absorbance was measured at 517 nm after 30 min.

Inhibition was calculated by using the following equation:

$$\% \text{ inhibition} = \left[A_{control} - A_{sample} / A_{control}\right] \times 100$$

IC50 values were calculated as the concentration of each sample required to give 50% DPPH radical scavenging activity with respect to absorbance of blank from the graph. The results were compared with BHT. The experiment was performed in triplicates and values are expressed in ± SD (Zhang et al. 2010).

Table 3 Antifungal activity measured as zone of inhibition at 50 µl (100 mg/ml) of solvent extracts of _Toona ciliata_ (leaf and flower) and standard antibiotics

SL no.	Pathogenic fungi	Solvent extracts (50 µl) zone of inhibition in mm(MIC in mg/ml)				Antibiotics
	Human pathogens	Petroleum ether	Chloroform	Ethyl acetate	Methanol	Miconozole
1	_Candida albicans_	0.00 ± 0.00	0.00 ± 0.00	0.00 ± 0.00	0.00 ± 0.00	13.66 ± 0.57
		(ND)	(ND)	(ND)	(ND)	(ND)
2	_Microsporum gypseum_	0.00 ± 0.00	0.00 ± 0.00	0.00 ± 0.00	0.00 ± 0.00	18.66 ± 0.57
		(ND)	(ND)	(ND)	(ND)	(ND)
3	_Microsporum canis_	0.00 ± 0.00	0.00 ± 0.00	0.00 ± 0.00	16.66 ± 0.57	18.00 ± 0.00
		(ND)	(ND)	(ND)	(1.25)	(1.25)
	Plant pathogens					Bavistin
4	_Aspergillus niger_	0.00 ± 0.00	0.00 ± 0.00	0.00 ± 0.00	0.00 ± 0.00	18.00 ± 0.00
		(ND)	(ND)	(ND)	(ND)	(ND)
5	_Aspergillus flavus_	0.00 ± 0.00	0.00 ± 0.00	0.00 ± 0.00	0.00 ± 0.00	17.66 ± 0.57
		(ND)	(ND)	(ND)	(ND)	(ND)
6	_Drechslera_	0.00 ± 0.00	0.00 ± 0.00	0.00 ± 0.00	0.00 ± 0.00	15.00 ± 0.00
		(ND)	(ND)	(ND)	(ND)	(ND)
7	_Fusarium verticillioides_	0.00 ± 0.00	0.00 ± 0.00	0.00 ± 0.00	0.00 ± 0.00	14.66 ± 0.57
		(ND)	(ND)	(ND)	(ND)	(ND)

Values are the mean of triplicates ± SD p < 0.05.
_ND_ Not Done.

ABTS assay

The antioxidant activity of _Toona ciliata_ test solvent extracts was determined for radical scavenging ability by ABTS method (Adedapo et al. 2009). The stock solution was prepared by using 7 mM ABTS solution and 2.4 mM potassium per sulfate solution separately. The working solution was made by mixing the two stock solutions in equal quantities and allowed them to react for 12 h. at room temperature in dark condition. 1.0 ml of _Toona ciliata_ test solvent extracts of differing concentrations (50–250 µgml^{-1} was added to 1.0 ml of ABTS solution and diluted with methanol. The absorbance was taken at 734 nm after 7 min using the spectrophotometer. The ABTS scavenging capacity of the extract was compared with that of BHT and percentage inhibition calculated as

$$\text{ABTS radical scavenging activity }(\%)= [(A_{control}-A_{sample})]/[(A_{control})]\times 100$$

Where, $A_{control}$ is the absorbance of ABTS radical + methanol; A_{sample} is the absorbance of ABTS radical + sample extract/standard. The experiment was performed in triplicates and values are expressed in ± SD.

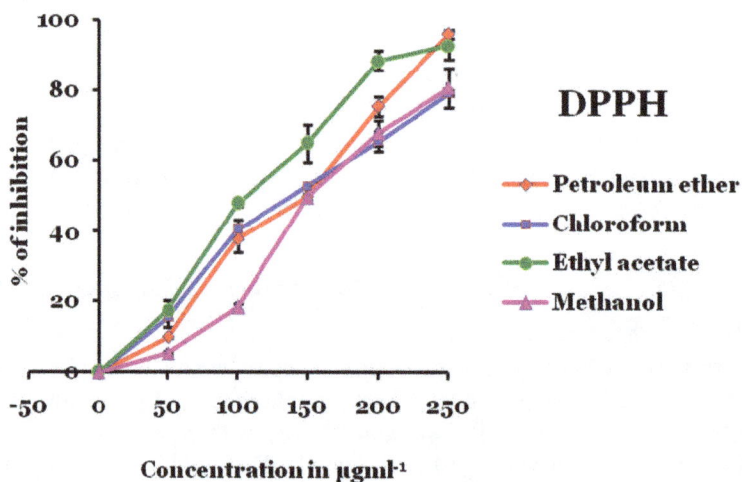

Figure 1 DPPH scavenging activity of test solvent extracts of _Toona ciliata_. Data expressed at p < 0.05.

Results and discussion

In the present study, preliminary chemical analysis of *Toona ciliata* leaf and flower extracts of different solvent extracts viz. petroleum ether, chloroform, ethyl acetate, methanol and water revealed the presence of carbohydrates, proteins, alkaloids and phytosterols (Table 1). Saponins was absent in all the extracts whereas flavonoids, tannins and phenolic compounds were found to be present in ethyl acetate and methanol extracts and glycosides was present in petroleum ether and aqueous extracts. Tables 2 and 3 shows the antibacterial and antifungal activities of test solvent extracts of *Toona ciliata* at 50 µl (100 mgml^{-1}) concentrations against human and phytopathogenic bacteria and fungi. Ethyl acetate extract showed moderate antibacterial activity against *Proteus mirabilis* and least activity against *Salmonella typhi* and *Staphylococcus epidermis*. Similarly, methanol extract also showed moderate antibacterial activity against *Proteus mirabilis* and least activity against *Klebsiella pneumoniae, Salmonella typhi, Staphylococcus aureus* and *Staphylococcus epidermidis*, whereas other test bacteria did not show any inhibition zone. Antibacterial activity against human and phytopathogenic bacteria was not observed with petroleum ether and chloroform extracts. Ethyl acetate and methanol extracts exhibited moderate activity against *Xanthomonas axonopodis* pv. *malvacearum* and least activity against *Xanthomonas campestris* pv. *vesicatoria, Xanthomonas oryzae* pv. *oryzae* and *Erwinia carotovora* in comparison with the standard tetracycline. Petroleum ether, chloroform and ethyl acetate solvent extracts did not exhibited any zone of inhibition against *Aspergillus niger, Aspergillus flavus, Drechslera, Fusarium verticillioides, Candida albicans* and *Microsporum gypsum* whereas, methanol extract showed significant activity only against *Microsporum canis* compared to the standard miconozole. Ethyl acetate and

methanol extracts of *Toona ciliata* exhibited lowest MIC varied from 10–2.5 mgml^{-1} against *Proteus mirabilis, Salmonella typhi, Staphylococcus epidermidis, Erwinia carotovora, Xanthomonas campestris* pv. *vesicatoria, Xanthomonas oryzae* pv. *oryzae* and *Xanthomonas axonopodis* pv. *malvacearum* wherein, standard antibiotics gentamicin and tetracycline showed MIC ranging between 0.625-0.156 mgml^{-1} and 0.156 mgml^{-1} against human pathogenic and phytopathogenic bacteria respectively. Methanolic extract of *Toona ciliata* showed MIC of 1.25 mgml^{-1} against *Microsporum canis*.

Antioxidant activity using DPPH and ABTS method is illustrated in the Figures 1 and 2. The test solvent extracts viz., petroleum ether, chloroform, ethyl acetate and methanol exhibited significant free radical scavenging activity wherein, as the concentration increases the percentage inhibition of free radical also increases. Petroleum ether, chloroform, ethyl acetate and methanol extracts of *Toona ciliata* showed DPPH significant activity with IC$_{50}$ value of 150, 135.5, 105 and 92.5 µgml^{-1} whereas ABTS scavenging activity showed IC$_{50}$ value of about 145, 120, 120.5 and 95 µgml^{-1} respectively. Standard BHT showed significant DPPH and ABTS scavenging activity with IC$_{50}$ value of 8 µgml^{-1} and 11.5 µgml^{-1} respectively.

In the present study, methanol extract of *Toona ciliata* revealed the presence of carbohydrates, alkaloids, flavonoids, phytosterols, tannins and phenolic compounds which justifies the earlier findings of Gautam et al. (2010). Scientific literatures perused by far suggest antimicrobial activity against human pathogens whereas no reports confer the evaluation of *Toona ciliata* against phytopathogens which has been reported in the present investigation with ethyl acetate and methanol extracts of *Toona ciliata* against *Erwinia carotovora, Xanthomonas axonopodis* pv. *malvacearum, Xanthomonas campestris* pv. *vesicatoria* and *Xanthomonas oryzae* pv. *oryzae* along

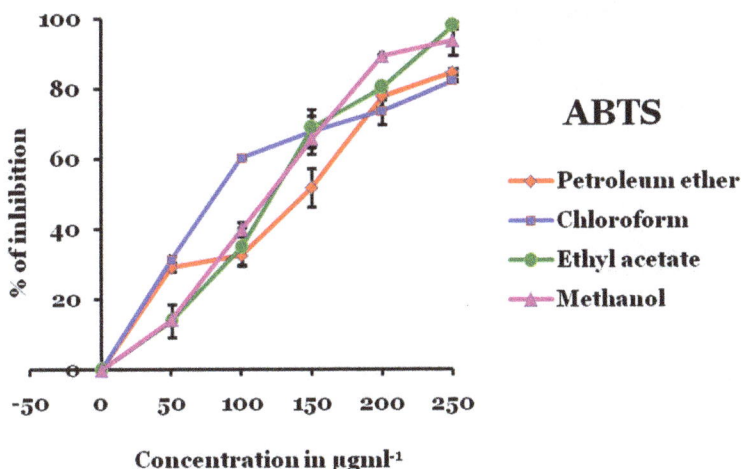

Figure 2 ABTS scavenging activity of test solvent extracts of *Toona ciliata*. Data expressed at p < 0.05.

human pathogen viz *Salmonella typhi, Staphylococcus epidermis and Klebsiella pneumoniae* suggests its antimicrobial potential which has also been studied previously against various pathogenic bacteria (Bibi et al. 2011; Kiladi 2012). Crude extracts upon evaluation of antioxidant activity using DPPH and ABTS method showed IC_{50} value ranging from 100–150 μgml^{-1} and 80–150 μgml^{-1} respectively with different test solvent extracts. Earlier report of antioxidant compound has been well described with potent DPPH activity resulting IC_{50} value 1.02 μgml^{-1} (Ekaprasada et al. 2009).

Conclusions

The present investigation concludes that methanol extract of *Toona ciliata* exhibited maximum inhibition against test human and phytopathogens. Crude extracts displayed significant antioxidant activity, thus results obtained in the present investigation are promising enough for further isolation and characterization to reveal any novel metabolite of pharmaceutical importance.

Competing interests

Both authors declare that they have no competing interests.

Authors' contributions

KSK design the experiment, carried out the experiment, and contributed in framing the article. SS supervised the work. Both authors read and approved the final manuscript.

Acknowledgments

The authors are thankful to the University of Grant Commission (UGC) – RGNFs, Govt. of India for providing financial support and the Department of Studies in Microbiology, University of Mysore, Manasagangotri, Mysore.

References

Adedapo AA, Jimoh FO, Koduru S, Masika PJ, Afolayan AJ (2009) Assessment of the medicinal potentials of the methanol extracts of the leaves and stems of *Buddleja saligna*. BMC Complem Altern Med 9:21, doi:10.1186/1472-6882-9-21

Bauer AW, Kirby WMM, Sherries SC, Tunk M (1966) Antibiotic susceptibility of testing by a standard single disc method. Amer J Clin Pathol 36:492–496

Bibi Y, Nisa S, Chaudhary MF, Zia M (2011) Antibacterial activity of some selected medicinal plants of Pakistan. BMC Complem Altern Med 11:52, doi:10.1186/1472-6882-11-52

Buatong J, Phongpaichit S, Rukachaisirikul V, Sakayaroj J (2011) Antimicrobial of crude extracts from mangrove fungal endophytes. World J Microbiol Biotechnol 27(12):3005–3008

Chen HD, Yang SP, Wu Y, Dong L, Yue JM (2009) Terpenoids from *Toona ciliata*. J Nat Prod 72(4):685–689

Chowdhury R, Rashid RB, Hasan CM (2002) Steroids and C-methyl coumarins from *Toona ciliata*. J Bang Acad Sci 26:219–222

Chowdhury R, Hasan CM, Rashid MA (2003) Antimicrobial activity of *Toona ciliata* and *Amoora rohituka*. Fitoterapia 74(1–2):155–158

Ekaprasada TM, Nurdin H, Ibrahim S, Dachriyanus (2009) Antioxidant activity of methyl gallate isolated from the leaves of *Toona sureni*. Indo J Chem 9(3):457–460

Gautam AD, Ahirwar DJ, Sujane M, Sharma GN (2010) Pharmacognostic evaluation of *Toona ciliata* bark. J Adv Pharm Technol Res 1(2):216–220

Ghani A (2003) Medicinal plants of Bangladesh, 2nd edn. The Asiatic society of Bangladesh, Dhaka, p 603

Gopalakrishnan G, Singh NDP, Kasinath V, Malathi R, Rajan SS (2000) Photo oxidation of cedrelone, a tetranortriterpenoid from *Toona ciliata*. Photochem Photobiol 72:464–466

Harborne JB (1973) Phytochemical methods a guide to modern techniques of plant analysis. Chapman and Hall, London, p 49e188

Kiladi CP (2012) Evaluation of antibacterial potential of *Toona ciliata* Roemer against ten different pathogenic bacteria. Int J Univ Pharm Life Sci 2(3):2249–6793

Liao SG, Yang SP, Yuan T, Zhang CR, Chen HD, Wu Y, Xu YK, Yue JM (2007) Limonoids from the leaves and stems of *Toona ciliata*. J Nat Prod 70:1268–1273

Lin Y, Wang C, Chen I, Jheng J, Li J, Tung C (2013) TIPdb: A database of anticancer, antiplatelet, and antituberculosis phytochemicals from indigenous plants in Taiwan. ScientificWorld J, doi:10.1155/2013/736386

Mahesh B, Satish S (2008) Antimicrobial activity of some important medicinal plant against plant and human pathogens. World J Agricult Sci 4:839–843

Malairajan P, Gopalakrishnan G, Narasimhan S, Jessi KVK (2006) Analgesic activity of some Indian medicinal plants. J Ethnopharmacol 106:425–428

Nagumanthri V, Rahiman S, Tantry BA, Nissankararao P, Phani kumar M (2012) *In vitro* antimicrobial activity of *Acacia nilotica, Ziziphus mauritiana, Bauhinia variegate* and *Lantana camara* against some clinical isolated strains. Iran J Sci Technol A2:213–217

Negi SJ, Bisht VK, Bhandari KA, Bharti MK, Sundriyal RC (2011) Chemical and pharmacological aspects of *Toona*(Meliaceae). Res J Phytochem 5:14–21

Neto JO, Agostinho SMM, Silva DMF, Vieira PC, Fernandes JB, Pinheiro AL, Vilela EF (1995) Limonoids from seeds of *Toona ciliata* and their chemosystematic significance. Phytochemistry 38:397–401

Ning J, He HP, Li SF, Geng ZL, Fang X (2010) Triterpenoids from leaves of *Toona ciliata*. J Asian Natl Prod Res 12:448–452

Satish S, Mohana DC, Ranhavendra MP, Raveesha KA (2007) Antifungal activity of some plant extracts against important seed borne pathogens of *Aspergillus* sp. J Agricult Technol 3:109–119

Sette LD, Passarini MRZ, Delarmelina C, Salati F, Duarte MCT (2006) Molecular characterization and antimicrobial activity of endophytic fungi from coffee plants. World J Microbiol Biotechnol 22:1185–1195

Shaohong C, Pengkang R, Yuntao Z (2010) Inhibitory effects of ethanol extract from *Toona sinensis* leaves on the formation of protein non-enzymatic. J Anhui Agricult Sci 11:5642

Sofowara A (1993) Medicinal plants and traditional medicine in Africa. Spectrum Book Ltd, Ibadan, Nigeria, p 289

Wang J, Liu H, Kurt'an T, M'andi A, S'andor A, Jia L, Zhang H, Guo Y (2011) Protolimonoids and norlimonoids from the stem bark of *Toona ciliata* var. *pubescens*. Org Biomol Chem 9(22):7685–7696

Zhang L, Gao Y, Zhang Y, Liu J, Yu J (2010) Change in bioactive compounds and antioxidant activities in pomegranate leaves. Sci Hortic 123:543–546

19

Proteomic and bioinformatic analysis of recurrent anaplastic oligodendroglioma

Yeonhee Hong[1,2†], Edmond Changkyun Park[1,2†], Eun-Young Shin[1,2,3], Sang-Oh Kwon[1], Young-Taek Oh[4], Byung-Ock Choi[5], Giwon Kim[4,5*] and Gun-Hwa Kim[1,2,3*]

Abstract

Background: Anaplastic oligodendroglioma (AO) is a type of glioma that is believed to originate from oligodendrocytes in the brain or from glial precursor cells. Recurrence of AO reduces the overall survival rate of patients and causes meningeal or even systemic spread/metastasis more frequently than other types of gliomas. We performed proteomic analysis of recurrent AO tumors to identify the proteins significantly expressed in recurrent AO and to understand biological characteristics of recurrent AO.

Findings: Using human brain tissues, we identified 401 proteins that were significantly expressed in recurrent AO. Through bioinformatic analysis, we determined that the majority of the identified proteins are involved in anti-apoptotic pathway and cell proliferation. In addition, our findings suggest that epidermal growth factor (EGF) signaling may be responsible for the development of recurrent AO.

Conclusions: These results will aid researchers in understanding the pathology of recurrent AO and identifying the therapeutic targets for the treatment of recurrent AO.

Findings

The goal of proteomics is a comprehensive, quantitative description of protein expression and its changes under the influence of biological perturbations such as disease or drug treatment (Anderson & Anderson 1998; Blackstock & Weir 1999). Proteins are vital parts of living organisms, as they are the main components of the physiological metabolic pathways of cells (James 1997). Recently, a number of proteomic studies have focused on the metabolic and signaling pathways associated with the formation and progression of tumors. The information obtained from proteomic analysis can be used to identify biomarkers or therapeutic targets of tumors (Alterovitz et al. 2008).

Anaplastic oligodendroglioma [AO, World Health Organization (WHO) grade III] is a type of glioma that is believed to originate from oligodendrocytes in the brain or from glial precursor cells. AO occurs primarily in adults (9.4% of all primary brain tumors), but also sometimes occurs in children (4% of all primary brain tumors) (Allison et al. 1997). AO is distinguished from other brain tumors by a unique constellation of molecular genetic alterations, including the coincident loss of chromosomal arms 1p and 19q in 50–70% of tumors (Gregory et al. 1998). AO also has a genetic profile similar to that of other primary gliomas, such as amplification of epidermal growth factor receptor (EGFR) expression, 10q loss, p16 deletion, and phosphatase and tensin homolog (PTEN) mutation without 1p loss or tumor protein 53 (TP53) mutation (Anthony et al. 2003). The standard treatment for AO is surgical resection followed by radiation and chemotherapy with procarbazine, lomustine, and vincristine (PCV) (Kyritsis et al. 1993; Allison et al. 1997; Cairncross et al. 1992; Cairncross & Macdonald 1988; Cairncross et al. 1998; Glass et al. 1992; Kleinberg et al. 1993; Mason et al. 1996). Despite high response rates (≥ 70%) to treatment, the median survival period is relatively short (Glass et al. 1992). Accumulation of diverse genetic and epigenetic alterations that occur due to interventions for treatment of the tumor may cause extensive changes in the expression of genes involved in oncogenesis leading to treatment-related effects such as aggressive transformation (Kleinberg et al. 1993).

* Correspondence: giwonkim@catholic.ac.kr; genekgh@kbsi.re.kr
†Equal contributors
4Department of Radiation Oncology, Ajou University School of Medicine, Suwon 443-721, Republic of Korea
1Division of Life Science, Korea Basic Science Institute, Daejeon 305-333, Republic of Korea
Full list of author information is available at the end of the article

Despite advances in the treatment and understanding of primary AO, recurrent AO is significantly less well understood. Although a recent study reported the microRNA expression profile in patients with recurrent AO (Kim et al. 2011), the biological characteristics of recurrent AO have never been studied. In this study, we conducted proteomic analysis of recurrent AO to examine protein profiles in recurrent AO and identify biological and molecular characteristics of recurrent AO.

Materials and methods

Patient samples and protein extraction

Institutional Review Board of the Catholic University of Korea approved of this study. The three specimens were obtained from patients with recurrent AO treated with total excision and adjuvant radiotherapy with total dose of 55.8 Gray (Additional file 1: Table S1) (Kim et al. 2011). The specimens were separated during surgery at the Seoul St. Mary's Hospital and stored in liquid nitrogen at -80°C.

Protein extraction was conducted previously described (Cao & Liang 2012) with minor modification. The specimens is added to RIPA buffer (89900, Pierce Biotechnology, IL, USA) containing protease inhibitor cocktail (78410, Pierce), and sonication. Centrifuge 14,000 rpm, 4°C, 20 min, and store supernatant at -80°C. After, the supernatant was used for SDS-PAGE and LC-MS/MS.

SDS-PAGE and in-gel digestion with trypsin

Protein samples were separated by 12% SDS-PAGE (mini-PROTEAN, BIO-RAD). A 100 ug of protein sample was applied to each lane, and the gels were stained with Coomassie Brilliant Blue R-250. In-gel digestion was conducted in accordance with the previously described method (Kim et al. 2006). Gels were fractionated into 10 parts according to molecular weight. Each part was digested with trypsin (1.2 ug) for 16 hours at 37°C after reduction and alkylation of cysteines of the proteins. Digested peptides were extracted by extraction solution (50 mM ammonium bicarbonate, 50% acetonitrile, and 5% trifluoroacetic acid). Digested peptides were resolved in 10 ul of sample solution containing 0.02% formic acid and 0.5% acetic acid, and stored at -80°C until required.

LC-MS/MS analysis and protein identification

The peptide samples (5 ul) were concentrated on a Easy-column™ (L 2cm, ID 100um, 120Å, C18-A1) trapping column (PROXEON, Denmark). Peptides were eluted from the column and directed onto a Easy-column™ (L 10cm, ID 75um, 120Å, C18-A2) reversephase column (PROXEON, Denmark) at a flow rate of 200 nl/min. Peptides were eluted by a gradient of 0~65% acetonitrile for 120 min. All MS and MS/MS spectra in the LTQ-

Velos ESI ion trap mass spectrometer (Thermo Scientific, USA) were acquired in a data-dependent mode. Each full MS (m/z range of 300 to 2,000) scan was followed by seven MS/MS scans of the most abundant precursor ions in the MS spectrum with dynamic exclusion enabled. For protein identification, MS/MS spectra were searched by MASCOT (Matrix science, www.matrixscience.com). The proteome sequence database of IPI human ver. 3.82 (http://www.ebi.ac.uk/IPI) was used as the database. Mass tolerance of parent ion and fragment ion was 1.5 Da and 1.3 Da, respectively. Cabamidomethylation of cysteine and oxidation of methionine were considered in MS/MS analysis as variable modifications of tryptic peptides. Mascot score was recalculated by percolator function in the mascot for increasing sensitivity of correct identification and false discovery ratios (FDR) of the search results was adjusted to below 2%.

Bioinformatic analysis

Using Ingenuity Pathway Analysis (IPA) tool, all of the identified proteins in recurrent AO were subjected analysis of indicated metabolic and signaling pathway. The identified proteins in recurrent AO were subjected to query global protein network analysis and direct/indirect interactions on proteins that are involved in pathways associated with epidermal growth factor receptor (EGFR). For functional enrichment analysis of Gene Ontology (GO) categories, GOfact online tool (http://61.50.138.118/gofact) was used. p-value of Fisher's exact test determines the probability that the association between the proteins in the dataset and the biological function/pathways explained by chance alone.

Results and discussion

Proteomic analysis of recurrent AO

First, we performed proteomics analysis using tumor sample from patient with recurrent AO who were treated with postoperative chemoradiotherapy. From LC-MS/MS, 401 proteins were shown significantly to express in recurrent AO (Additional file 2: Table S2) and then we analyzed the 401 identified proteins using the Ingenuity Pathway Analysis (IPA) tool (http://www.ingenuity.com). The results revealed the majority of the metabolically associated proteins that was involved in biosynthesis of carbohydrates, such as glyoxylate and dicaboxylate metabolism, oxidative phosphorylation and glycolysis/gluconeogenesis (Figure 1A). With respect to the signaling pathways active in recurrent AO, the identified proteins were associated with pathways responsible for cell growth (Figure 1B). These findings

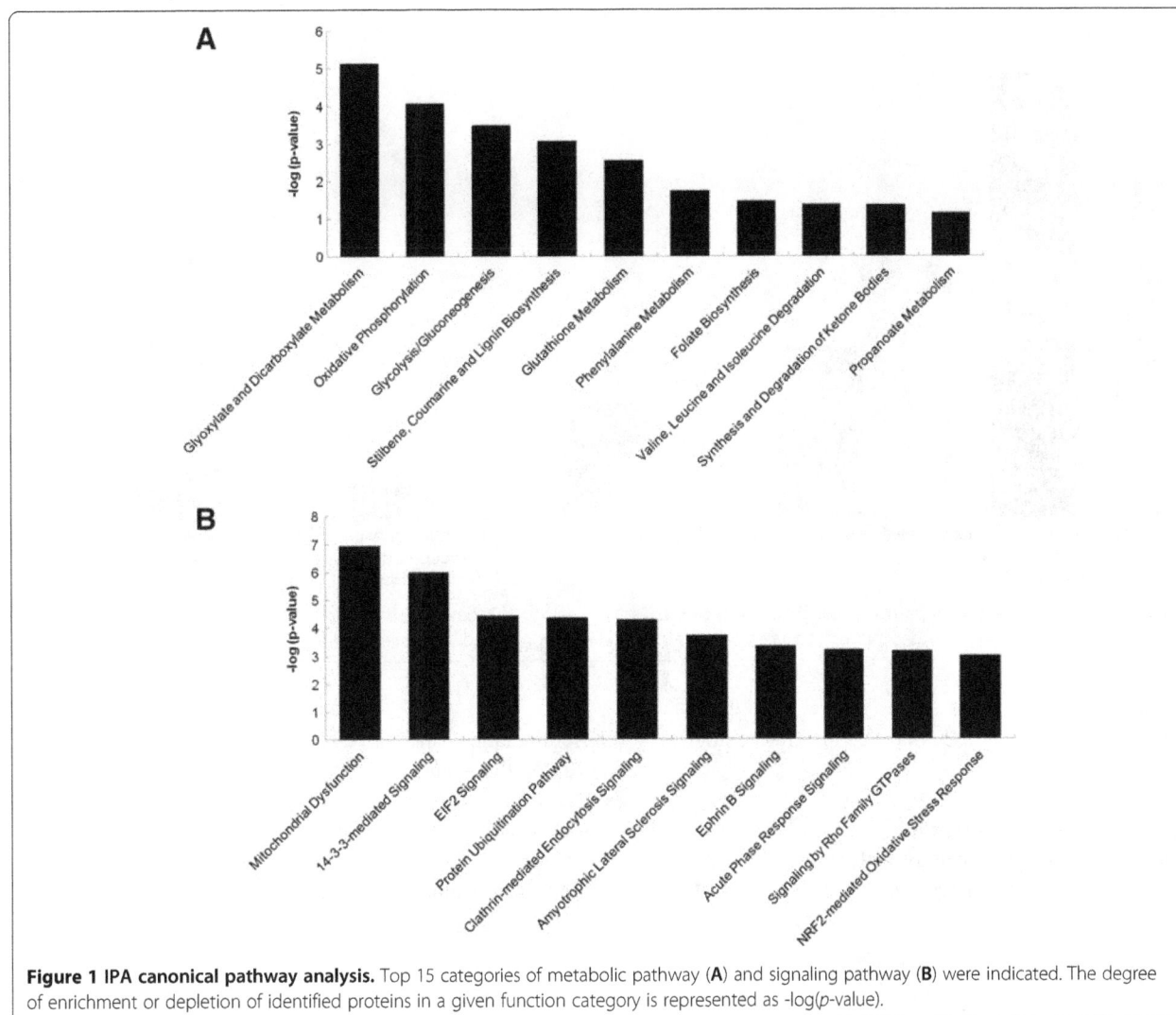

Figure 1 IPA canonical pathway analysis. Top 15 categories of metabolic pathway (**A**) and signaling pathway (**B**) were indicated. The degree of enrichment or depletion of identified proteins in a given function category is represented as -log(*p*-value).

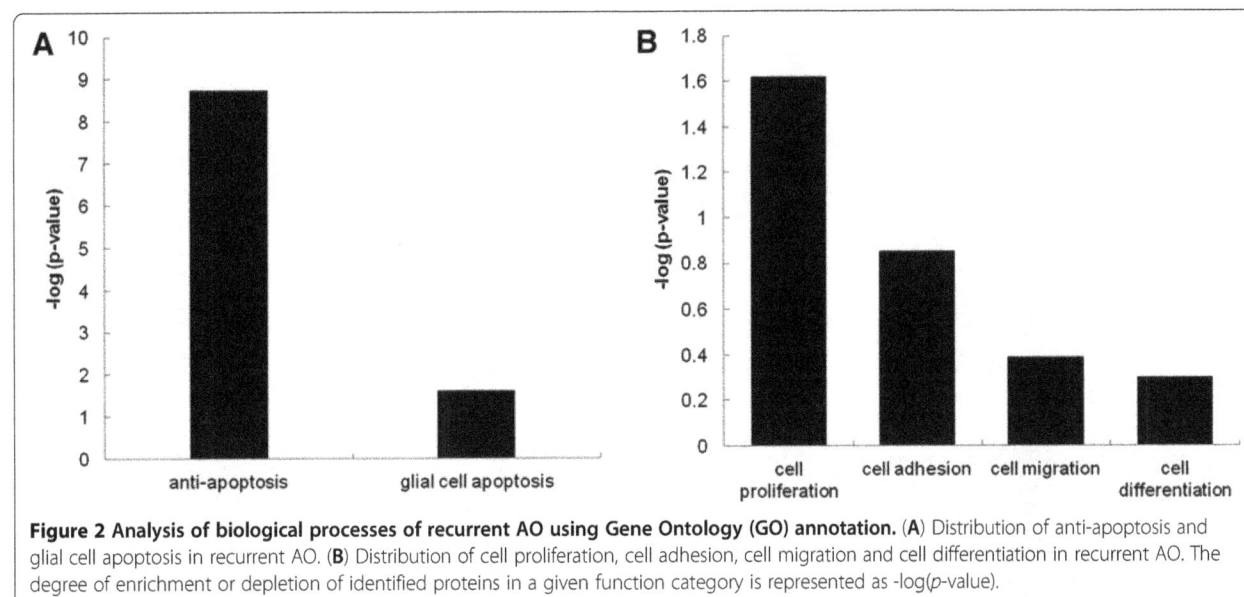

Figure 2 Analysis of biological processes of recurrent AO using Gene Ontology (GO) annotation. (**A**) Distribution of anti-apoptosis and glial cell apoptosis in recurrent AO. (**B**) Distribution of cell proliferation, cell adhesion, cell migration and cell differentiation in recurrent AO. The degree of enrichment or depletion of identified proteins in a given function category is represented as -log(*p*-value).

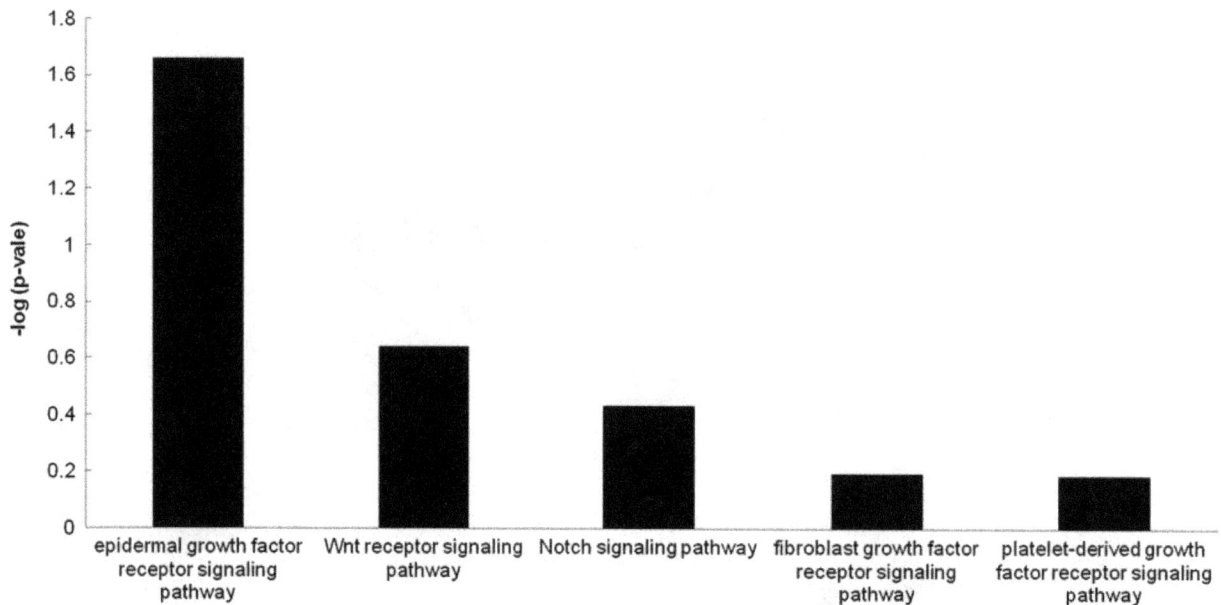

Figure 3 Distribution of signaling pathway associated with cancer. Gene Ontology (GO)-determined signaling pathways associated with cancer. The degree of enrichment or depletion of identified proteins in a given function category is represented as -log(p-value).

indicate that recurrent AO may be in the biological state of actively growth.

Biological process of recurrent AO

To confirm the biological state of actively growth and discover the specific cause of recurrent AO, Gene Ontology (GO) annotation analysis was performed. As shown in Figure 2A, the results indicated that the identified proteins were connected to anti-apoptotic functions, while only a few were associated with glial cell apoptosis. This may indicate why recurrent AO is resistant to chemoradiotherapy. Anti-apoptotic effects often contribute to cancer cell survival and

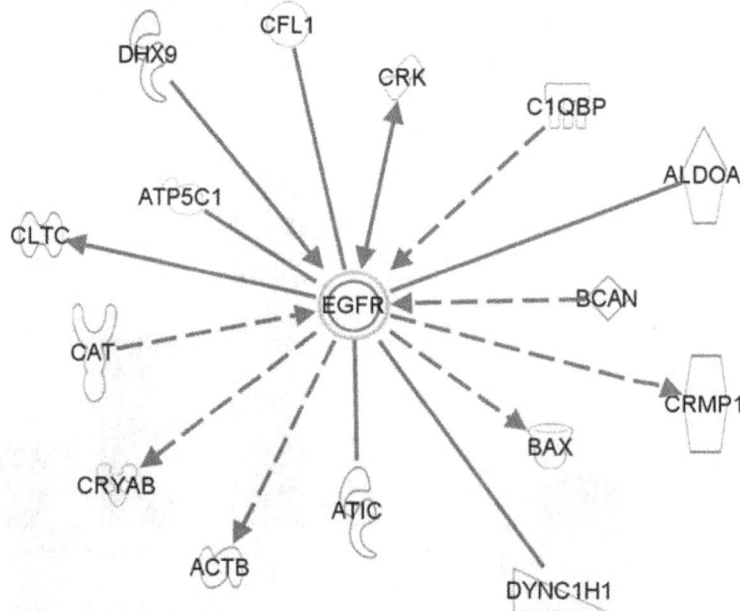

Figure 4 Proteins related to EGFR in recurrent AO. IPA-determined network of identified proteins in the recurrent AO that may be involved in EGFR signaling. Dotted line, indirect; Solid line, direct.

chemoresistance in cancer cells (Williams et al. 2005). The 18 proteins that are involved in anti-apoptosis pathways are listed in Additional file 1: Table S3.

Multiple biological processes, including cell proliferation, adhesion, and migration, have been shown to be involved in tumor development (Lee 1992). Base on our analyses, the 14 proteins appear to be linked to cell proliferation processes (Figure 2B). Fourteen of these proteins are presented in Additional file 1: Table S4. Tumor cell is related to cell adhesion and motility. Cell motility is necessary to moves within tissues during invasion and metastasis by their own motility (Ymazaki et al. 2005; Zang et al. 2006). The 11 proteins and 9 proteins appear to be linked to cell adhesion and migration, respectively (Additional file 1: Table S5 and Table S6). These findings indicate that the expression of proteins related to anti-apoptotic and cell proliferation pathways are closely related to the development of recurrent AO. Understanding the functions of these proteins in recurrent AO would contribute to the development of therapeutics for the prevention or treatment of this disease.

Signaling pathways associated with tumor development in recurrent AO

Recent studies claimed that growth factor signaling is involved in tumorigenesis and the development of malignancy (Lee 2001). EGF signaling is well known to be involved in the autonomous growth of cancer cells (Normanno et al. 2001), and FGF signaling plays a pivotal role in cancer development (Cao et al. 2011). Wnt and Notch signaling are also associated with tumor development (Allenspach et al. 2002; Paul 2000). Therefore, we investigate whether the identified proteins is involved in various signaling pathway. The results showed that EGF signaling is highly associated with recurrent AO compared to other signaling pathways (Figure 3). This is in agreement with a previous report showing that amplification of EGFR, which permits evasion of cancer cell death, has been observed in cases of recurrent AO that develop resistance to treatment (Paul 2000). Additionally, Wnt and Notch signaling were also associated with recurrent AO. Further studies regarding the role of Wnt and Notch signaling during the development of AO recurrence will be necessary.

Proteins related to EGFR in recurrent AO

Based on the fact that EGF signaling is highly related to recurrent AO (Figure 3) and activation of EGFR is responsible for the resistance of recurrent AO to treatment, novel regulators or effectors that mediate EGFR signaling in recurrent AO were sought. EGFR signaling plays roles in cell proliferation, angiogenesis, and inhibition of apoptosis processes that are essential for cancer

development (Normanno et al. 2001). By additional bioinformatics analysis of the identified proteins identified using IPA tool, we indeed found that 15 proteins are closely associated with EGFR (Figure 4 and Additional file 1: Table S7), implying that these proteins may play an important role in the development of recurrent AO.

Conclusion

In this study, we identified proteins significantly expressed in recurrent AO. Extensive bioinformatics analysis revealed that these proteins are involved in biological processes and signaling pathways that are associated with the development of recurrent AO. Cells in recurrent AO are highly resistant to apoptosis and actively proliferating. Analysis of signaling pathway enrichment also showed that EGF signaling is active in recurrent AO. This is the first analysis to identify and characterize the proteins associated with recurrent AO. Further studies, investigating the role of these identified proteins in recurrent AO will contribute to a more thorough understanding of the molecular mechanisms that mediate the development of this disease.

Additional files

Additional file 1: Supplementary tables.

Additional file 2: Table S2.

Competing interests
The authors declare that they have no competing interests.

Authors' contributions
YH and ECP designed experiments, analyzed data, and wrote the manuscript. EYS performed proteomic analysis and SOK carried out bioinformatic analysis. YTO and BOC shared evaluation of patients and sampling. GK and GHK supervised the study. All authors read and approved the final manuscript.

Acknowledgments
This work was supported by Seoul St. Mary's Clinical Medicine Research Program year of 2009 and the Catholic Medical Center Research Foundation year of 2010 through the Catholic University of Korea, Korea Basic Science Institute (T32414), and the Pioneer Research Center Program of the National Research Foundation of Korea funded by the Ministry of Education, Science and Technology (2012-0000432).

Author details
[1]Division of Life Science, Korea Basic Science Institute, Daejeon 305-333, Republic of Korea. [2]Pioneer Research Center for Protein Network Exploration, Korea Basic Science Institute, Daejeon 305-333, Republic of Korea. [3]Department of Functional Genomics, University of Science and Technology (UST), Daejeon 305-333, Republic of Korea. [4]Department of Radiation Oncology, Ajou University School of Medicine, Suwon 443-721, Republic of Korea. [5]Department of Radiation Oncology, Seoul St. Mary's Hospital, The Catholic University of Korea, Seoul 137-701, Republic of Korea.

References
Allenspach EJ, Maillard I, Aster C, Warren S (2002) Notch Signaling in Cancer. Cancer Biol Ther 5:466–476

Allison RR, Schulsinger A, Vongtama V, et al. (1997) Radiation and chemotherapy improve outcome in oligodendroglioma. Int J Radiat Oncol Biol Phys 37:399–403

Alterovitz G, Xiang M, Liu J, Chang MF (2008) System-wide peripheral biomarker discovery using information theory. P.S.biocomuting 13:231–242

Anderson NL, Anderson NG (1998) Proteome and proteomics: new technologies, new concepts, and new words. Electrophoresis 19:1853–1861

Anthony B, Khe HX, Antoine FC, Delattre JV (2003) Primary brain tumours in adults. Lancet 361:323–31

Blackstock WP, Weir MP (1999) Proteomics: quantitative and physical mapping of cellular proteins. Trends Biotechnol 17:121–127

Cairncross JG, Macdonald DR (1988) Successful chemotherapy for recurrent malignant oligodendroglioma. Ann Neurol 23:360–364

Cairncross JG, Macdonald DR, Ramsay DA (1992) Aggressive oligodendroglioma: A chemosensitive tumor. Neurosurgery 31:78–82

Cairncross JG, Ueki K, Zlatescu MC, et al. (1998) Specific genetic predictors of chemotherapeutic response and survival in patients with anaplastic oligodendrogliomas. J Natl Cancer Inst 90:1473–1479

Cao R, Liang S (2012) Liver plasma membranes: an effective method to analyze membrane proteome. Methods Mol Biol 909:113–123

Cao XM, Meng HS, Hong F, Ying QS (2011) Fibroblast Growth Factor Receptor 4 Regulates Proliferation and Antiapoptosis During Gastric Cancer Progression. Cancer 1:5340–5313

Glass J, Hochberg FH, Gruber ML, et al. (1992) The treatment of oligodendrogliomas and mixed oligodendroglioma-astrocytomas with PCV chemotherapy. J Neurosurg 76:741–745

Gregory J, Keisuke U, Magdelena CZ, David KL (1998) Specific Genetic Predictors of Chemotherapeutic Response and Survival in Patients With Anaplastic Oligodendrogliomas. J Natl Cancer Inst 90:1473–1479

James P (1997) Protein identification in the post-genome era: the rapid rise of proteomics. Q Rev Biophys 30:279–331

Kim YH, Cho K, Yun SH, Kim JY, Kwon KH, Yoo JS, Kim SI (2006) Analysis of aromatic catabolic pathways in Pseudomonas putida KT 2440 using a combined proteomic approach: 2-DE/MS and cleavable isotope-coded affinity tag analysis. Proteomics 6:1301–1318

Kim G, Park EC, Chung HR, Jeon SS, Kim SI, Jang HS, Kim GH, Choi BO (2011) MicroRNA expression profiling in recurrent anaplastic oligodendroglioma treated with postoperative radiotherapy. JAST 2:97–104

Kleinberg L, Wallner K, Malkin MG (1993) Good performance status of long-term disease-free survivors of intracranial gliomas. Int J Radiat Oncol Biol Phys 26:129–133

Kyritsis AP, Yung WK, Bruner J, et al. (1993) The treatment of anaplastic oligodendrogliomas and mixed gliomas. Neurosurgery 32:365–370

Lee AS (1992) Mammalian stress response: induction of the glucose-regulated protein family. Curr Opin Cell Biol 4:267–273

Lee AS (2001) The glucose-regulated proteins: stress induction and clinical application. Trends Biochem Sci 26:504–511

Mason WP, Krol GS, DeAngelis LM (1996) Low-grade oligodendroglioma responds to chemotherapy. Neurology 46:203–207

Normanno N, Bianco C, De Luca A, Salomon DS (2001) The role of EGF related peptides in tumor growth. Front Biosci 6:d685–d707

Paul P (2000) Wnt signaling and cancer. Genes Dev 14:1837–1851

Williams J, Lucas PC, Griffith KA, et al. (2005) Expression of BclxL in ovarian carcinoma is associated with chemoresistance and recurrent disease. Gynecol Oncol 96:287–295

Ymazaki D, Kurisu S, Takenawa T (2005) Regulation of cancer cell motility through actin reorganization. Cancer Sci 96:379–386

Zang XP, Bullen EC, Manjeshwar S, Jupe ER, Howard EW, Pneto JT (2006) Enhanced motility of KGF-transfected breast cancer cells. Anticancer reseatch 26:961–966

20

Removal of fluoride from polluted waters using active carbon derived from barks of *Vitex negundo* plant

Mekala Suneetha[1], Bethanabhatla Syama Sundar[1] and Kunta Ravindhranath[2*]

Abstract

Background: Deleterious effects of fluoride contamination in ground waters on the health of human beings are well known and intensive research on developing de-fluoridation methods is globally pursued. Of the various methodologies, increasing interest is being envisaged in using the adsorption methods based on active carbons derived from plant material. In the present investigation, Nitric acid activated carbon derived from barks of Vitex negundo plant (NVNC) is probed for its de-fluoridation abilities.

Methods: The activated carbon is characterized adopting various physicochemical methods and surface morphological studies are carried out using FT-IR and SEM-EDX techniques. The effect of various parameters such as pH, sorbent dosage, agitation time, initial concentration of fluoride, temperature, particle size and presence of foreign ions on the extraction of the fluoride is studied adopting Batch methods. The adsorption process is analyzed with Freundlich, Langmuir, Temkin and Dubinin-Radushkevich (D-R) isotherms and kinetics of adsorption is studied using pseudo first-order, pseudo second-order, Weber and Morris intraparticle diffusion, Bangham's pore diffusion and Elovich equations. The methodology developed is applied to real ground water samples.

Results: De-fluoridation is maximum at the pH: 7.0, adsorbent dosage: 4.0g/lit; equilibrium time: 50 min, Particle size: 45μ and temperature: $30 \pm 1°C$. The correlation coefficient values for the adsorption isotherms: Freundlich, Langmuir, Temkin and Dubinin-Radushkevich are 0.929, 0.998, 0.980 and 0.946 respectively and for kinetic models: pseudo-first-order, pseudo-second-order, Weber and Morris intraparticle diffusion, Bangham's pore diffusion and Elovich equations are 0.989, 0.994, 0.874, 0.902 and 0.912 respectively. The Temkin heat of sorption, B, and the Dubinin-Radushkevich mean free energy, E, for the activated carbon adsorbent are 0.196 J/mol and 7.07 kJ/mol respectively.

Conclusions: Nitric acid activated carbon derived from barks of Vitex negundo (NVNC) plant is found to be an effective adsorbent for the de-fluoridation of waters. The adsorption process is satisfactorily fitted with Langmuir adsorption isotherm with good correlation coefficient value and it indicates monolayer adsorption. The adsorption kinetics is found to follow pseudo-second-order kinetics. The Dubinin-Radushkevich mean free energy and Temkin heat of sorption confirm the physisorption nature as these are lower than 20kJ/mol. The procedure developed is remarkably successful in de-fluoridation of real ground water samples.

Keywords: Fluoride removal; Batch adsorption technique; Inexpensive adsorbents; Surface characterization; Adsorption isotherms; Kinetic models; Applications

* Correspondence: ravindhranath.kunta@gmail.com
[2]Department of Chemistry, K L University, Vaddeswaram, 522 502 Guntur Dt., AP, India
Full list of author information is available at the end of the article

Introduction

Fluoride in drinking water has both beneficial and harmful effects on human health (Maheshwari and Meenakshi 2006). Though fluoride in minute amounts is an essential component for bones and for the formation of dental enamel in animals and humans (Grynpas et al. 2000; Jackson et al. 1973; Fawell et al. 2006; Kumar and Moss 2008; Underwood 1997), its high concentration cause irreversible demineralization of bones and tooth tissues which is known as dental and skeletal fluorosis, damage to the brain, liver, and kidney, headache, skin rashes, bone cancer, and even death in extreme cases (Susheela 2001; Barbier et al. 2010; Gazzano et al. 2010; Ayoob and Gupta 2006; Chaturvedi et al. 1990; Wang and Reardon 2001; Lounici et al. 1997; Srimurali et al. 1998; Savinelli and Black 1958; Ganvir and Das 2011; Chinoy 1991). The maximum permissible limit of fluoride in water is 1.5 mg/l as per WHO standards (WHO 1984, 2004; BIS 1991).

A variety of treatment procedures have been reported for the removal of excess fluoride from polluted waters based on precipitation (Aldaco et al. 2007; Akbar et al. 2008; Reardon and Wang 2000; Cengeloglu et al. 2002; Yadav et al. 2006; Nawlakhe et al. 1975; Saha 1993), ion exchange (Meenakshi and Viswanathan 2007; Castel et al. 2000; Feng Shen et al. 2003; Chubar et al. 2005; Apambire et al. 1997; Singh et al. 1999), reverse osmosis (Sehn 2008; Simons 1993), Donnan dialysis (Tor 2007; Garmes et al. 2002; Hichour et al. 1999), electrodialysis (Lahnid et al. 2008; Menkouchi et al. 2007; Hichour et al. 2000; Adikari et al. 1989; Amer et al. 2001), nanofiltration (Liu et al. 2007), membrane-based methods (Dieye et al. 1998; Mjengera and Mkongo 2003; Lhassani et al. 2001; Mameri et al. 2001), electrocoagulation (Hu et al. 2003), and adsorption on to various adsorbents (Shihabudheen et al. 2006; Onyango et al. 2006; Tripathy et al. 2006; Mohapatra et al. 2004; Raichur and Jyoti Basu 2001). The choice of the method depends on conditions like area, concentration, availability of resources, etc. Among all these techniques, adsorption methods have more advantages because of their greater accessibility, economical, ease of operation, and effectiveness in removing fluoride from water to the maximum extent (Chauhan et al. 2007; Ayoob and Gupta 2008; Venkata Mohan et al. 2007).

Activated carbons prepared from various raw materials exhibit good fluoride uptake capacity (McKee and Jhonston 1934). Activated coconut shell carbon (Arulanantham et al. 1989), zirconium-impregnated ground nut shell carbon (Alagumuthu and Rajan 2010a), zirconium-impregnated coconut shell carbon (Sai sathish et al. 2007), *Dolichos lablab* carbon (Rao et al. 2009), zirconium-impregnated cashew nut shell carbon (Alagumuthu and Rajan 2010b), *Phyllanthus emblica*-activated carbon (Alagumuthu and Veeraputhiran 2011), *Acacia farnesiana* carbon (Kishore and Hanumantharao 2011), *Moringa indica*-activated carbon (Karthikeyan and Siva Elango 2007), *Cynodon dactylon*-activated carbon (Alagumuthu et al. 2011), *Typha angustata*-activated carbon (Hanumantharao et al. 2012a), Pine wood and Pine bark chars (Mohan et al. 2012), and activated carbon derived from steam pyrolysis of rice straw (Daifullah et al. 2007) have been explored for their adsorption nature towards fluoride. These techniques suffer from one or the other drawbacks, and a universally accepted, simple, eco-friendly and economical methods are still alluding researchers.

Hence, in this work, we searched for active carbons derived from plant materials belonging to different classes of plant kingdom. Our primary investigations indicated that there is strong affinity between fluoride and active carbons derived from barks of *Vitex negundo* plant. So, this work is devoted to study in depth the sorption characteristics of the said active carbon towards fluoride with respect to various physicochemical parameters such as adsorbent dosage, particle size, temperature, contact time, effect of pH, presence of foreign ions, and initial fluoride concentration. Further, the mechanism of sorption is probed in order to establish theoretical grounds for the observed sorption phenomenon by making morphological studies adopting such techniques like Fourier transform infrared spectroscopy (FTIR), scanning electron microscopy (SEM), and energy-dispersive X-ray spectroscopy (EDX). Kinetics of the adsorption process has been also investigated. Further, the adoptability of the methodologies developed in this work has been tested with diverse groundwater samples collected from different places in Guntur District of Andhra Pradesh.

Methods

Plant description

Of the various classes of plant materials tested for their sorption abilities towards fluoride, it has been noted that the active carbon derived from the barks of *V. negundo* plant shows affinity for fluoride.

The *V. negundo* plant (Figure 1), known as five-leaved chaste tree, or Vavili or Nalla-vavili belongs to the Lamiaceae family in the plant kingdom. It is a small tree growing from 2 to 8 m in height with reddish-brown barks and is commonly found near bodies of water, grasslands, and mixed open forests. All parts of the plant like leaves, roots, bark, fruits, flowers, and seeds can be used medicinally in the form of powder, decoction, juice, oil, tincture, sugar/water/honey paste, dry extract.

Preparation of active carbon

The barks of *V. negundo* plant were collected in bulk, crushed into small pieces, washed with fluoride-free water, and completely dried under sunlight for 2 days. The dried material was carbonized in muffle furnace in the absence of air at 500°C for about 4 h. After carbonization, the

Figure 1 *Vitex negundo* plant.

carbon was washed with fluoride-free water; the process was repeated for several times and then filtered. After that, the carbon was dried in air oven at 110°C, and it was sieved into desired particle sizes and then the carbon was subjected to liquid phase oxidation by mixing with 0.1 N HNO_3 and boiled for 2 to 3 h on flame. Then, the carbons were washed with double distilled water to remove the excess acid and dried at 150°C for 12 h. This activated carbon prepared from *V. negundo* was named as NVNC.

Reagents and chemicals

All the chemicals used were of analytical reagent grade purchased from Merck. India Pvt. Ltd. (Bengaluru, India) and Sd. Fine Chemicals (Mumbai, Maharashtra, India), and all solutions were prepared by using double distilled water throughout this study. Stock solution of fluoride (100 ppm) was prepared and was suitably diluted as per the need. SPADNS solution, Zirconyl acid reagent, Acid Zirconyl-SPADNS reagent, and reference solution were prepared as per the literature.

Experimental procedure

Batch mode adsorption studies were adopted. Test solution of 5 mg/l of fluoride was prepared by diluting appropriate quantity of standard fluoride (100 mg/l) solution with double distilled water because the maximum concentration of fluoride reported in groundwater of most of the fluoride affected areas is around 5 mg/l. From this 5 mg/l of fluoride simulated solution, 100 ml of solution was pipetted out into a 250-ml conical flask at room temperature 30°C ± 1°C, and to it, weighed quantity of the prepared active carbon adsorbent was added. Then, the conical flask along with test solution and adsorbent was shaken in horizontal shaker at 120 rpm. At the end of the desired contact time, the conical flask was removed from shaker and allowed to stand for 2 min for settling the adsorbent, and the adsorbent was filtered using Whatman No.42 filter paper. The filtrate was analyzed for residual fluoride concentration by SPADNS method using UV-visible spectrophometer (Model No: Elico U.V-2600, ELICO, Hyderabad, India) as described in Standard Methods of Water and Waste Water Analysis (APHA (American Public Health Association) 1985) at λ_{max} 570 nm.

Fluoride ion analysis

The percentage removal of fluoride ion and amount adsorbed (in mg/g) were calculated using the following equations:

$$\% \text{ Removal } (\%R) = \frac{(C_i\text{-}C_e)}{C_i} \times 100$$

$$\text{Amount adsorbed } (q_e) = \frac{(C_i\text{-}C_e)}{m} V$$

where C_i = initial concentration of the fluoride solution in mg/l, C_e = equilibrium concentration of the fluoride solution in mg/l, m = mass of the adsorbent in grams, and V = volume of the fluoride test solution in liters.

The same procedure has been adopted for the experiments carried out by varying the physicochemical parameters such as adsorbent dosage, pH of the fluoride solution, initial concentration of the standard fluoride solution, particle size, temperature, presence of foreign ions, and agitation time. Further, the same method was adopted in monitoring the concentration of fluoride in the characterization and adsorption studies.

Methodologies adopted for the adsorbent characterization

Physicochemical parameters

Various properties of the bio-sorbent, NVNC, were explored using standard methods, and the most important features of the results obtained were presented in Table 1. The pH for the activated carbon adsorbent was determined using the Elico pH meter, model LI-120, and the pH_{ZPC} was determined using the pH equilibrium method (Kadirvelu et al. 2000; Marsh and Rodriguez-Reinoso 2006; Newcombe et al. 1993). Particle size was determined using American Standard Test Method (ASTM) sieves (El-Hendawy et al. 2001). Iodine number (ASTM D4607-94 2006; Hill and Marsh 1968), decolorizing power (Girgis and El-Hendawy 2002; Rozada et al. 2005), and other parameters such as apparent density, moisture, loss on ignition, ash, water soluble matter, and acid soluble matter were analyzed by using standard test methods (Namasivayam and Kadirvelu 1997; BIS (Bureau of Indian Standards) 1989).

Boehm titration

The surface functional groups of oxygen were determined according to Boehm titration (Meldrum and Rochester 1990a; Boehm 1994; Bandosz et al. 1992). One gram of carbon sample was placed in 50 ml of the solution containing 0.05 N of sodium hydroxide, 0.05 N of sodium carbonate, and 0.05 N of sodium bicarbonate. The bottles were sealed and shaken for 24 h, and the mixture was filtered consequently. The excess base was titrated with 0.05 N HCl solution. The number of acidic sites was determined under the assumptions that NaOH

Table 1 Physicochemical properties of NVNC

Serial No.	Parameter		Value
1	Apparent density (g/ml)		0.288
2	Moisture content (%)		6.98
3	Loss on Ignition (LOI) (%)		90.26
4	Ash content (%)		4.89
5	Water soluble matter (%)		0.67
6	Acid soluble matter (%)		0.81
7	Decolorizing power (mg/g)		373
8	pH		7.14
9	pH_{ZPC}		9.35
10	Iodine number (mg/g)		684
11	Particle size (μ)		45
12	BET analysis - surface area (m²/g)	Before	262.6
		After	194.8
13	Surface functional groups (meq/g) - Boehm titration		
I	Carboxyl		1.247
II	Lactonic		1.024
III	Phenolic		1.330
IV	Carbonyl		1.242
V	Total basic groups		6.876

neutralizes carboxylic, lactonic, and phenolic groups; that Na_2CO_3 neutralizes carboxylic and lactonic groups; and that $NaHCO_3$ neutralizes only carboxylic groups. The number of basic sites was calculated from the amount of hydrochloric acid reacted with the carbon.

BET surface area

The Brunauer-Emmett-Teller (BET) surface area was determined using computer-controlled nitrogen gas adsorption analyzer at 77 K by Quantachrome Nova-Win - Data Acquisition and Reduction for NOVA instruments version 10.01 (Quantachrome Instruments, Boynton Beach, FL, USA). The BET-N_2 surface area was obtained by applying the BET equation to the adsorption data (Kadirvelu et al. 2000; Hashim 1994; Brunauer et al. 1938).

FTIR analysis

The surface functional groups on activated carbon adsorbent were examined using FTIR. The spectra were measured from 4,000 to 500 cm^{-1} on a BRUKER VERTEX 80/80v FTIR spectrometer (Bruker AXS, Inc., Madison, WI, USA), optical resolution of <0.06 cm^{-1}, with automatic and vacuum compatible beam splitter changer (BMS-c) option. Anhydrous KBr was used as a pellet material.

SEM

The SEM is one of the most versatile instruments available for the examination and analysis of the microstructure morphology and chemical composition characterizations. The microphotographs of these carbons were recorded using LEO 1420 VP compact variable pressure digital SEM, manufactured by Leo Electron Microscopy Ltd. (Cambridge, UK; beam voltage 500 to 2,000 V, magnifications ×250 to ×65,000, resolution 3 nm at 1,000 V).

EDS

EDS, EDX, or XEDS is an analytical technique used for the elemental analysis to identify what those particular elements are and their relative proportions (atomic %) of a sample. Each element has a unique atomic structure allowing unique set of peaks on its X-ray spectrum and was recorded using BRUKER EDX two-dimensional VANTEC-500 detector.

Results and discussions

The effect of various parameters such as adsorbent dosage, pH, contact time, initial fluoride ion concentration, particle size, and presence of foreign ions on the adsorption of the fluoride ion from aqueous solution by activated carbon adsorbent, NVNC, have been represented by plots as shown in Figure 2.

Effect of adsorbent dosage

The percentage removal of the fluoride ion by adsorption onto activated carbon adsorbent was studied by

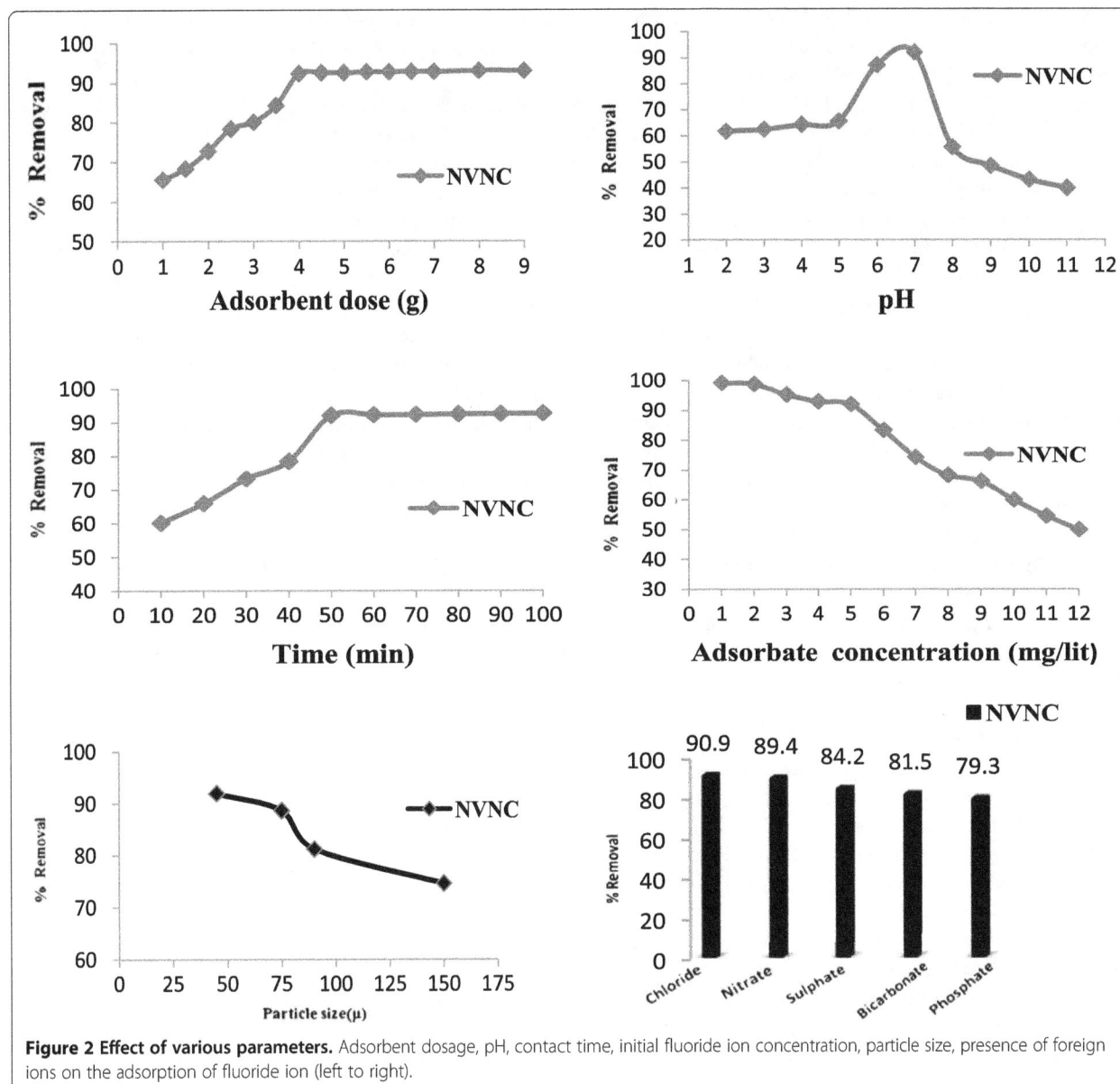

Figure 2 Effect of various parameters. Adsorbent dosage, pH, contact time, initial fluoride ion concentration, particle size, presence of foreign ions on the adsorption of fluoride ion (left to right).

varying the dosage of the activated carbon adsorbent, NVNC, in the range from 1.0 to 9.0 g/l, with a solution of 5 mg/l of fluoride ion concentration, at agitation time of 50 min, at a fixed pH = 7 and at a temperature of $30°C \pm 1°C$. The results obtained were plotted as percentage removal of the fluoride ion versus adsorbent dosage as shown in Figure 2.

As can be inferred from the graphs that the percentage removal of the fluoride ion increases rapidly with increase in adsorbent dosage and after optimum dosage of 4.0 g/l, there has been no significant change in the percentage removal of the fluoride ion.

Effect of pH

The effect of pH of the fluoride ion solution on the removal of the fluoride ion from aqueous solution was studied by varying the pH from 2 to 11 under optimum parameters, i.e., with a solution of 5 mg/l of fluoride ion concentration, contact times of 50 min, dose of 4.0 g/l for NVNC, and at a temperature of $30°C \pm 1°C$. The results obtained were plotted as percentage removal of the fluoride ion versus pH as shown in Figure 2.

As pH is less than pH_{ZPC} (Table 1), the net charge on surface of the solid activated carbon adsorbents is positive due to adsorption of excess H^+, which favors adsorption due to columbic attraction. So, it is expected that acidic conditions favor the fluoride adsorption on to the surface of the solid activated carbon adsorbent. But according to fluoride speciation, neutral hydrofluoride which has less affinity towards active carbon adsorbents is predominant in the pH at less than 3 (Lagergren 1898), and so, less adsorption is observed at low pH conditions. At low pH values, the fluoride ion converts into neutral HF, and thereby anion exchanging nature is lost.

When pH increases from 2 to 11, the percentage removal of the fluoride ion increases up to 7, and then onwards, the percentage removal of the fluoride ion decreases. The optimum pH range has been found to be 6 to 7 and below, and above this range, the percentage removal of the fluoride ion is less (Figure 2).

Effect of agitation time

In adsorption system, contact time plays an important role, irrespective of the other experimental parameters affecting the adsorption kinetics. In order to study the effect of contact time on kinetics of adsorption of the fluoride ion, the adsorption experiments were conducted and the extent of removal of the fluoride ion was known by varying the contact time from 10 to 100 min, with a solution of 5 mg/l of fluoride ion concentration, at a fixed pH = 7 and at a temperature of $30°C \pm 1°C$. The results obtained were plotted as percentage removal of the fluoride ion versus contact time (min) as shown in Figure 2.

As contact time increases, initially, the percentage removal of the fluoride ion is increased rapidly and after a certain time, approached an almost constant value indicating an attainment of equilibrium condition at which the rate of adsorption of fluoride onto the surface of the sorbent is equal to the rate of desorption. The rate of removal of the fluoride ion with time is higher at initial stages because of the availability of more active sites on the surface of the adsorbent, and with increase in contact time, the availability of active sites on the surface of the adsorbent decreases and this result in the decrease of the fluoride ion removal rate by the adsorbent. The decreased removal rate indicates the possible monolayer of the fluoride ions on the outer surface and pores of the adsorbent leading to pore diffusion onto inner surface of adsorbent particles (Namasivayam and Kadirvelu 1994; Yadav et al. 2006). From the observed results, the optimum contact time of activated carbon adsorbent NVNC is 50 min.

Effect of initial concentration

The effect of initial concentration of the fluoride ion solution on the extent of removal of the fluoride ion from aqueous solution was studied by varying the initial concentration of the fluoride ion solution from 1 to 12 mg/l under constant parameters, i.e., at a fixed pH = 7, contact times of 50 min with a dose of 4.0 g/l, and at a temperature of $30°C \pm 1°C$. The results obtained were plotted as percentage removal of the fluoride ion versus initial concentration of the fluoride ion solution as shown in Figure 2.

With an increase in initial concentration of the fluoride ion solution, the percentage removal of the fluoride ion has been decreased due to insufficient number of active sites that are available on the activated carbon adsorbents to adsorb the fluoride ions from highly concentrated solution of the fluoride ions. At low concentrated solution of the fluoride ions, sufficient numbers of active sites are available on activated carbon adsorbents, and hence, most of the fluoride ions interact with the active sites on the activated carbon adsorbent, and thus, percentage removal of fluoride is more.

With an increase in the initial concentration of the fluoride ion solution from 1 to 12 mg/l, the percentage removal (%R) of the fluoride ion is decreased from 99.2% to 49.9%, and adsorption capacity (q_e) increased from 0.248 to 1.497 mg/g for NVNC.

Effect of particle size

The effect of particle size on fluoride removal by NVNC was explored to have a better understanding of the adsorption process with 45 to 150 μ mesh sized particles under constant parameters, i.e., with a solution of 5 mg/l of fluoride ion concentration, at a fixed pH = 7, contact time of 50 min with a dose of 4.0 g/l, and at a temperature of $30°C \pm 1°C$. The results obtained were plotted as percentage removal of the fluoride ion vs. particle size as shown in Figure 2.

With an increase in particle size of activated carbon adsorbent, the percentage removal of the fluoride ion has been decreased due to insufficient number of surface active sites that are available on activated carbon adsorbent to adsorb fluoride ions. The rate of fluoride adsorption on smaller particles was significantly greater than that on larger sized particles. This is attributed to the fact that the lesser the particle size, the more will be the surface area and the more will be the number of active sites available for adsorption processes for a given amount of adsorbent. The percentage removal (%R) of the fluoride ion increases with decrease in particle size of activated carbon adsorbent, and hence, 45 μ is fixed as the optimum size of the particles of activated carbon adsorbent.

Interfering ions

The effect of interfering ions on fluoride removal by NVNC was explored to have a better understanding of the adsorption process with 50 mg/l concentration of interfering ions such as chloride, nitrate, sulfate, bicarbonate, and phosphate under constant parameters, i.e., with a solution of 5 mg/l of fluoride ion concentration, at a fixed pH = 7, contact time of 50 min with a dose of 4.0 g/l, and at a temperature of $30°C \pm 1°C$. The results obtained were presented in Table 2 and plotted as shown in Figure 2.

The impact of interfering ions present in water on fluoride adsorption by the activated carbon adsorbent, NVNC, follows the order $PO_4^{3-} > HCO_3^- > SO_4^{2-} > NO_3^- > Cl^-$.

Previous research indicates that chloride and nitrate ions form outer sphere surface complexes while sulfate ions form both outer sphere and inner sphere surface complexes (Onyango et al. 2004). Hence, chloride ions as well as nitrate ions have less interference with fluoride removal while sulfate ions have some significant effect on fluoride removal efficiency. The most important factor affecting fluoride removal efficiency was the bicarbonate alkalinity of the water, and it reduces the positive charge on the active sites of the active carbon and thereby reduces the affinity of the active sites of the adsorbent for fluoride adsorption. This results in the decrease in the percentage removal of the fluoride ion from water.

Phosphate ion which is having high negative charge compared to other anions needs three close surface groups and adsorbed on adsorbents as inner sphere surface complex. This inner spherically adsorbing phosphate ion (Goldberg and Sposito 1984a, 1984b; Zhang and Spark 1990) can significantly interfere with the fluoride ion and hence decrease in the percentage removal of the fluoride ion from water. In the presence of interfering ions chloride, nitrate, sulfate, bicarbonate, and phosphate, the percentage removal of the fluoride ion decreases from 92% to 90.9%, 89.4%, 84.2%, 81.5%, and 79.3%, respectively.

Effect of temperature

The effect of solution temperature was studied by conducting the experiment at different temperatures 303, 313, and 323 K and at optimum conditions, i.e., with a solution of 5 mg/l of fluoride ion concentration, at a fixed pH = 7, contact time of 50 min, dosage of 4.0 g/l for NVNC, and results obtained were plotted as $\ln K_d$ vs. $1/T$ as shown in Figure 3. Thermodynamic parameters of the adsorption process such as change in free energy (ΔG) (kJ/mole), change in enthalpy (ΔH) (kJ/mole), and change in entropy (ΔS) (kJ/mole) were determined at different temperatures by using the equations (Alagumuthu and Rajan 2010a; Karthikeyan and Siva Elango 2007)

$$\Delta G = -RT \ln K_d$$

$$\ln K_d = \Delta S/R - \Delta H/RT$$

$$K_d = q_e/C_e \text{ and}$$

$$\Delta G = \Delta H - T\Delta S$$

where K_d is the distribution coefficient for the adsorption, q_e is the amount of fluoride ion adsorbed on the activated carbon adsorbent per liter of solution at equilibrium, C_e is the equilibrium concentration of fluoride ion solution, T is the absolute temperature in Kelvin, R is the gas constant. ΔG is the change in free energy, ΔH is the change in enthalpy, and ΔS is the change in entropy.

The values of ΔH and ΔS were obtained from the slope and intercept of a plot between $\ln K_d$ and $1/T$ and ΔG values were obtained from the equation $\Delta G = \Delta H - T\Delta S$ and tabulated (Horsfall and Spiff 2005; Viswanathan and Meenakshi 2010).

Table 2 Effect of interfering ions on fluoride ion removal from aqueous solution by the activated carbon: NVNC

Serial No.	Adsorbent	Maximum extractability at optimum conditions	Extractability of fluoride ion in the presence of 50 mg/l of interfering ions at optimum (pHs)				
			Cl^-	NO_3^-	$S O_4^{2-}$	HCO_3^-	PO_4^{3-}
1	NVNC	92.0%, pH:7, 50 min	90.9%, pH: 7, 50 min	89.4%, pH: 7, 50 min	84.2%, pH: 7, 50 min	81.5% pH: 7, 50 min	79.3%, pH: 7, 50 min

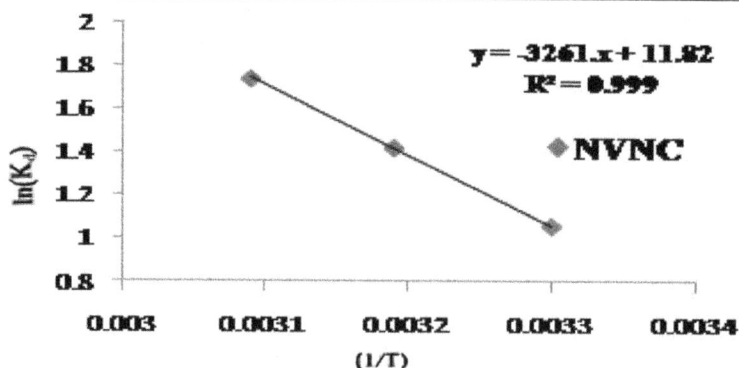

Figure 3 Effect of temperature on the adsorption of fluoride ion.

It is observed that with an increase in the temperature from 303 to 323 K (30°C to 50°C), the percentage removal of fluoride ion increases from 92% to 95.8% for NVNC.

As the temperature increases, the thickness of the outer surface of the activated carbon adsorbent decreases and the kinetic energy of the fluoride ion increases, and hence, the rate of diffusion of the fluoride ion increases across the external boundary layer and internal pores of the activated carbon adsorbent.

As can be inferred from the Table 3, the values of ΔH are positive, which indicates the physisorption and endothermic nature of adsorption (Bouberka et al. 2005). The R^2 values close to one also indicates that adsorption process is endothermic nature. The positive values of ΔS indicate the increased disorder and randomness at the solid solution interface of the fluoride ion with the adsorbent (Sairam Sundaram et al. 2009). The negative values of ΔG indicate the spontaneous nature of adsorption process.

Adsorption isotherms

The adsorption isotherms are one of the most significant methods for representing the adsorption capacity of the adsorbent and the mechanism of the adsorption system. The purpose of an adsorption isotherm is to evaluate the relation between the fluoride concentrations remaining in the bulk solution to the amount of fluoride adsorbed at the solid/solution interface. Four well-known models, Freundlich, Langmuir, Dubinin-Radushkevich, and Temkin adsorption isotherms have been selected for describing adsorption isotherms at a constant temperature for water and wastewater treatment application.

Linear form of Freundlich isotherm equation is as follows:

$$\log (q_e) = \log k_f + \left(\frac{1}{n} \right) \log C_e$$

Linear form of Langmuir isotherm equation is as follows:

$$(C_e/q_e) = (a_L/k_L)C_e + 1/k_L$$

The significant features of the Langmuir isotherm model can be defined by the dimensionless constant separation factor R_L which is expressed by the following equation:

$$R_L = 1/ (1 + a_L C_i)$$

where, k_F and $1/n$ are the Freundlich constants, C_i is the initial fluoride ion concentration, q_e (mg/g) is the amount of fluoride ions adsorbed per unit weight of the adsorbent (mg/g), and k_L and a_L are the Langmuir constants related to capacity and energy of adsorption, respectively.

In defluoridation of water using activated carbon adsorbents, linear plot of log (q_e) vs. log (C_e) at different fluoride ion concentrations were found to be linear as shown in Figure 4 and confirmed the applicability of Freundlich isotherm model. When C_e/q_e is plotted against C_e, a straight line was observed as shown in Figure 4. The slope, a_L/k_L, and intercept, $1/k_L$, of the straight line showed that the adsorption followed the Langmuir isotherm.

The Freundlich constants k_F and $1/n$ were calculated from the intercept and slope of the plots, respectively. The Langmuir constants a_L and k_L were calculated from the slope and intercept of the straight line, respectively. The observed linear relationships were statistically

Table 3 Thermodynamic parameters of fluoride ion adsorption on NVNC

Parameter	ΔH (kJ/mol)	ΔS (J/mol/K)	ΔG (kJ/mol)			R^2
Temperature (K)			303	313	323	0.999
NVNC	27.11	98.27	−2.6658	−3.6485	−4.6312	

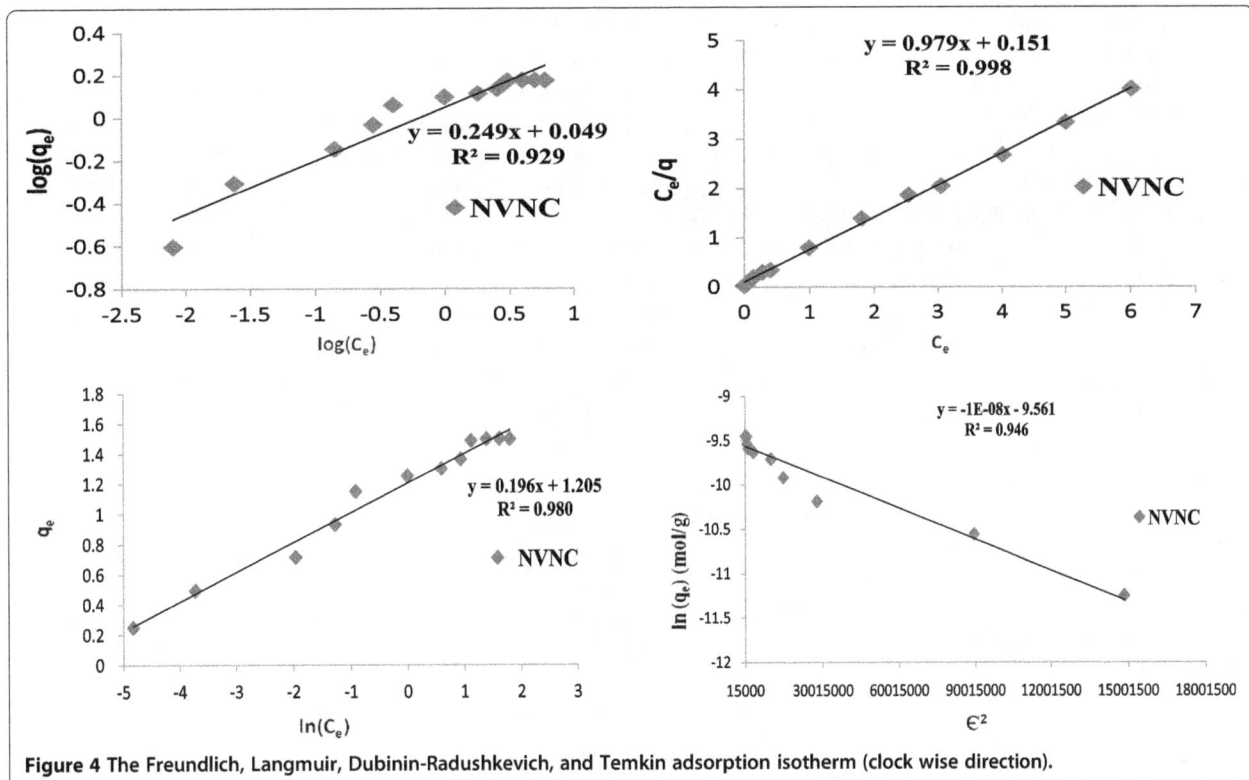

Figure 4 The Freundlich, Langmuir, Dubinin-Radushkevich, and Temkin adsorption isotherm (clock wise direction).

important as evidenced from the correlation coefficients (R^2 values) close to unity, which indicated that the applicability of these two adsorption isotherms confirmed the heterogeneous surface of the adsorbent and the monolayer coverage of the fluoride ion on the active carbon surface.

According to Hall et al. (1966), the separation factor R_L indicates the isotherm's shape and the nature of the adsorption process as unfavorable ($R_L > 1$), linear ($R_L = 1$), favorable ($0 < R_L < 1$), and irreversible ($R_L = 0$). The correlation coefficient (R^2) value was higher for the activated carbon adsorbent NVNC in the Langmuir isotherm than in the Freundlich isotherm, and the value of dimensionless separation factor (R_L), 0.0299, which was found to be fraction in the range of 0 to 1 indicates the favorability of the Langmuir isotherm than the Freundlich isotherm.

Linear form of Temkin isotherm equation is as follows:

$q_e = B \ln C_e + B \ln A$ where $RT/b = B$

Linear form of Dubinin-Radushkevich isotherm equation is as follows:

$\ln q_e = -B\varepsilon^2 + \ln Q_m$ where $\varepsilon = RT \ln(1 + 1/C_e)$ where B is the Temkin constant related to heat of sorption (J/mol), A is the Temkin isotherm constant (L/g), b is the Temkin isotherm constant, q_m (mol/g) is the Dubinin-Radushkevich monolayer adsorption capacity, β is a constant related to energy, $E = 1\sqrt{2\beta}$, ε is the polanyi potential, R is a gas constant (8.314 J/mol K), T is the absolute temperature, C_e is the equilibrium concentration of the fluoride solution in mg/l, and q_e is the amount of fluoride ions adsorbed per unit weight of the adsorbent (mg/g).

In defluoridation of water using activated carbon adsorbents, linear plot of (q_e) vs. ln (C_e) at different fluoride ion concentrations were found to be linear as shown in Figure 4 and confirmed the applicability of Temkin ($R^2 = 0.980$) isotherm model. When ln (q_e) (mol/g) is plotted against ε^2, a straight line with slope β and intercepts ln q_m showed that the adsorption followed the Dubinin-Radushkevich ($R^2 = 0.946$) isotherm as shown in Figure 4.

The Temkin constants A and B were calculated from the intercept and slope of the plots. The Dubinin-Radushkevich constant, β, can be calculated from the slope of the straight line. The observed linear relationships were statistically important as evidenced from the correlation coefficients (R^2 values) close to unity, which indicated the applicability of these two adsorption isotherms and confirmed the heterogeneous surface of the adsorbent. And the Dubinin-Radushkevich mean free energy, E, can be calculated from the relation $E = 1\sqrt{2\beta}$. The mean free energy, E, was found to be 7.07 kJ/mol for the activated carbon adsorbent NVNC. This indicates that the mechanism of adsorption is 'physisorption'. When the $E < 8$ kJ/mol, it is an indication of physisorption (Monika et al. 2009) (nonspecific adsorption) dominating the chemisorptions and ion exchange.

Physisorption is also called nonspecific adsorption which occurs as a result of long-range weak van der Waals forces between adsorbates and adsorbents. According to Atkins (Atkins 1999) also, if the free energy, E and Temkin heat of sorption (B) are less than 20 kJ/mol, the physisorption is predominant.

The values of the Freundlich and Langmuir Temkin and Dubinin-Radushkevich adsorption isotherm constants together with the correlation coefficients values were presented in Table 4.

Kinetic study of adsorption

In the present work, some kinetic models, namely, pseudo-first-order model, pseudo-second-order model, Weber and Morris intraparticle diffusion model, Bangham's pore diffusion model, and Elovich equations are discussed to study the rate and kinetics of adsorption of the fluoride ion onto the activated carbon adsorbent NVNC. The kinetics of adsorption describes the solute uptake rate, which in turn governs the residence time of adsorption reaction.

The pseudo-first-order model

On adsorption of the fluoride ion onto the activated carbon adsorbent, linear plot of log ($q_e - q_t$) vs. t at different contact times for the extent of removal of the fluoride ion from water is applied to confirm the applicability of pseudo-first-order model. The pseudo-first-order equation is log ($q_e - q_t$) = log $q_e - k_1 t/2.303$

It was found that the plot of log ($q_e - q_t$) vs. t should give a linear relationship from which values of the k_1 and q_e could be determined from the slope and intercept of the plot, respectively.

The pseudo-second-order model

On adsorption of fluoride ion on to the activated carbon adsorbent, linear plot of (t / q_t) vs. t at different contact times for the extent of removal of the fluoride ion from water is applied to confirm the applicability of pseudo-second-order model. The pseudo-first-order equation is $t/q_t = 1/k_2 q_e^2 - (1/q_e)t$.

The plot of (t / q_t) vs. t should give a linear relationship from which the values of the k_2 and q_e could be determined from the intercept and slope of the plot, respectively.

Weber and Morris intraparticle diffusion model

On adsorption of the fluoride ion onto the activated carbon adsorbent, linear plot of (q_t) vs. $t^{1/2}$ at different contact times for the extent of removal of fluoride ion from water is applied to confirm the applicability of Weber and Morris intraparticle diffusion model. The Weber and Morris intraparticle diffusion equation is as follows:

$$q_t = k_{ip} t^{1/2} + c$$

The plot of (q_t) vs. $t^{1/2}$ should give a linear relationship from which the value of the Weber and Morris intraparticle diffusion rate constant, k_{ip}, could be determined from the slope of the plot.

Bangham's pore diffusion model

On adsorption of the fluoride ion onto the activated carbon adsorbent, linear plot of log [log ($C_i/(C_i - q_t m)$)] vs. log(t) at different contact times for the extent of removal of the fluoride ion from water is applied to confirm the applicability of Bangham's pore diffusion model. Bangham's pore diffusion equation is as follows:

$$\log [\log(C_i/C_i - q_t m)] = \log (k_o/2.303\ V) + \alpha \log(t)$$

The linear plots of log [log ($C_i/(C_i - q_t m)$)] vs. log (t) should give a linear relationship from which the value of the Bangham's pore diffusion rate constant, k_o, and a constant, α, could be determined from the intercept and slope of the plot, respectively.

Elovich equation

On adsorption of the fluoride ion onto the activated carbon adsorbent, linear plot of (q_t) vs. ln(t) at different

Table 4 Adsorption and kinetic parameters

Serial No.	Adsorption isotherms and kinetic models		Slope	Intercept	R^2
1	Freundlich isotherm		0.249	0.490	0.929
2	Langmuir isotherm	$R_L = 0.0299$	0.979	0.151	0.998
3	Temkin isotherm	$B = 0.196$ J/mol	0.196	1.205	0.980
4	Dubinin-Radushkevich isotherm	$E = 7.07$ kJ/mol	−1E−08	−9.561	0.946
5	Pseudo-first-order model		−0.012	−0.258	0.989
6	Pseudo-second-order model		0.777	7.320	0.994
7	Bangham's pore diffusion model		0.538	−0.976	0.902
8	Elovich model		0.207	0.249	0.912
9	Weber and Morris intraparticle diffusion model		0.066	0.567	0.874

contact times for the extent of removal of the fluoride ion from water is applied to confirm the applicability of Elovich equation. The Elovich equation is as follows:

$$q_t = 1/\beta \ln(\alpha\beta) + 1/\beta \ln(t)$$

The linear plot of (q_t) vs. $\ln(t)$ should give a linear relationship from which the value of the constants α and β could be determined from the intercept and slope of the plot, respectively.

The plots of all these five kinetic models are as shown in Figure 5.

The values of pseudo-first-order, pseudo-second-order, Weber and Morris intraparticle diffusion, Bangham's pore diffusion, and Elovich model constants together with the correlation coefficients values were presented in Table 4.

The experimental data revealed that of the five kinetic models, namely, pseudo-first-order, pseudo-second-order, Weber and Morris intraparticle diffusion, Bangham's pore diffusion, and Elovich models when correlated with the linear forms, the correlation coefficient value of $R^2 = 0.994$ for the pseudo-second-order model is greater than the other kinetic models, and this indicates that the pseudo-second-order model is the best fit to the experimental data of the present studied adsorption system. The next to follow the order is the pseudo-first-order model with $R^2 = 0.989$, Elovich model with $R^2 = 0.912$, Bangham's pore diffusion model with $R^2 = 0.902$, and the least is Weber and Morris intraparticle diffusion model ($R^2 = 0.874$).

The correlation coefficient of $R^2 = 0.912$ value for the Elovich equation suggest that the diffusion of the

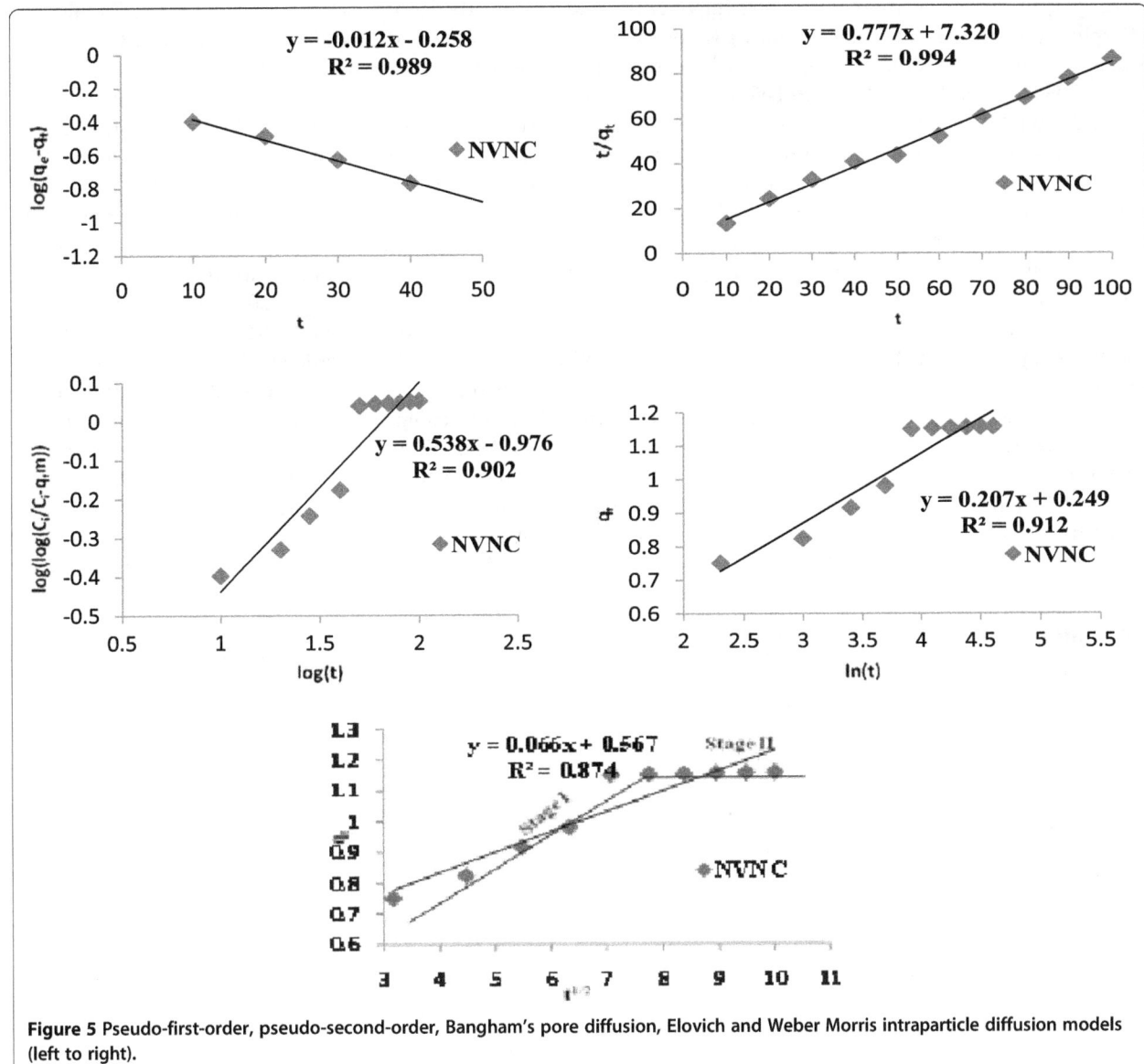

Figure 5 Pseudo-first-order, pseudo-second-order, Bangham's pore diffusion, Elovich and Weber Morris intraparticle diffusion models (left to right).

fluoride ion follows the Elovich kinetics pattern and the rate determining step is diffusion in nature, and this diffusional rate limiting is more prominent in fluoride ion adsorption onto the activated carbon adsorbent. The correlation coefficient of $R^2 = 0.902$ value indicates that the adsorption kinetics follows the Bangham's equation and the adsorption of fluoride ion onto the activated carbon adsorbent is Bangham's pore diffusion controlled.

The low correlation coefficient value of $R^2 = 0.874$ for the Weber and Morris intraparticle diffusion model indicates that the intraparticle diffusion model does not apply to the present studied adsorption system.

Characterization of the adsorbent NVNC

The different physicochemical properties of the active carbon, NVNC, were presented in Table 1. The surface characteristics make it fit for adsorption of the fluoride ion. Zero point charge (pH_{ZPC}) of the adsorbent is more important for adsorption processes. In the present study, adsorption of anion is favored because for the NVNC, the obtained result is $pH < pH_{ZPC}$. The BET surface area before defluoridation was $262.6\ m^2/g$, and the value reduces to $194.8\ m^2/g$ after defluoridation. The presence of surface functional groups of oxygen like phenol, carbonyl, hydroxyl, and lactones were determined according to Boehm titration, and from Table 1, it was clear that for activated carbon, the total basic groups were greater than the total acidic groups. The presence of these groups was confirmed by FTIR results.

FTIR spectroscopy of NVNC before and after defluoridation

It gives confirmation for the presence of specific functional groups on the surface of carbon materials. Several characteristic bands were observed in the FTIR spectrum of NVNC (vide Table 5 and Figure 6), and each of the bands can be assigned to specific functional group based on the previous assignments made in literature.

Even though a cluster of functional groups were present on the carbon surface, the prominent among them was a sharp and intense band centered around $1,691.51\ cm^{-1}$ which was attributed to the carbonyl (-C=O) stretching vibration of quinine or quinone or conjugated ketone (Ji et al. 2007; Biniak et al. 1997; Yu et al. 2008; Ishizaki and Marti 1981; Shin et al. 1997; Moreno-Castilla et al. 1998; Starsinic et al. 1983; Zawadzki 1989). The carbon surface is oxidized by treatment with concentrated HNO_3 leading to the generation of such quinone type carbon functional groups which bear significance in the redox chemistry of carbon materials. Carbonyl functional groups are known to be pronounced in oxidized carbon materials rather than the original parent carbon material (Budinova et al. 2006). In addition, stretching vibration bands of surface and hydrogen bonded -O-H group of alcohols, phenols, and chemisorbed water (Daifullah et al. 2003; Ibrahim et al. 1980; Yang and Lua 2003; Puziy et al. 2003), the peaks pertaining to asymmetric -C-H stretching vibration of aliphatic $-CH_3$ or $-CH_2$ groups (Biniak et al. 1997; Yu et al. 2008; Puziy et al. 2003; Rajeshwari et al. 2001); peaks due to the in plane bending vibration of -C-H of methylene group (Budinova et al. 2006; Rajeshwari et al. 2001; Ozgul et al. 2007); peaks due to -C-O stretching in alcohols, phenols, ethers, esters, acids, epoxides, lactones, and carboxylic anhydrides (Shin et al. 1997; Budinova et al. 2006; Rajeshwari et al. 2001; Gomez-Serrano et al. 1994; Figueiredo et al. 1999; El-Hendawy 2003; Park et al. 1997; Attia et al. 2006; Lapuente et al. 1998); the peaks pertaining to the -C=O stretching in carbonyl and carboxyl groups and in lactones (Zawadzki 1989; Nageswara Rao et al. 2011; Fanning and Vannice 1993; Painter et al. 1985; Zhuang et al. 1994); peaks due to the out-of-plane deformation vibrations of -C-H group in aromatic structures (Nageswara Rao et al. 2011; Meldrum and

Table 5 Bands assigned to the surface functional groups of NVNC before and after

| Serial No. | Wave number (cm^{-1}) | | Bond stretching |
	NVNC (before)	NVNC (after)	
1	3855.00, 3743.37, 3676.37, 3617.38 3563.84, 3386.65, 3289.23	3843.03, 3745.86, 3679.14, 3614.13, 3562.86, 3334.89, 3227.25	-O-H in alcohols, acids, phenols, and -N-H in amines and amides
2	2932.15, 2876.53, 2822.24, 2786.84	2957.30, 2866.67, 2828.66, 2747.80	-C-H in $-CH_3$ and $-CH_2$
3	1866.07, 1832.92, 1743.93	1868.79, 1832.60, 1743. 59	-C = O in carbonyl, carboxyl groups, and lactones
4	1691.47	1691.51	-C = O in quinine or quinone
5	1644.47, 1517.28, 1425.74, 1393.18	1644.14, 1516.82, 1426.50, 1391.95	-C = C- in aromatic rings, -C = O in highly conjugated carbonyl groups, and -C-H deformations in Alkanes
6	1339.81, 1187.65, 1146.36, 1078.53	1360.57, 1223.45, 1079.16, 1040.23	-C-O- in alcohols, phenols, ethers, esters, acids, epoxides, lactones, and carboxylic anhydrides
7	875.55, 830.19, 744.37, 695.57	871.09, 831.03, 758.74, 695.74	-C-C- deformations and out of plane -C-H deformations in aromatic rings

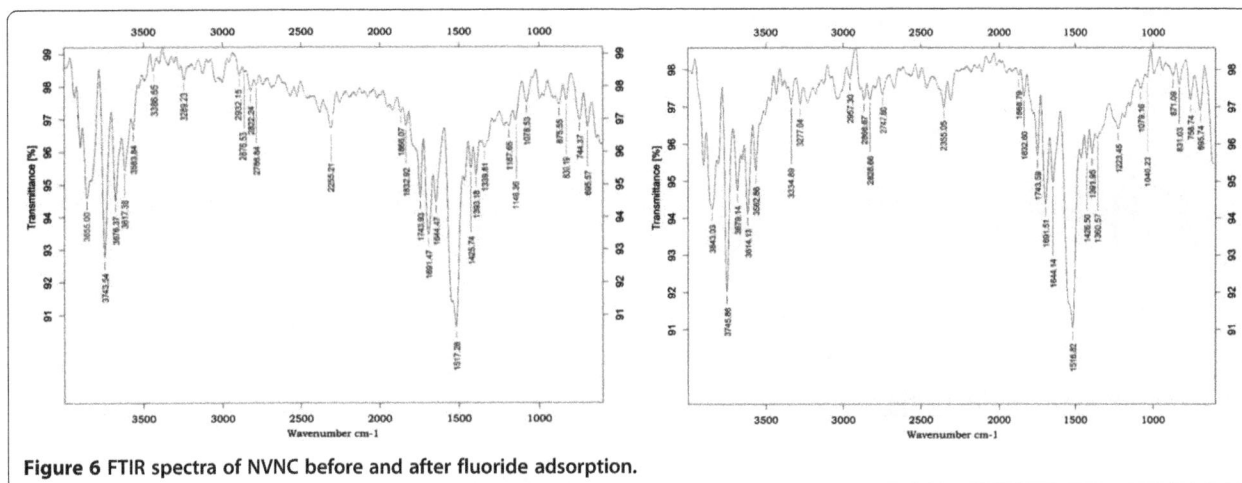

Figure 6 FTIR spectra of NVNC before and after fluoride adsorption.

Rochester 1990b, 1990c) have been noted. The absence of specific peak pertains to -C-F, suggesting that the adsorption process was physisorption but not chemisorptions. The shifts or decrease in the percentage of transmittance in FTIR spectra before and after deflouidation in the range 4,000 to 500 cm^{-1} (Figure 6 and

Table 5) indicates that the sorption of fluoride on the surface of the adsorbent is through physisorption.

SEM analysis of NVNC before and after defluoridation
SEM micrographs were studied for surface and morphological characteristics of activated carbon material

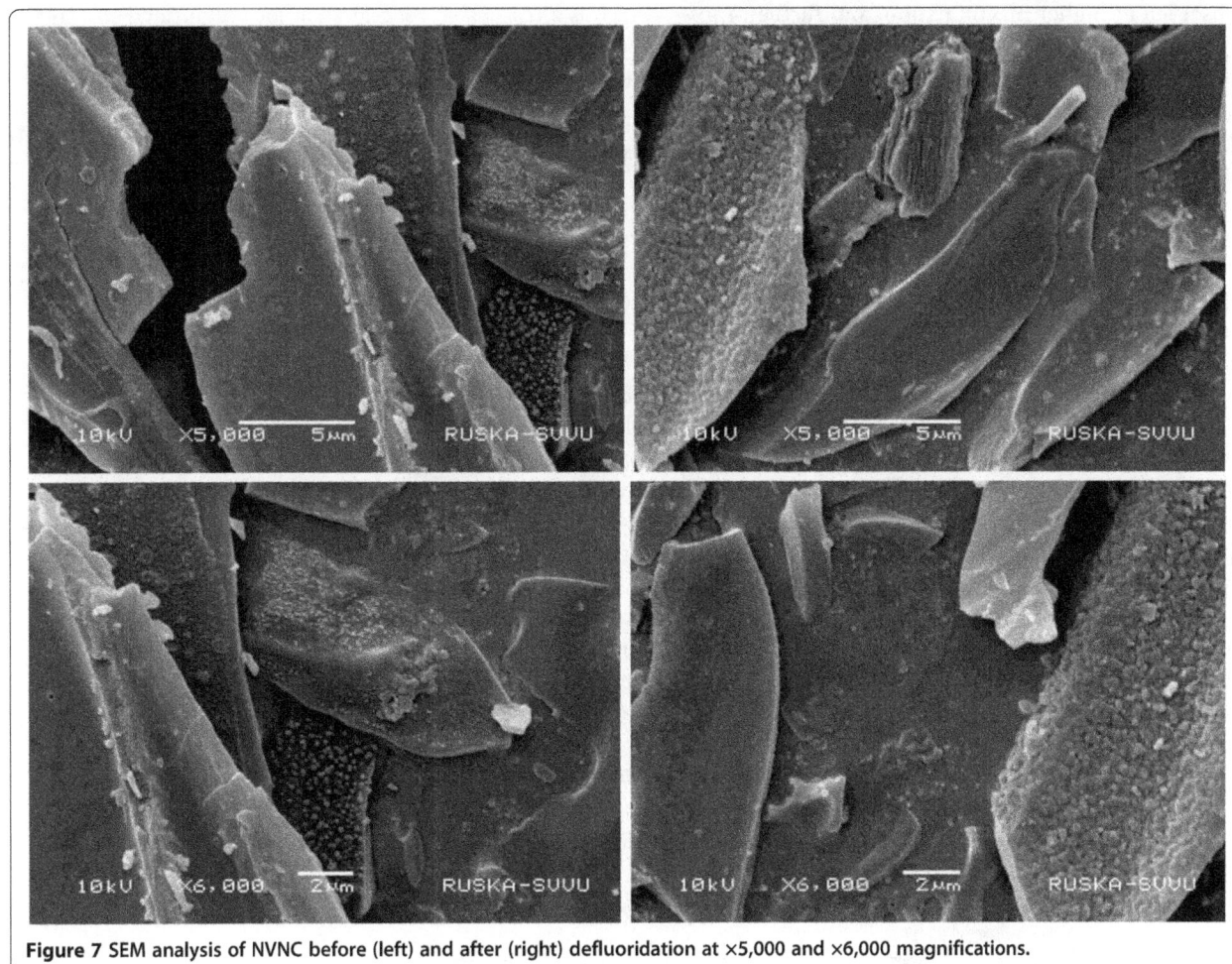

Figure 7 SEM analysis of NVNC before (left) and after (right) defluoridation at ×5,000 and ×6,000 magnifications.

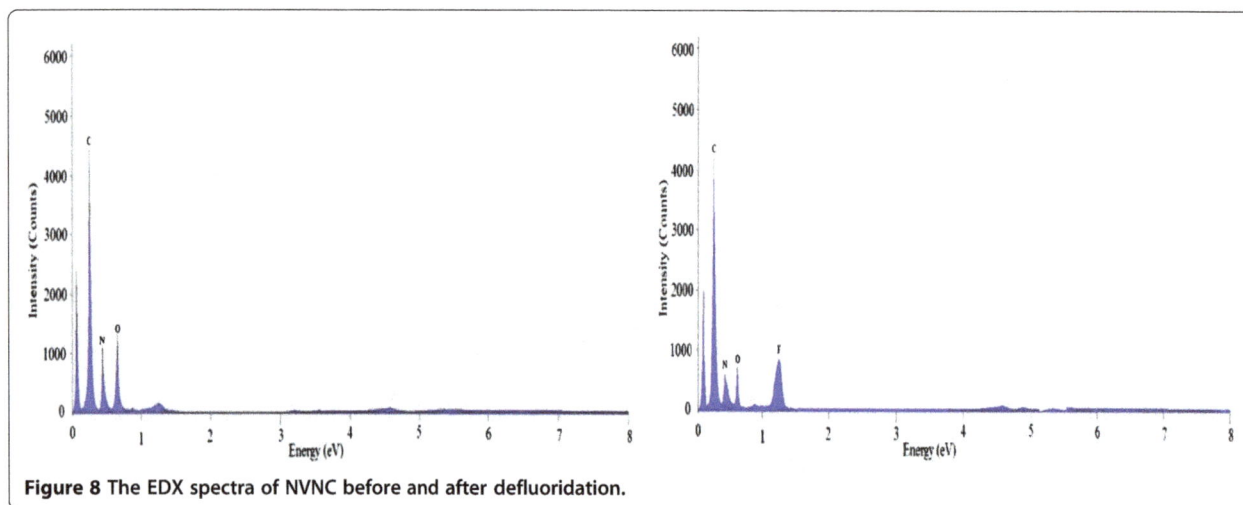

Figure 8 The EDX spectra of NVNC before and after defluoridation.

(vide Figure 7). The examination of the SEM micrographs of the activated carbon material showed that in micrographs, dark areas indicated pores and grey areas indicated the carbon matrix and showed rough surface of the adsorbent that provided large surface area for adsorption. Overall, a well-developed porous surface was observed at higher magnification, and further, randomly distributed pore size was observed in all micrographs. The grey surface area of every micrograph contained smaller microparticles (nm to μm) which might be due to the activated sites or surface functional groups of the carbon. The adsorption of fluoride might be due to the presence of pores and active groups on the surface. SEM micrographs proved the fluoride adsorption.

EDX elemental analysis

The EDX elemental analysis highlighted the presence of carbon and oxygen ions in untreated and fluoride treated sample (Figure 8 and Table 6). Nitrogen is also present but in less mass amount compared with carbon and oxygen. The intensity of the fluoride signals was higher in samples treated with fluoride and was below the limit of detection on the untreated samples. It can clearly conclude that the oxygen element concentration has been reduced after adsorption process, but simultaneously, fluoride concentration is increased in the adsorbent and hence the fluoride ion may replace the ion-containing oxygen atom (OH⁻) on the surface of adsorbent.

Comparison with other carbon adsorbents

The efficiency of the adsorbent developed in this work, viz., NVNC has been compared with the already existing adsorbents reported in the literature based on the adsorbent mass needed to reduce the fluoride concentration from 4 mg/l (average of fluoride concentration in potable water) to 1 mg/l which is a reasonable concentration for health standards. In this concern, the amount of adsorbent has been estimated using a mass balance equation (Ramos et al. 1999):

$$VC_i = mq + VC_e$$

because q and C_e must be in equilibrium, q for adsorbents can be calculated using Langmuir equation, and C_i, C_e, and V are 4, 1 mg/l, and 1 m³ (1,000 liters), respectively. The results and also the values pertaining to the other active carbons reported in literature have been presented in the Table 7.

From the table, it may be noted that the active carbon developed in this work, namely, NVNC is more effective than commercially available F-4000 carbon and also than many other active carbons reported in literature such as *A. farnesiana* fruits carbon, *T. angustata* plant carbon, *Lagenaria siceraria* shell carbon, and aluminum-impregnated carbon. Though KMnO₄-modified carbon prepared from rice straw, aluminum-supported carbon nanotubes, and carbons loaded with specific chemical moieties pecan nut shells (CMPNS) seem to be scoring better than NVNC, they have their inherent disadvantages. In the case of

Table 6 Elemental analysis of NVNC by EDX spectra before and after defluoridation

Composition	Elemental analysis (EDX) (before)					Elemental analysis (EDX) (after)				
	CK	NK	OK	FK	Total	CK	NK	OK	FK	Total
Energy (eV)	0.3	0.5	0.6	1.1		0.3	0.5	0.6	1.1	
Wt.% (mass ratios)	84.3	3.0	12.7	-	100	84.1	1.1	10.1	4.7	100
At.% (atomic percentages)	84.6	2.9	12.5	-	100	85.1	0.9	9.8	4.2	100

Table 7 Comparison study with other carbon adsorbents

Serial No.	Adsorbent	pH	Ci (mg/l)	q_e (mg/g)	V/m (l/g)	m (kg)	Reference
1	KMnO4 modified carbon prepared from rice straw	2.0	5 to 20	3.40	1.13	0.89	Daifullah et al. (2007)
2	Aluminum impregnated carbon	4.0	0.5 to 15	1.006	0.34	2.98	Ramos et al. (1999)
3	Alumina supported carbon nanotubes	6.0	1 to 50	9.68	3.20	0.31	Li et al. (2001)
4	*Acacia farnesiana* fruits Carbon	7.0	1.5 to 15	0.333	0.121	4.50	Hanumantharao et al. (2012b)
5	CMPNS (carbons loaded with specific chemical moieties pecan nut shells)	7.0	5 to 40	2.3	0.7667	1.30	Montoya et al. (2012)
6	Commercially available F-400 carbon (produced from coconut shells by Calgon Corporation)	6.2-7.5	0.5 to 15	0.0741	0.025	40.50	Ramos et al. (1999)
7	*Typha angustata* plant carbon	7.0	1.5 to 15	0.429	0.121	3.50	Hanumantharao et al. (2012a)
8	*Lagenaria siceraria* shell carbon	7.0	1.5 to 15	0.375	0.122	4.00	Hanumantharao et al. (2012c)
9	NVNC	7.0	1 to 12	1.150	0.3833	2.61	Present work

$KMnO_4$-modified carbon from rice straw, the optimum pH condition is 2 which is a limiting factor for its application to defluoridation of drinking water. In view of the availability of plenty of raw material, the ease of deriving the active carbons and, further, the simplicity of the technique make the NVNC better than the rest two.

Further, it may be noted that the values of various physicochemical properties of NVNC have been found

Table 8 Fluoride ion concentration (before and after defluoridation) of ground water samples

Sample number	Village name	C_i (mg/l) (before defluoridation)	C_f (mg/l) (after defluoridation) With NVNC	% Removal
1	Sivapuram	3.75	0.454	87.9
2	Koppukonda	3.82	0.554	85.5
3	Thimmayapalem	3.56	0.495	86.1
4	Narasayapalem	3.48	0.432	87.6
5	Brahmanapalli	4.27	0.743	82.6
6	Mada manchipadu	3.88	0.442	88.6
7	Andugulapadu	3.69	0.432	88.3
8	Tsouta palem	4.09	0.483	88.2
9	Venkupalem	3.37	0.377	88.8
10	Nagulavaram	3.28	0.538	83.6
11	Peda kancherla	3.68	0.545	85.2
12	Narasarayani palem	3.52	0.538	84.7
13	Dondapadu	3.95	0.545	86.2
14	Vinukonda	3.62	0.391	89.2
15	Gokana konda	4.21	0.531	87.4
16	Enugupalem	3.69	0.523	85.7
17	Surepalli	3.45	0.521	84.9
18	Ummadivaram	3.59	0.582	83.8
19	Perumalla palli	4.12	0.655	84.1
20	Nayanipalem	3.68	0.592	83.9
21	Settupalli	3.72	0.513	86.2
22	Vithamrajupalli	3.43	0.579	83.1
23	Neelagangavaram	3.49	0.520	85.1

to be comparable with the other activated carbons, i.e., the moisture content value with the activated carbons of *Euphorbia antiquorum* L. (Palanisamy and Sivakumar 2008), loss on ignition value with the activated Kaza's carbons (Sreenivasa Rao et al. 2011), ash content value with the activated carbon of bituminous coal (Cuhadaroglu and Aydemir Uygun 2008), pH value with the activated carbons of *E. Antiquorum* L. (Palanisamy and Sivakumar 2008) and coconut shell (Gimba and Turoti 2006), and the basicity values with the activated carbon prepared from rice straw (Daifullah et al. 2007).

Moreover, the calculated BET surface area value for the adsorbent, NVNC, has been found to be 262.6 m^2/g which is more than some other active carbons such as *L. siceraria* shell carbon (198.5 m^2/g) (Hanumantharao et al. 2012c), peanut hull (208 m^2/g) (Periasamy and Namasivayam 1994), and cassava peel (200 m^2/g) (Rajeshwari et al. 2001), and hence, NVNC shows more adsorption capacity than the said active carbons.

It may be inferred from this discussion that the prepared carbon, NVNC, exhibits considerably greater fluoride adsorption potential when compared with some of the less attractive and low-cost materials.

Applications

The adoptability of the methodology developed with the new bio-sorbent in this work for removing fluoride has been tried with some real water samples collected from groundwaters in fluoride-affected areas in Vinukonda Mandal of Guntur District of Andhra Pradesh.

The samples were subjected to extraction for fluoride using the said bio-sorbents developed in this work at optimum conditions of pH, equilibration time, and sorbent concentration. The results obtained were presented in Table 8.

From Table 8, it is evident that the concentration of fluoride in all groundwater samples collected from various villages of Vinukonda Mandal, Guntur District have been varied from 3.37 to 4.27 mg/l. The data indicates that most of the samples contain excess of fluoride beyond the permissible World Health Organization limit 1.5 mg/l (WHO 1984, 2004; BIS 1991). Hence, in the present work, the defluoridation studies were carried out on these particular samples using the adsorbent, NVNC, in order to reduce the fluoride content below the permissible limit. The concentrations of fluoride ion in these samples after defluoridation were analyzed and reported in Table 8.

It may be inferred that NVNC effectively decrease the fluoride content in groundwater samples to below permissible limits under optimum experimental conditions.

It can be inferred from Table 8 that the methodology developed in this research work using active carbon NVNC is remarkably successful.

Conclusions

A new activated carbon adsorbent has been developed for fluoride removal from aqueous solution in this study. The results indicated that the maximum fluoride adsorption takes place at the optimum pH of 7.0, the adsorbent dosage of 4.0 g/l, equilibration time of 50 min, and at a temperature of 30°C ± 1°C.

It was observed that the adsorption process satisfactorily fitted with Langmuir adsorption isotherm which had good correlation coefficient value indicating monolayer adsorption and also confirmed the heterogeneous surface of the adsorbent due to Freundlich isotherm. The Temkin and Dubinin-Radushkevich isotherms also established the linear relationship which indicated the applicability of these two adsorption isotherms and confirmed the heterogeneous surface of the adsorbent. The Dubinin-Radushkevich mean free energy, *E*, was found to be 7.07 kJ/mol for the activated carbon adsorbent NVNC. An energy value of <8 $kJmol^{-1}$ was an indication of physisorption. Hence, adsorption is not restricted to monolayer coverage as purposed for chemisorption.

In kinetic studies of defluoridation, pseudo-first-order, pseudo-second-order, Weber and Morris intraparticle diffusion model, Bangham's pore diffusion model, and Elovich equations were applied to identify the rate and kinetics of adsorption process. The adsorption process had good correlation coefficient values with pseudo-second-order, Bangham's pore diffusion, and Elovich equations which indicated that the process fitted with pseudo-second-order model and pore diffusion played a very important role in controlling the rate of the reaction. Even though the plots of Weber and Morris intraparticle diffusion made straight lines with correlation coefficients, they fail to pass through origin. This indicated that the process of the mechanism of adsorption was complex in nature with the more than one mechanism limiting the rate of adsorption, i.e., particle diffusion of fluoride adsorption was more towards the rate controlling step than intraparticle diffusion model. The FTIR studies indicated the participation of the surface sites of the adsorbent in the adsorbent interaction. Characterization of the activated carbon adsorbent through FTIR and SEM-EDX techniques confirmed the adsorption of the fluoride ion on the adsorbent surface. The active carbon developed has been found to be effectively decreasing the fluoride content in real groundwater samples below the permissible limits, and hence, the active carbon NVNC can be successfully applied in wastewater treatment technologies in controlling the fluorides.

Author details
[1]Department of Chemistry, Acharya Nagarjuna University, 522 510 Guntur Dt., AP, India. [2]Department of Chemistry, K L University, Vaddeswaram, 522 502 Guntur Dt., AP, India.

References

Adikari SK, Tipnis UK, Harkare WP, Govindan KP (1989) Defluoridation during desalination of brackish water by electrodialysis. Desalination 71:301–312

Akbar E, Maurice SO, Aoyi O, Shigeo A (2008) Removal of fluoride ions from aqueous solution at low pH using schwertmannit. J Hazard Mater 152:571–579

Alagumuthu G, Rajan M (2010a) Kinetic and equilibrium studies on fluoride removal by zirconium (IV)-impregnated ground nutshell carbon. Hem Ind 64(4):295–304

Alagumuthu G, Rajan M (2010b) Equilibrium and kinetics of adsorption of fluoride onto zirconium impregnated cashew nut shell carbon. Chem Eng J 158:451–457

Alagumuthu G, Veeraputhiran V (2011) Sorption Equilibrium of fluoride onto Phyllanthus emblica activated carbon. Int J Res Chem Environ 1(1):42–47

Alagumuthu G, Veeraputhiran V, Venkataraman R (2011) Fluoride sorption using Cynodon dactylon based activated carbon. Hem Ind 65(1):23–35

Aldaco R, Garea A, Irabien A (2007) Calcium fluoride recovery from fluoride Waste water in a fluidized bed reactor. Water Res 41:810–818

Amer Z, Bariou B, Mameri N, Taky M, Nicolas S, Elimidaoui A (2001) Fluoride removal from brakish water by electro dialysis. Desalination 133:215–223

Apambire WB, Boyle DR, Michel FA (1997) Geochemistry, genesis and health implications of fluoriferous groundwaters in the upper regions of Ghana. Environ Geol 33:13–24

APHA (American Public Health Association) (1985) Standard methods for the Examination of Water and Waste water. APHA, Washington, DC

Arulanantham AJ, Krishna TR, Balasubramanium (1989) Studies on fluoride removal by coconut shell carbon. Ind J Environ Health 13:531

ASTM D4607-94 (2006) Standard test method for Determination of Iodine number of Activated Carbon

Atkins P (1999) Physical chemistry, 6th edn. Oxford University Press, London, pp 857–864

Attia AA, Rashwan WE, Khedr SA (2006) Capacity of activated carbon in the removal of acid dyes subsequent to its thermal treatment. Dyes Pigments 69:128–136

Ayoob S, Gupta AK (2006) Fluoride in drinking water: A review on the status and tress effects. Environ Sci Technol 36:433–487

Ayoob S, Gupta AK (2008) Insights into isotherm making in the sorptive removal of fluoride from drinking water. J Hazard Mater 152:976–985

Bandosz TJ, Jagiello J, Schwarz JA (1992) A comparison of methods to asses surface acidic groups on activated carbons. Anal Chem 64:891–895

Barbier O, Arreola-Mendoza L, Del Razo LM (2010) Molecular mechanisms of fluoride Toxicity. Chem Biol Interact 188:319–333

Biniak S, Szymanski G, Siedlewski J, Swiatkowski A (1997) The characterization of activated carbons with oxygen and nitrogen surface groups. Carbon 35 (12):1799–1810

BIS (1991) Indian Standards for Drinking Water-Specification. Bureau of Indian Standards, New Delhi

BIS (Bureau of Indian Standards) (1989) Activated Carbon, Powdered and Granular-Methods of sampling and its tests. BIS, New Delhi, p 877

Boehm HP (1994) Some aspects of the surface chemistry of carbon blacks and others carbons. Carbon 32:759–769

Bouberka Z, Kaoha S, Kamecha, Elmaleh S, Derriche Z (2005) Sorption study of an acid dye from an aqueous solutions using modified clays. J Hazard Mater 119:117–124

Brunauer S, Emmett PH, Teller E (1938) Adsorption of Gases in Multimolecular Layers. J Am Chem Soc 60:309–315

Budinova T, Ekinci E, Yardim F, Grimm A, Bjornbom E, Minkova V, Goranova M (2006) Characterization and application of activated carbon produced by H_3PO_4 and water vapor activation. Fuel Process Techonol 87:899–905

Castel C, Schweizer M, Simonnot MO, Sardin M (2000) Selective removal of fluoride ions by a two-way ion-exchange cyclic processes. Chem Eng Sci 55:3341–3352

Cengeloglu Y, Esengul K, Ersoz M (2002) Removal of Fluoride from aqueous solution by using red mud. Sep Pur Technol 28:81–86

Chaturvedi AK, Yadva KP, Yadava KC, Pathak KC, Singh VN (1990) Defluoridation of water by adsorption on fly ash. Water Air Soil Pollut 49(1–2):51–61

Chauhan VS, Dwivedi PK, Iyengar L (2007) Investigations on activated alumina based domestic defluoridation units. J Hazard Mater 139:103–107

Chinoy NJ (1991) Effects of fluoride on physiology of animals and human beings. Indian J Environ Toxicol 1:17–32

Chubar NI, Samanidou VF, Kouts VS, Gallios GG, Kanibolotsky VA, Strelko VV, Zhuravlev IZ (2005) Adsorption of fluoride, chloride, bromide and bromate ions on a novel ion exchange. J Colloid Interface Sci 291:67–74

Cuhadaroglu D, Aydemir Uygun O (2008) Production and characterization of activated carbon from a bituminous coal by chemical activation. African J Biotech 7(20):3703–3710

Daifullah AAM, Girgis BS, Gad HMH (2003) Utilization of agro-residues (rice husk) in small waste water treatment plans. Mater Lett 57:1723–1731, http://www.sciencedirect.com/science/article/pii/S0167577X02010583

Daifullah AAM, Yakout SM, Elreefy SA (2007) Adsorption of fluoride in aqueous solutions using $KMnO_4$ modified activated carbon derived from steam pyrolysis of rice straw. J Hazard Mater 147:633–643

Dieye A, Larchet C, Auclair B, Mar-Diop C (1998) Elimination des fluorures parla dialyse ionicque croisee. Eur Polym J 34:67–75

El-Hendawy ANA (2003) Influence of HNO_3 oxidation on the structure and adsorptive properties of corncob-based activated carbon. Carbon 41:713–722

El-Hendawy ANA, Samra SE, Girgis BS (2001) Adsorption characteristics of activated carbons obtained from corncobs. Colloids Surf A Physicochem Eng Aspects 180:209–221

Fanning PE, Vannice MA (1993) A DRIFTS study of the formation of surface groups on carbon by oxidation. Carbon 31(5):721–730

Fawell J, Bailey K, Chilton J, Dahi E, Fewtrell L, Magara Y (2006) Fluoride in drinking water. WHO IWA Publishing, London-Seattle

Feng Shen X, Chen PG, Guohua C (2003) Electrochemical removal of fluoride ions from industrial wastewater. Chem Eng Sci 58:987–993

Figueiredo JL, Perria MFR, Freitas MMA, Orfao JJM (1999) Modification of the surface chemistry of activated carbons. Carbon 37(9):1379–1389

Ganvir V, Das K (2011) Removal of Fluoride from Drinking Water Using Aluminum Hydroxide Coated Rice Husk Ash. J Hazard Mater 185(2–3):1287–1294

Garmes H, Persin F, Sandeaux J (2002) Defluoridation of groundwater by a hybrid process combining adsorption and Donnan dialysis. Desalination 145:287–291

Gazzano E, Bergandi L, Riganti C, Aldieri E, Doublier S, Costamagna C, Bosia A, Ghigo D (2010) Fluoride effects: the two faces of Janus. Curr Med Chem 17:2431–2441

Gimba A, Turoti A (2006) Adsorption efficiency of coconut shell-based activated carbons on colour of molasses. SWJ 1(1):21–26

Girgis BS, El-Hendawy ANA (2002) Porosity development in activated carbons obtained from date pits under chemical activation with phosphoric acid. Microporous Mesoporous Mater 52:105–117

Goldberg S, Sposito G (1984a) A chemical model of phosphate adsorption by soils: 1. Reference oxide minerals. Soil Sci Soc Am J 48:772–778

Goldberg S, Sposito G (1984b) A chemical model of phosphate adsorption by soils: II. Noncalcareous soils. Soil Sci Soc Am J 48:779–783

Gomez-Serrano V, Acedo-Ramos M, Lopez-Peinado AJ, Venezuela-Calahorro C (1994) Study of Surface Functional Groups by FT-IR. Fuel 73(3):387–395

Grynpas MD, Chachra D, Limeback H (2000) In: Henderson JE, Goltzman D (eds) The Action of Fluoride on Bone, The Osteoporosis Primer. Cambridge University Press, Cambridge, UK

Hall KR, Eagleton LC, Acrivos A, Vermevlem T (1966) Pore and solid diffusion kinetics in fixed bed adsorption under constant pattern conditions. Indian Eng Chem Fundam 5:212–219

Hanumantharao Y, Kishore M, Ravindhranath K (2012a) Characterization and Defluoridation Studies of Active Carbon Derived from Typha Angustata Plants. J Analytical Sci Technol 3(2):167–181

Hanumantharao Y, Kishore M, Ravindhranath K (2012b) Characterization and defluoridation studies using activated Acacia Farnesiana carbon as adsorbent. Ele J Environ Agric Food Chem 11(5):442–458

Hanumantharao Y, Kishore M, Ravindhranath K (2012c) Characterization and adsorption studies of Lagenaria siceraria shell carbon for the removal of fluoride. Inter J Chem Tech Res 4(4):1686–1700

Hashim MA (ed) (1994) Symposium on Bioproducts Processing-Technologies for the Tropics, Kuala Lumpur, Malaysia. INST Chemical Engineers, UK

Hichour M, Persin F, Molenat J, Sandeaux J, Gavach C (1999) Fluoride removal from diluted solutions by Donnan dialysis with anion exchange membranes. Desalination 122:53–62

Hichour M, Persin F, Sandeaux J, Gavach C (2000) Water defluoridation by Donann Dialysis and electro dialysis. Sep Purif Technol 18:1–11

Hill A, Marsh H (1968) A study of the adsorption of iodine and acetic acid from aqueous solutions on characterized porous carbons. Carbon 6(I):31–39

Horsfall M, Spiff A (2005) Effects of temperature on the sorption of Pb2+ and Cd2+ from aqueous solution by Caladium bicolor (Wild Cocoyam) biomas. J Biotechnol 8:162–169

Hu CY, Lo SL, Kuan WH (2003) Effect of co-exiting anions on fluoride removal in electrocoagulation process using aluminium electrodes. Water Res 37:4513–4523

Ibrahim DM, El-Hemaly SA, Abdel-Kerim FM (1980) Study of rice-husk ash silica by infrared spectroscopy. Thermo Chimica Acta 37:307–314

Ishizaki C, Marti I (1981) Surface oxide structures on a commercial activated carbon. Carbon 19:409–412, http://www.sciencedirect.com/science/article/pii/0008622381900233

Jackson D, Murray JJ, Fairpo CG (1973) Lifelong benefits of fluoride in drinking water. Br Dental J 134(10):419–422

Ji Y, Li T, Li Z, Wang X, Liu Q (2007) Preparation of activated carbons by microwave heating KOH activation. Appl Surf Sci 254(2):506–512

Kadirvelu K, Faur-Brasquet C, Le Cloirec P (2000) Removal of Cu (II), Pb(II), and Ni (II) by Adsorption onto Activated Carbon Cloths. Langmuir 16:8404–8409

Karthikeyan G, Siva Elango S (2007) Fluoride sorption using Morringa Indica-based activated carbon. Iran J Environ Health Sci Eng 4(1):21–28

Kishore M, Hanumantharao Y (2011) Preparation and development of adsorbent carbon from Acacia farnesiana for defluoridation. Int J Plant Animal Environ Sci 1(3):209–223

Kumar JV, Moss ME (2008) Fluorides in dental public health programs. Dent Clin North Am 52:387–401

Lagergren S (1898) About the theory of so-called adsorption of soluble substances. Kungliga Svenska, Vetenskapsakademiens, Handlingar 24(4):1–39

Lahnid S, Tahaikt M, Elaroui K, Idrissi I, Hafsi M, Laaziz I, Amor Z, Tiyal F, Elmidaoui A (2008) Economic evaluation of fluoride removal by electrodialysis. Desalination 230:213–219

Lapuente R, Cases F, Garces P, Morallon E, Vazquez JL (1998) A voltammeter and FTIR-ATR study of the electro polymerization of phenol on platinum electrodes in carbonate medium: Influence of sulfide. J Electroanal Chem 451:163–171

Lhassani A, Rumeau M, Benjelloun D, Pontie M (2001) Selective demineralization of water by nanofiltration application to the defluoridation of brackish water. Water Res 35:3260–3264

Li YH, Wang S, Cao A, Zhao D, Zhang X, Xu C, Luan Z, Ruan D, Liang J, Wu D, Wei B (2001) Adsorption of fluoride from water by amorphous alumina supported on carbon nanotubes. Chem Phys Lett 350(5&6):412–416

Liu J, Xu Z, Li X, Zhang Y, Zhou Y, Wang Z, Wang X (2007) An Improved process to prepare high separation performance PA/PVDF hollow fiber composite nano filtration membranes. Sep Purif Technol 58:53–60

Lounici H, Addour L, Belhocine D, Grib H, Nicolas S, Bariou B (1997) Study of a new technique for fluoride removal from water. Desalination 114:241–251

Maheshwari RC, Meenakshi (2006) Fluoride in drinking water and its removal. J Hazard Mater 137(1):456–463

Mameri N, Lounici H, Belhocine D, Grib H, Prion DL, Yahiat Y (2001) Defluoridation of Sahara Water by small electro coagulation using bipolar Aluminium Electrodes. Sep Purif Technol 24:113–119

Marsh H, Rodriguez-Reinoso F (2006) Activated Carbon, Elsevier Science & Technology Books., pp 401–462

McKee RH, Jhonston WS (1934) Removal of fluorides from drinking water. Ind Eng Chem 26(8):849–850

Meenakshi S, Viswanathan N (2007) Identification of selective ion-exchange resin for fluoride sorption. J Colloid Interface Sci 308:438–450

Meldrum BJ, Rochester CH (1990a) In Situ Infrared Study of the Surface Oxidation of Activated Carbon Dispersed in Potassium Bromide. J Chem Soc Faraday Trans 86:2997–3002

Meldrum BJ, Rochester CH (1990b) In situ infrared study of the surface oxidation of activated carbon in oxygen and carbon dioxide. J Chem Soc Faraday Trans 86(5):861–865

Meldrum BJ, Rochester CH (1990c) In situ infrared study of the modification of the surface of activated carbon by ammonia, water and hydrogen. J Chem Soc Faraday Trans 86(10):1881–1884

Menkouchi SMA, Annouar S, Tahaikt M (2007) Fluoride Removal for Underground Brackish Water by Adsorption on the Natural Chitosan and by Electrodialysis. Desalination 212:37

Mjengera H, Mkongo G (2003) Appropriate defluoridation technology for use in fluorotic areas in Tanzania. Phys Chem Earth 28:1097–1104

Mohan D, Sharma R, Vinod K, Singh PS, Pittman CU Jr (2012) Fluoride Removal from Water using Bio-Char, a Green Waste, Low-Cost Adsorbent: Equilibrium Uptake and Sorption Dynamics Modeling. Ind Eng Chem Res 51:900–914

Mohapatra D, Mishra D, Mishra SP, Roy Chaudhury G, Das RPJ (2004) Use of oxide minerals to abate fluoride from water. Colloid Interface Sci 275:355–359

Monika J, Garg V, Kadirvelu K (2009) Chromium (VI) removal from aqueous solution, using sunflower stem waste. J Hazard Mater 162:365–372

Montoya VH, Montoya LAR, Petriciolet AB, Moran MAM (2012) Optimizing the removal of fluoride from water using new carbons obtained by modification of nut shell with a calcium solution from egg shell. Biochem Eng J 62:1–7

Moreno-Castilla C, Carrasco-Marin F, Maldonado-Hodar FJ, Rivera-Utrilla J (1998) Effects of non-oxidant and oxidant acidtreatments on the surface properties of an activated carbonwith very low ash content. Carbon 36(1–2):145–151

Nageswara Rao M, Chakrapani CH, Rajeswara Reddy BV, Suresh Babu CH, Hanumantha Rao Y, Somasekhara Rao K, Rajesh K (2011) Preparation of activated kaza's carbons from bio-materials and their characterization. Int J Appl Biol Pharm Tech 2(3):610–618

Namasivayam C, Kadirvelu K (1994) Coir pith, an agricultural waste by-product, for the treatment of dyeing waste water. Biores Technol 48:79–81

Namasivayam C, Kadirvelu K (1997) Agricultural solid wastes for the removal of heavy metals: Adsorption of Cu(II) by coir pith carbon. Chemosphere 34:377–399

Nawlakhe WG, Kulkarni DN, Pathak BN, Bulusu KR (1975) De-fluoridation of water by Nalgonda technique. Ind J Environ Health 17:26–65

Newcombe G, Hayes R, Drikas M (1993) Granular activated carbon: importance of surface properties in the adsorption of naturally occurring organics. Colloids Surf A:Physicochem Eng Aspects 78:65–71

Onyango MS, Kojima Y, Aoyi O, Bernardo EC, Matsuda H (2004) Adsorption equilibrium modeling and solution chemistry dependence of fluoride removal from water by trivalent-cation-exchanged zeolite F-9. J Colloid Interface Sci 279(2):341–350

Onyango MS, Matsuda H, Alain T (2006) Chapter 1: Fluoride Removal from Water Using Adsorption Technique. Adv Fluorine Sci 2:1

Ozgul G, Adnan O, Safa Ozcan A, Ferdi Gercel H (2007) Preparation of activated carbon from a renewable bio-plant of Euphorbia rigida by H2SO4 activation and its adsorption behavior in aqueous solutions. Appl Surf Sci 253:4843–4852, http://www.sciencedirect.com/science/article/pii/S0169433206013900

Painter P, Starsinic M, Coleman M (1985) Determination of functional groups in coal by Fourier transform interferometry. In: Fourier transform infrared spectroscopy, 4th edn. Academic Press, New York, pp 169–189

Palanisamy PN, Sivakumar P (2008) Production and Characterization of a novel non-conventional low-cost adsorbent from Euphorbia Antiquorum L. Rasayan J Chem 1(4):901–910

Park SH, McClain S, Tian ZR, Suib SL, Karwacki C (1997) Surface and bulk measurements of metals deposited on activated carbon. Chem Mater 9:176–183

Periasamy K, Namasivayam C (1994) Process development for removal and recovery of cadmium from wastewater by a low-cost adsorbent: adsorption rates and equilibrium studies. Ind Eng Chem Res 33:317–320

Puziy I, Poddubnaya O, Martinez-Alonso A, Suarez-Garcia F, Tascon J (2003) Synthetic carbons activated with phosphoric acid III. Carbons prepared in air. Carbon 41:1181–1191

Raichur AM, Jyoti Basu M (2001) Adsorption of fluoride onto mixed rare earth Oxides. Sep Purif Technol 24:121–127

Rajeshwari S, Sivakumar S, Senthilkumar P, Subburam V (2001) Carbon from cassava peel, an agricultural waste, as an adsorbent in the removal of dyes and metal ions from aqueous solutions. Biores Technol 81:1–3

Ramos RL, Ovalle-Turrubiartes J, Sanchez-Castillo MA (1999) Adsorption of fluoride from aqueous solution on aluminum-impregnated carbon. Carbon 37:609–617

Rao VB, Mandava, Subba Rao M, Prasanthi V, Muppa R (2009) Characterization and defluoridation studies of activated Dolichos Lab Lab carbon. Rasayan J Chem 2(2):525–530

Reardon EJ, Wang Y (2000) A limestone reactor for fluoride removal from waste waters. Environ Sci Technol 34:3247–3253

Rozada F, Otero M, Garcia I (2005) Activated carbons from sewage sludge and discarded tyres, Production and optimization. J Hazard Mater 124(1–3):181–191

Saha S (1993) Treatment of aqueous effluent for fluoride removal. Water Res 27:1347–1350

Sai sathish R, Raju NSR, Raju GS, Nageswara Rao G, Anil Kumar K, Janardhana C (2007) Equilibrium and Kinetic Studies for Fluoride Adsorption from Water on Zirconium impregnated Coconut Shell Carbon. Sep Sci Technol 42:769–788

Sairam Sundaram C, Viswanathan N, Meenakshi S (2009) Defluoridation of water using magnesia/chitosan composite. J Hazard Mater 163(2–3):618–624

Savinelli EA, Black AP (1958) Defluoridation of water with activated alumina. J Am Water Works Assoc 50:34–44

Sehn P (2008) Fluoride removal with extra low energy reverse osmosis membranes: three years of large scale field experience in Finland. Desalination 223:73–84

Shihabudheen MM, Atul KS, Ligy P (2006) Water Res 40:3497–3506

Shin S, Jang J, Yoon SH, Mochida I (1997) A study on the effect of heat treatment on functional groups of pitch based activated carbon fiber using FTIR. Carbon 35(12):1739–1743

Simons R (1993) Trace element removal from ash dam waters by nanofiltration and diffusion dialysis. Desalination 89:325–341

Singh G, Kumar B, Sen PK (1999) Removal of fluoride from spent pot liner leach ate using ion exchange. Water Environ Res 71:36–42

Sreenivasa Rao V, Somasekhara Rao K, Nageswara Rao M, Upasana Sinha B (2011) Studies on the surface characterisation of newly prepared activated Kaza's carbons. Asian J Biochem Pharmacol Res 2(1):567–584

Srimurali M, Pragathi A, Karthikeyan J (1998) A study on removal of fluoride from drinking water by adsorption onto low cost materials. Environ Pollut 99:285–289

Starsinic M, Taylor RL, Walker PL Jr, Painter PC (1983) FTIR Studies of Saran chars. Carbon 1(1):69–74

Susheela AK (2001) Treatise on Fluorosis, Fluorosis Research and Rural Development Foundation, India, Fluoride, vol 34., pp 181–183

Tor A (2007) Removal of fluoride from water using anion-exchange membrane under Donnan dialysis condition. J Hazard Mater 141:814–818

Tripathy SS, Bersillon J-L, Gopal K (2006) Removal of fluoride from drinking water by adsorption onto alum-impregnated activated alumina. Sep Purif Technol 50:310–317

Underwood EJ (1997) Trace Elements in Human and Animal Nutrition. Academic Press, New York, p 545

Venkata Mohan S, Ramanaiah SV, Rajkumar B, Sarma PN (2007) Removal of fluoride from aqueous phase by biosorption onto algal biosorbent Spirogyra Sp. I02: sorption mechanism elucidation. J Hazard Mater 141:465–474

Viswanathan N, Meenakshi S (2010) Enriched fluoride sorption using alumina/chitosan composite. J Hazard Mater 178:226–232

Wang Y, Reardon EJ (2001) Activation and regeneration of a soil sorbent for defluoridation of drinking water. Appl Geochem 16:531–539

WHO (1984) Guidelines for drinking water quality. Health criteria and other supporting information. Geneva, Switzerland, vol 2

WHO (2004) Guidelines for Drinking Water Quality. World Health Organization, Geneva

Yadav AK, Kaushik CP, Haritash AK, Kansal A, Neetu R (2006) Defluoridation of drinking water using brick powder as an adsorbent. J Hazard Mater 128:289–293

Yang T, Lua A (2003) Characteristics of activated carbons prepared from pistachio-nut shells by physical activation. J Colloid Interface Sci 267:408–417

Yu C, Qiu JS, Sun YF, Li XH, Chen G, Zhao ZB (2008) Adsorption removal of thiophene and dibenzothiophene from oils with activated carbon as adsorbent: effect of surface chemistry. J Por Mater 15:151–157

Zawadzki J (1989) Chemistry and Physics of Carbon. Marcel Dekker, New York, pp 147–380

Zhang PC, Spark DL (1990) Kinetics and mechanisms of sulfate adsorption/desorption on goethite using pressure-jump relaxation. Soil Sci Soc Am J 54:1266–1273

Zhuang QL, Kyotani T, Tomita A (1994) DRIFT and TK/TPD analyses of surface oxygen complexes formed during carbon gasification. Energy Fuels 8:714–718

21

Optical characteristics and photothermal conversion of natural iron oxide colloid

Tae Yeon Kang[1], Ki Soo Chang[2], Jae Young Kim[2], Seon-Kang Choi[3] and Weon-Sik Chae[1*]

Abstract

Background: Chemical compositions and spectroscopic characteristics of the natural floating colloids in brine mineral water were investigated in this study.

Methods: The natural colloidal materials were investigated using electron microscopy, X-ray crystallography, elemental analysis, and absorption and emission spectroscopies.

Results: The natural colloidal particles have a spherical shape, with average diameter of 200 nm, and amorphous crystalline structure. The colloids are mostly composed of iron and oxygen atoms; they also contained small amounts of trace elements and rare earth minerals. In particular, the colloids show remarkable absorption and emission characteristics in the wide spectral region from ultraviolet (UV) to near infrared (NIR), which could make it useful in photoconversion and hyperthermal applications.

Conclusion: From the photothermal conversion efficiency measurement using an infrared thermography under irradiation of visible and NIR light, interestingly, it was found that the natural colloids have higher photothermal conversion efficiency, as compared with those of several different-typed minerals.

Keywords: Natural colloid; Geumjin spring water; Rare earth mineral; Near-infrared emission; Photothermal conversion

Background

Brine mineral water (BMW) is defined as any spring water that is gushed out from the bedrock located within about 1 km from the coast generally. BMW is known to include more abundant mineral ingredients, such as calcium (Ca), magnesium (Mg), strontium (Sr), manganese (Mn), zinc (Zn), nickel (Ni), and iron (Fe) in comparison with other ocean deep water. Moreover, BMW has an excellent mineral balance similar to that of human body fluids (Kim et al. 2008; Moon et al. 2004). Recently, BMW drawn from 1,100 m below the coast terrace of Geumjin (GJ) area (Gangneung City, Republic of Korea) has attracted special attention because it contains several functional minerals such as selenium (Se) and vanadium (V). Moreover, it is confirmed that this BMW has the most suitable mineral balance for Ca and Mg particularly because it is helpful for absorbing Ca to the human

body. Using this unique BMW from the GJ area (GJ BMW), many studies have been recently performed in the various fields of industry, like functional food, cosmetics, and medicine. Kim et al. (2010) reported for the effect of the GJ BMW on atopic dermatitis *in vivo* with atopic dermatitis model. They have shown that the GJ BMW can not only suppress the ear swelling induced by trimellitic anhydride (TMA) but also attenuate hyperactivated lymph nodes stimulated by TMA. Moreover, they reported that the growth of several kinds of cancer cells was inhibited by GJ BMW through a dose-dependent manner (Kim et al. 2009). Contrary to the various studies and the practical uses of the GJ BMW, however, the floating colloidal particles, which are observed in the GJ BMW, are still not well known. These colloidal particles are suspended in high concentration in the GJ BMW, and they cause the GJ BMW to have a unique opaque color like red wine. However, after several hours, the floating colloids are mostly deposited on the bottom by self-aggregation. By now, there has been no systematic study on these sediments. In this work, we report on the

* Correspondence: wschae@kbsi.re.kr
[1]Gangneung Center, Korea Basic Science Institute (KBSI), Gangneung 210-702, Republic of Korea
Full list of author information is available at the end of the article

chemical compositions and spectroscopic properties of the natural colloidal particles in the GJ BMW. The optical properties of the colloidal particles were investigated by ultraviolet–visible (UV–vis) absorption and near-infrared photoluminescence (NIR-PL) spectroscopies and then their photothermal conversion characteristics were compared with those of the other mineral materials by infrared (IR) thermography. Here, we first report the unique optical characteristics of broad vis-NIR absorption and intense NIR emissions from the natural colloidal material. Interestingly, the colloidal material shows notable photothermal conversion property.

Methods

Sampling

BMW were collected at the Geumjin spring area located in the Gangneung City in the east region of the Republic of Korea (Figure 1). The BMW is originally a clear solution, but its color gradually turns to red orange with time under ambient condition. The red orange colloidal particles were typically sedimented within a day. The sedimented colloidal particles in the spring water were

collected by centrifugation and then rinsed thoroughly with deionized (DI) water. The samples were dried at 100°C for 24 h in an oven, which were further used for all analyses and measurements.

Morphology and composition characterization

To study the particle formation kinetics, dynamic light scattering (DLS) measurement values were examined using an electrophoretic light scattering spectrophotometer (ELS-8000, Otsuka Electronics Co., Ltd., Osaka, Japan). The shape and size of the GJ colloidal particles were analyzed using a field emission scanning electron microscope (FESEM, SU-70, Hitachi Ltd. Tokyo, Japan) operating at 15-kV accelerating voltage. Powder X-ray diffraction crystallography (XRD) patterns were obtained with an X-ray diffractometer (X'Pert Pro, PANalytical, Almelo, The Netherlands). Elemental composition was determined using an energy dispersive X-ray spectrometer (EDS) which is attached to the FESEM. More quantitative analysis of trace elements in the GJ colloidal particles was performed by inductive coupled plasma mass spectrophotometry (ICP-MS) (Elan DRCII, Perkin

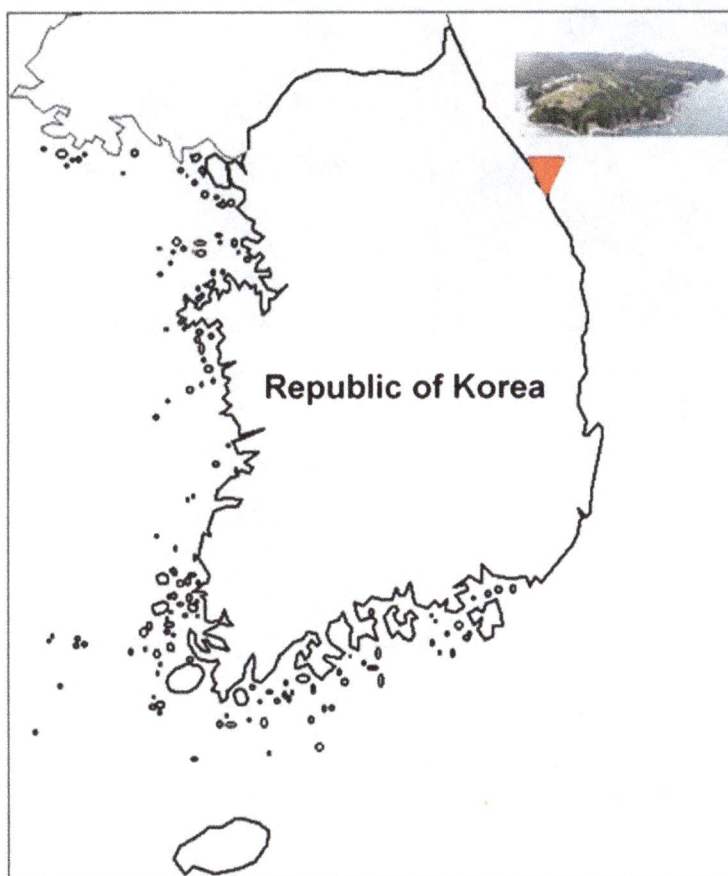

Figure 1 Geologic location of the Geumjin spring waters sampled in the Gangneung City, Republic of Korea.

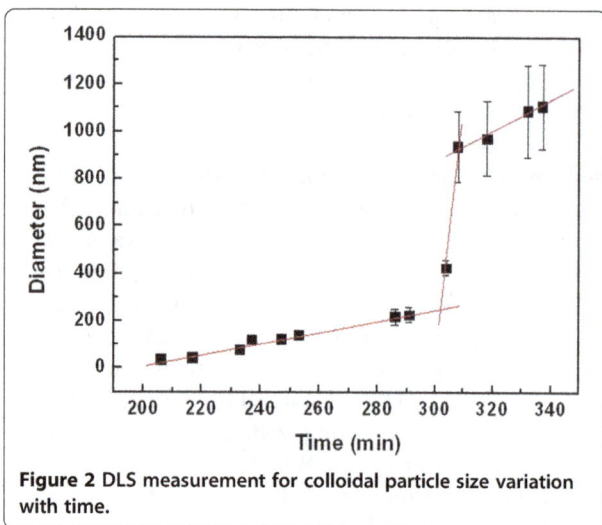

Figure 2 DLS measurement for colloidal particle size variation with time.

Elmer, Santa Clara, CA, USA) and inductively coupled plasma atomic emission spectrometry (ICP-AES) (JY Ultima2C, Jobin Yvon, Paris, France).

Spectroscopic characterization

UV–vis absorption spectrum of the GJ colloidal particles which are dispersed in ethylene glycol was recorded immediately using an UV–vis spectrophotometer (S-3100, Scinco Co., Ltd., Seoul, South Korea) at room temperature. A conventional quartz cuvette of 1-cm optical path was used for the measurements.

NIR photoluminescence (PL) spectrum of the GJ colloidal particles was also measured on a spectrophotometer (iHRA-330 PL, Jobin Yvon, Horiba) equipped with a liquid nitrogen-cooled InGaAs photodetector in a wavelength range of 800 to 1500 nm with a monochromatic 580-nm light (Xe lamp) as an excitation source.

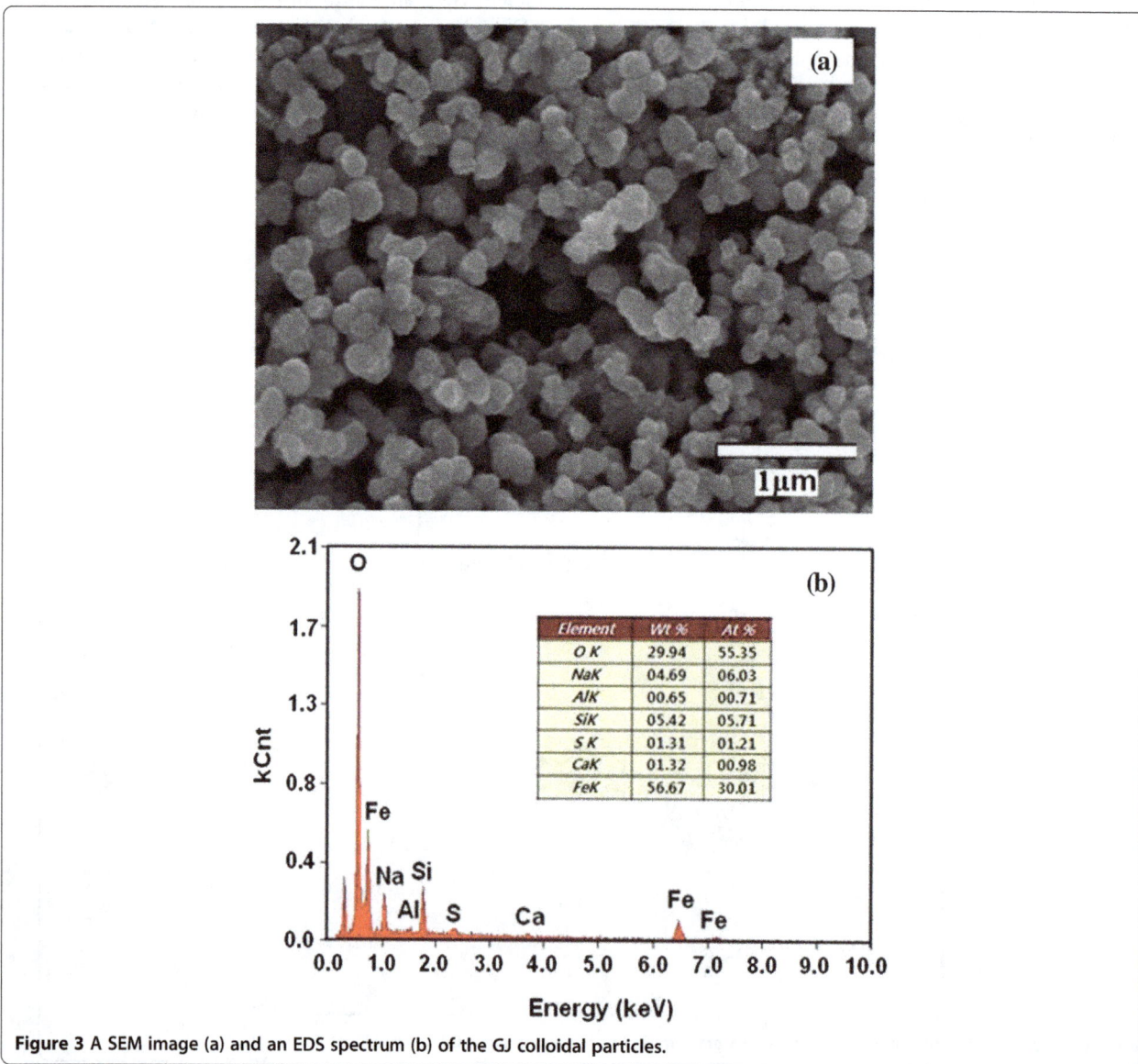

Element	Wt %	At %
O K	29.94	55.35
NaK	04.69	06.03
AlK	00.65	00.71
SiK	05.42	05.71
S K	01.31	01.21
CaK	01.32	00.98
FeK	56.67	30.01

Figure 3 A SEM image (a) and an EDS spectrum (b) of the GJ colloidal particles.

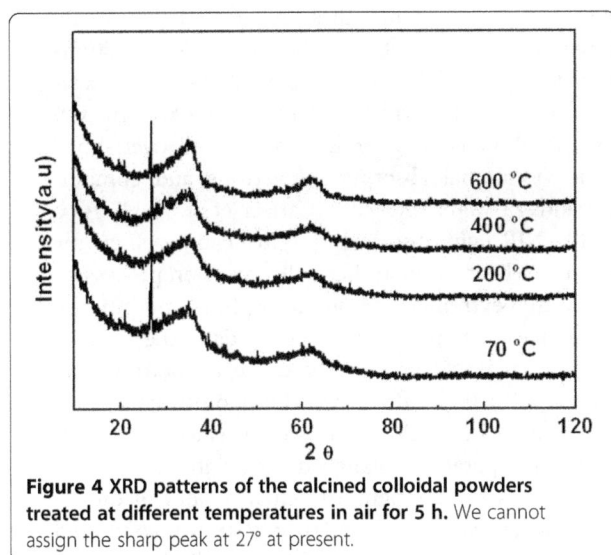

Figure 4 XRD patterns of the calcined colloidal powders treated at different temperatures in air for 5 h. We cannot assign the sharp peak at 27° at present.

Photothermal conversion measurements

Halogen illuminator (FHL-101, 100 W, Asahi-Spectra Co., Ltd., Tokyo, Japan) was used as a visible-NIR light source. The illuminating light from FHL-101 was delivered to the surface of the sample by fiber optic ring light guide (MRG53-1000S, Moritex Corp., Saitama, Japan). The distance between the sample surface and the fiber optic ring was about 8 cm. The illumination power was fixed at 200 mW/cm^2 at the sample surface. After light illumination, the temperature change of the samples was recorded using an infrared thermal camera (SC7600, FLIR Systems, Croissy-Beaubourg, France), every 10 s for 30 min. All the dispersed samples (10 mg/10 ml DI water) were placed in Teflon bath (10-mm diameter, 500 μl).

Results and discussion

Red orange colloidal particles were first formed in the GJ BMW, further grew, and sedimented with time. The

Table 1 Analysis of the GJ colloid powder by ICP-AES

Element	Concentration (ppm)
Al	14,050
Ca	32,090
Fe	411,100
K	2,873
Mg	3,263
P	115.3
Ti	468.9
Mn	86.38
Na	873.7
Pb	133.0

Resolution: 0.005 nm (UV) and 0.05 nm (visible). 0.07 g of the powder sample was used.

sedimentation kinetics of the GJ colloidal particles were investigated by DLS measurements. Figure 2 shows the particle size variation for the GJ colloidal particles as a function of time. The red orange colloidal particles appeared after around 3 h and further increased in size as time goes on. After about 5 h, the size of the colloidal particles increased to around 200 nm and then the particles rapidly agglomerated to larger ones with an average size of 1 μm.

A FESEM image and an EDS spectrum of the GJ colloidal particles are shown in Figure 3. The colloidal particles have spherical shape and with about 200 nm in diameter; the primary colloidal particles aggregated by tens, which well agreed with the DLS results. The EDS analysis reveals that the colloidal particles are mainly composed of Fe and oxygen (O) with a minor number of minerals, such as sodium (Na), silicon (Si), and aluminum (Al).

The crystal structure of the colloidal particles was investigated by XRD analysis. The colloidal powder samples were calcined at 70°C, 200°C, 400°C, and 600°C in air for 5 h, respectively, which were further used for the XRD analysis. As shown in Figure 4, the

Table 2 Analysis of the GJ colloid powder by ICP-MS

Element	Concentration (ppb)
Ti	551,829
V	2,384
Cr	148,264
Ni	9,030
Cu	902,819
Zn	33,141
As	49,894
Rb	17,418
Sr	728,497
Y	5,453
Zr	13,756
Ag	801
Ba	107,025
Ce	7,452
Pr	815
Nd	4,355
Sm	716
Gd	1,107
Dy	616
Er	380
Yb	291
Pb	156,200
Th	1,724
U	470

Resolution: 0.005 nm (UV) and 0.05 nm (visible). 0.07 g of the powder sample was used.

measured XRD patterns do not show any distinct diffraction peak except two broad diffractions at approximately 35° and 62° for all colloidal samples. This means that the GJ colloidal particles have amorphous structural characteristics in crystallography, and this unique structural feature does not be much changed even for the heat-treated colloids at high temperature. A very similar XRD pattern was previously observed for the amorphous iron oxide/hydroxide materials (Kwon et al. 2005). In the case of the GJ colloid, the amorphous structure may be additionally attributed to the number of co-doped chemical compositions to the iron oxide matrix, as described below.

The chemical compositions of the GJ colloidal powder were determined by ICP-AES and ICP-MS (Tables 1 and 2). It shows very high concentrations of Fe (approximately above 40%) and some trace elements, such as Ca, Al, Mg, and K. Moreover, the GJ colloid also contains quite some amount of functional minerals (Se and V) as well as various rare earth minerals (Y, Ce, Nd, etc.). These functional minerals and rare earth minerals would be extremely useful in a variety of applied fields.

Absorption and photoluminescence spectrum of the GJ colloid, which is dispersed in ethylene glycol, are shown in Figure 5. The absorption spectrum (Figure 5a) shows broad absorption features from 350 to 900 nm with a maximum peak at 580 nm. This wide-range absorption means that the GJ colloidal particles can be utilized as a photoabsorber for the visible and infrared light. Figure 5b shows the photoluminescence spectrum of the GJ colloid at room temperature under the excitation wavelength of 580 nm. Interestingly, intense characteristic emissions are observed at the near-infrared regions (800 to 1,400 nm). Recently, NIR irradiation combined with NIR absorbing agents play a crucial role in the biomedical applications of anticancer treatment, hyperthermia therapy, and *in vivo* molecular imaging (Park et al. 2009; Lee et al. 2010; Park et al. 2010; Park et al. 2008). This is because the NIR light can penetrate deeply into the skin tissue and then this enables local heating to destroy target cancer cells by the assistance of NIR absorber and heating complex agents. This method enables an efficient anticancer treatment route without undesired side effects to normal organs and tissues (Sherlock et al. 2011). Moreover, NIR fluorophores offer better advantages than visible light fluorophores in molecular imaging applications. By now, quantum dots and rare earth metal reagents have been widely developed and studied in order to be used as NIR emitting probes (Ogawa et al. 2009; Sharma et al. 2008; Weissleder 2001). However, for the first time, at the best of our knowledge, we observed the presence of intense NIR emission from the natural colloidal product of iron oxide as main constituents. In the case of the GJ colloid, although it contains various co-doped elements, these broad and intense PL emissions may be attributed to the electronic transitions between the partly filled d electronic states (d-d transitions) of iron coupled with oxygen (Ronda 2007). In particular, the observed NIR emissions are well overlapped with the vibrational absorption (overtone and combination) bands of water molecule (Carleer et al. 1999). As a result, the NIR light absorbed by water can be efficiently transferred to heat by photothermal conversion process.

Hence, the extremely broad absorption and intense emission characteristics of the GJ colloids in the vis-NIR range could make it possible to cause a photothermal conversion with high efficiency. The photothermal conversion ability of the GJ colloid was measured by monitoring the temperature change during light irradiation. Figure 6 shows the temperature enhancement of the GJ colloid powder/water mixture (10 mg/10 ml), compared to pure DI water as a reference. The overall illumination power was approximately 200 mW/cm². The temperature of the GJ colloid powder/water mixture (0.5 ml) and pure water (0.5 ml) was increased to 30.3°C and 29.1°C, respectively, as shown in Figure 6a. From the repetitive

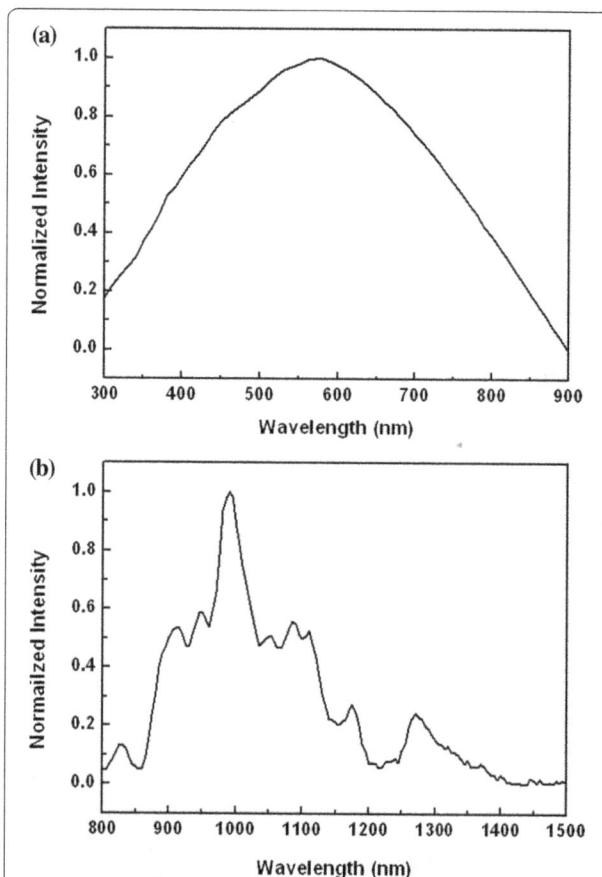

Figure 5 Absorption (a) and NIR emission (b) spectra of the GJ colloidal material dispersed in ethylene glycol. The excitation wavelength is 580 nm.

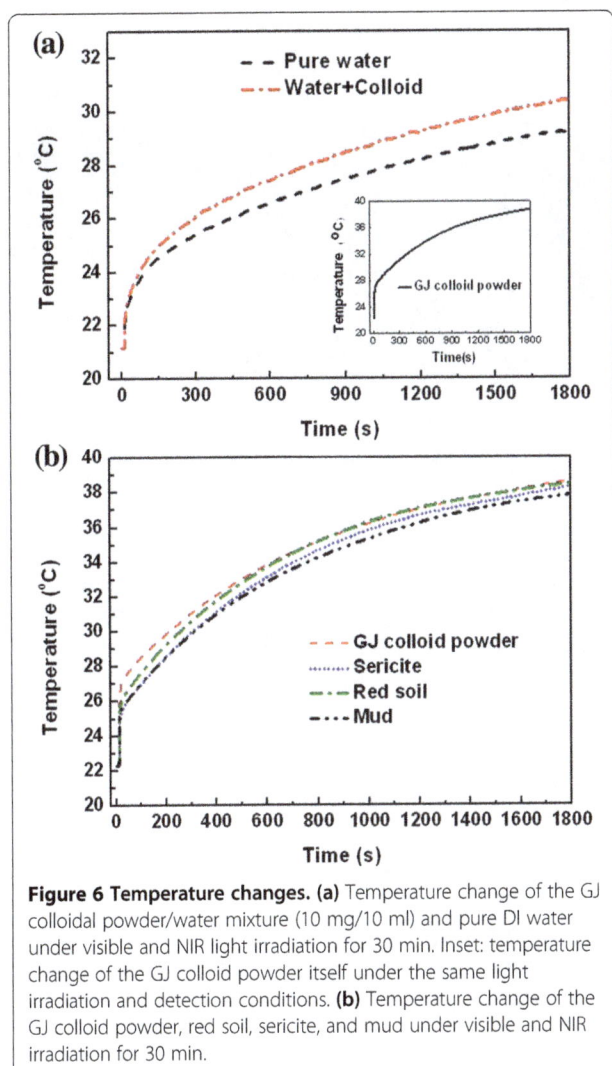

Figure 6 Temperature changes. (a) Temperature change of the GJ colloidal powder/water mixture (10 mg/10 ml) and pure DI water under visible and NIR light irradiation for 30 min. Inset: temperature change of the GJ colloid powder itself under the same light irradiation and detection conditions. **(b)** Temperature change of the GJ colloid powder, red soil, sericite, and mud under visible and NIR irradiation for 30 min.

soil, sericite, and mud. Under the same conditions, the GJ colloidal powder showed the best photothermal conversion efficiency for all measurement time, as compared to those of the other similar natural materials (Figure 6b). Approximating similar self-heating effects of the natural materials themselves through nonradiative vibrational relaxations, it is considered that this superior photothermal conversion ability of the GJ colloid is due to the wide absorption characteristics for visible-NIR light and subsequent efficient energy transfer processes of the excited electronic energy of the colloid to excite vibrational modes of surrounding water.

Conclusion

We have investigated for the composition, structure, and optical characteristics, and photothermal conversion efficiency of the floating colloidal particles in brine mineral water at the Geumjin area. In this study, we first observed unique optical characteristics of broad vis-NIR absorption and intense NIR emissions from the natural colloidal material including additional co-doped elements. The colloidal material shows notable photothermal conversion efficiency compared to the other natural products. The obtained results show that the GJ colloidal particles would have distinctive promise for use in various fields, such as therapeutic and biomedical diagnosis applications in addition to conventional photothermal therapy.

Competing interests
The authors declare that they have no competing interests.

Authors' contributions
WSC designed the study. TYK and WSC performed the research. KSC and JYK carried out the photothermal analysis. SKC did the sampling of the materials. TYK and WSC wrote the manuscript. All authors read and approved the final manuscript.

Acknowledgements
This work was supported by a grant (no. PGB066) from the Catholic University of Korea (Seoul St. Mary's Hospital), a grant (no. K3208F) from the Korea Basic Science Institute (KBSI), and partly by a KRF grant (no. 2011–0008671). We thank the Korea Basic Science Institute, Seoul Center (Seoul, Korea) for the help with the ICP-AES and ICP-MS measurements.

Author details
[1]Gangneung Center, Korea Basic Science Institute (KBSI), Gangneung 210-702, Republic of Korea. [2]Center for Analytical Instrumentation Development, Korea Basic Science Institute (KBSI), Daejeon 305-806, Republic of Korea. [3]Tongyang Life Science Corp., Seoul 135-995, Republic of Korea.

measurements, it is consistently observed that the temperature of the GJ colloid powder/water mixture was more enhanced than that of pure DI water under the same conditions. This enhanced temperature elevation can be assigned to the direct energy transfer of the emissions of the GJ colloid by the vibrational absorptions of water, as mentioned above. In addition, light energy absorbed by the GJ colloidal powder itself can be also converted to heat and then be transferred to water. The inset in Figure 6a depicts the temperature change of the GJ colloidal powder itself under visible and NIR irradiation for 30 min. The temperature of the GJ colloidal powder was drastically increased up to approximately 38.5°C. This indicates that the absorbed light energy in the excited states of iron oxide moiety can be transferred to thermal energy through nonradiative intramolecular vibrational relaxations and/or transitions (Jackson et al. 1981). Hence, photothermal conversion efficiency measurements were also performed for similar natural materials, such as red

References
Carleer M, Jenouvrier A, Vandaele AC, Bernath PF, Me´rienne MF, Colin R, Zobov NF, Polyansky OL, Tennyson J, Savin VA (1999) The near infrared, visible, and near ultraviolet overtone spectrum of water. J Chem Phys 111:2444–2450
Jackson WB, Amer NM, Boccara AC, Fournier D (1981) Photothermal deflection spectroscopy and detection. Appl Optics 20:1333–1344

Kim YJ, Jung IS, Song HJ, Choi EY, Choi IS, Choi YJ (2008) Study of deep ground sea-like water on antioxidant activity and the immune response in RAW264.7 macrophages. J Life Sci 18:329–335

Kim WJ, Li H, Yoon TJ, Sim JM, Choi SK, Lee KH (2009) Inhibitory activity of brine mineral water on counter cell growth, materials and angiogenesis. Korean J Food Nut 22:542–547

Kim JJ, Kim WJ, Sim JM, Choi SK, Kwon SS, Kim JD, Lee KH (2010) Effect of brine mineral water on TMA-induced contact hypersensitivity reaction in the mouse model. Korean J Food Nut 23:440–445

Kwon SK, Kimijima K, Kanie K, Muramatsu A, Suzuki S, Matsubara E, Waseda Y (2005) Inhibition of conversion process from $Fe(OH)_3$ to β-FeOOH and α-Fe_2O_3 by the addition of silicate ions. ISIJ Int 45:77–81

Lee SM, Park H, Yoo KH (2010) Synergistic cancer therapeutic effects of locally delivered drug and heat using multifunctional nanoparticles. Adv Mater 22:4049–4053

Moon DS, Jung HJ, Kim HJ, Shin PK (2004) Comparative analysis on resources characteristics of deep ocean water and brine groundwater. J Kor Soc Mar Environ Engi 7:42–46

Ogawa M, Kosaka N, Choyke PL, Kobayashi H (2009) In vivo molecular imaging of cancer with a quenching near-infrared fluorescent probe using conjugates of monoclonal antibodies and indocyanine green. Cancer Res 69:1268–1272

Park H, Yang J, Seo S, Kim K, Suh J, Kim D, Haam S, Yoo KH (2008) Multifunctional nanoparticles for photothermally controlled drug delivery and magnetic resonance imaging enhancement. Small 4:192–196

Park H, Yang J, Lee J, Haam S, Choi IH, Yoo KH (2009) Multifunctional nanoparticles for combined doxorubicin and photothermal treatments. ACS Nano 3:2919–2926

Park JH, Von Maltzahn G, Xu MJ, Fogal V, Kotamraju VR, Ruoslahti E, Bhatia SN, Sailor MJ (2010) Cooperative nanomaterial system to sensitize, target, and treat tumors. Proc Natl Acad Sci USA 107:981–986

Ronda C (2007) Luminescence: from theory to applications. Wiley, New York

Sharma R, Wendt JA, Rasmussen JC, Adams KE, Marshall MV, Sevick-Muraca EM (2008) New horizons for imaging lymphatic function. Ann N Y Acad Sci 1131:13–36

Sherlock SP, Tabakman SM, Xie L, Dai H (2011) Photothermally enhanced drug delivery by ultrasmall multifunctional FeCo/graphitic shell nanocrystals. ACS Nano 5:1505–1512

Weissleder R (2001) A clearer vision for in vivo imaging. Nat Biotechnol 19:316–317

22

Simultaneous quantification of major bioactive constituents from Zhuyeqing Liquor by HPLC-PDA

Hong-ying Gao[1], Shu-yun Wang[1], Hang-yu Wang[2], Guo-yu Li[2], Li-fei Wang[3], Xiao-wei Du[3], Ying Han[3], Jian Huang[2] and Jin-hui Wang[1,2]*

Abstract

Background: Zhuyeqing Liquor (ZYQL) is a famous traditional Chinese functional liquor. For quality control of ZYQL products, quantitative analysis using high-performance liquid chromatography coupled with photodiode array detector (HPLC-PDA) was undertaken.

Methods: Eighteen compounds from ZYQL were simultaneously detected and used as chemical markers in the quantitative analysis, including 3-hydroxy-4,5(R)-dimethyl-2(5H)-furanone (M1), isobiflorin (M2), vanillic acid (M3), biflorin (M4), genipin 1-O-β-D-gentiobioside (M5), 1-sinapoyl-β-D-glucopyranoside (M6), geniposide (M7), epijasmnoside A (M8), ferulic acid (M9), luteolin 8-C-β-glucopyranoside (M10), isoorientin (M11), narirutin (M12), hesperidin (M13), 6'-O-sinapoylgeniposide (M14), 3,5-dihydroxy-3',4',7,8-tetramethoxyl flavones (M15), 3',4',3,5,6,8-hexamethoxyl flavone (M16), kaempferide (M17), and tangeretin (M18).

Results: The separation by gradient elution was achieved on SHIMADZU VP-ODS column (4.6 × 150 mm, 5 μm) at 30°C with methanol (A)/0.1% phosphoric acid (B) as the mobile phase. The detection wavelengths were 254, 278, and 335 nm. The optimized HPLC method provided a good linear relation ($r \geq 0.9991$ for all the target compounds), satisfactory precision (RSD values less than 1.47%) and good recovery (97.40% to 103.44%). The limits of detection ranged between 0.20×10^{-4} and 64.90×10^{-4} μg/μL for the different analytes. Furthermore, the optimum sample preparation was obtained from HPD_{100} column eluted with water and 95% ethanol, respectively.

Conclusions: Quality control of ZYQL products, in total seven samples and twelve parent plants, was examined by this method, and results confirmed its feasibility and reliability in practice.

Keywords: Zhuyeqing Liquor; Bioactive constituent; Quantitative analysis; HPLC-PDA

Background

Zhuyeqing Liquor (ZYQL), authorized as a functional health liquor in 1998 by the Ministry of Public Health in China, is a famous traditional Chinese functional liquor. The history of ZYQL could be traced back to the Warring States Period and became popular among people in the South and North Dynasties. In the Tang Dyansty and Song Dynasty, it had reached its climax (Yang 2007). ZYQL was designed based on the principles of traditional Chinese medicine (TCM) and comprises 12 herbs: *Lophatherum gracile* Brongn. (Zhuye), *Gardenia jasminoides* Ellis (Zhizi), *Lysimachia capillipes* Hemsl. (Paicao), *Angelica sinensis* (Oliv.) Diels (Danggui), *Kaempferia galanga* L. (Shannai), *Citrus reticulata* Blanco (Chenpi), *Chrysanthemum morifolium* Ramat. (Juhua), *Amomum villosum* Lour. (Sharen), *Santalum album* L. (Tanxiang), *Eugenia caryophyllata* Thunb. (Gongdingxiang), *Aucklandia lappa* Decne. (Guangmuxiang), and *Lysimachia foenum-graecum* Hance (Linglingxiang). According to its long-term history use, ZYQL has various biological properties such as anti-oxidant, anti-fatigue, and immunoenhancement (Han 2007).

Up to now, many studies show solicitude for the color, smell, and taste of the health functional liquor; few studies pay close attention to its chemical constituents and quality control. Currently, chemical analytical methods for the quality control of ZYQL have not been established. Therefore, it is necessary to establish a rapid and effective method

* Correspondence: wjh.1972@aliyun.com
[1]School of Traditional Chinese Materia Medica 49#, Shenyang Pharmaceutical University, Wenhua Road 103, Shenyang 110016, People's Republic of China
[2]School of Pharmacy, Shihezi University, Shihezi 832002, People's Republic of China
Full list of author information is available at the end of the article

for the quantitative analysis of the health functional liquor. In this study, the system of high-performance liquid chromatography coupled with photodiode array detector (HPLC-PDA) was used for analyzing the chemical profile of ZYQL. This method includes many advantages like high speed detection, excellent peak shapes, less solvent usage, well-defined chemical constituents, and simultaneous detection of multi-constituents, which is better than finger-printing. Thus, simultaneous determination by RP-HPLC method is suitable for quantitative analysis and can be used as an effective tool to evaluate herbal medicine products.

Methods

Chemicals and materials

Methanol (HPLC-grade) was purchased from Fisher Scientific Co. (Franklin, MA, USA). Water for HPLC analysis was purified by a Milli-Q water purification system (Millipore, Billerica, MA, USA). Phosphoric acid (analytical grade) was purchased from Tianjin Guangfu Chemical Reagent Co. Ltd. (Tianjin, China). Other solvents from Tianjin Guangfu Chemical Reagent Co. Ltd. (Tianjin, China) were all of analytical grade.

Reference compounds of **M1** to **M18** (Figure 1) were isolated previously from ZYQL by author, structures of which were elucidated by comparison of spectral data (UV, MS, ^{1}H NMR, and ^{13}C NMR) with the literature data (Lin et al. 2006; Okamura et al. 1998; Huang et al. 2012; Zhang and Chen 1997; Ma et al. 2009; Miyake et al. 2007; Liu et al. 2011; Chen et al. 2008; Rayyan et al. 2005; Kumarasamy et al. 2004; Ke et al. 1999; Yoo et al. 2002; Dinda et al. 2011; Esteban et al. 1986; Ballester et al. 2013; Wang et al. 2010; Hòrie et al. 1998). The purity of each reference standard

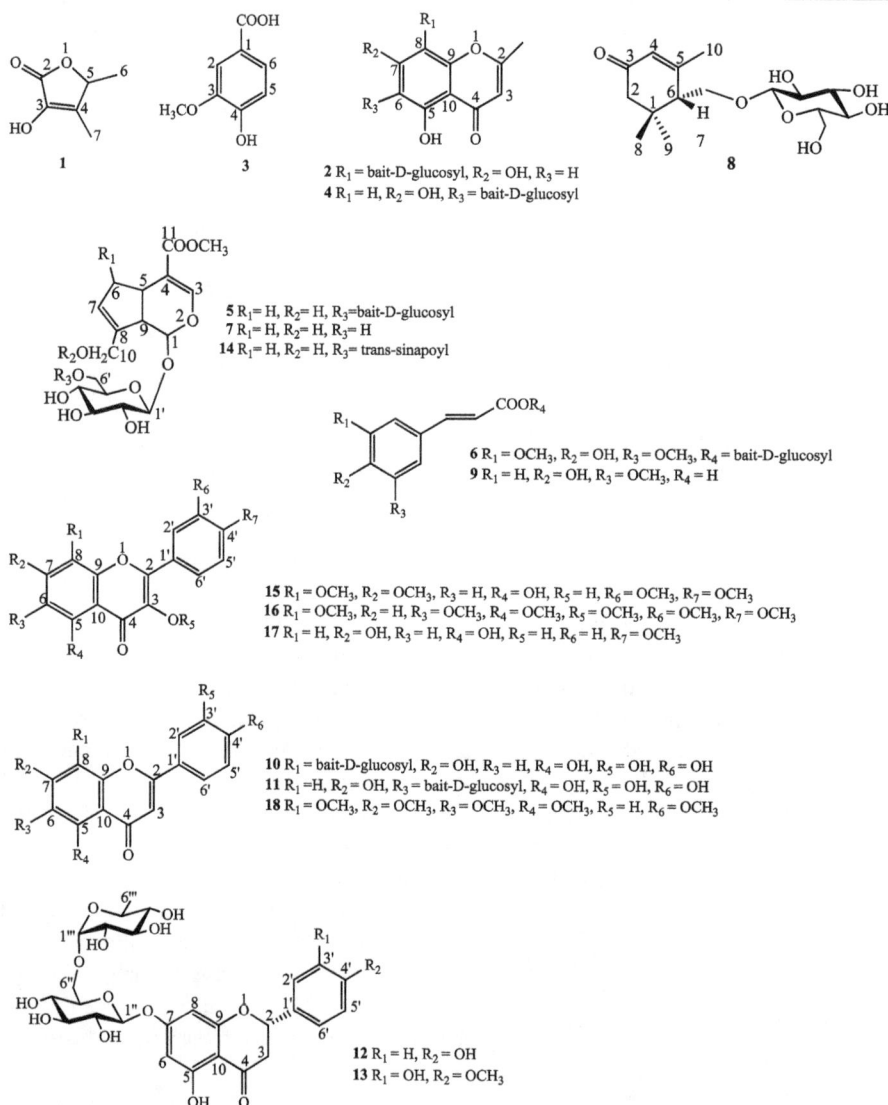

Figure 1 Structures of compounds M1 to M18.

was determined to be above 98% by HPLC analysis based on a peak area normalization method, detected by HPLC-PDA and confirmed by HR-ESI-TOF-MS and NMR spectroscopy.

The samples of different batch and different alcoholicity of ZYQL and the 12 parent plants were provided by Shanxi XinghuaCun Fen Jiu Group Co., Ltd. (Shanxi, China). The 12 parent plants were identified by Professor Jincai Lu (Shenyang Pharmaceutical University, Shenyang, China). The voucher specimen was deposited at Shenyang Pharmaceutical University (Shenyang, China) and registered under the number ZYQL 2011050101.

Instrumentation and chromatographic conditions

Chromatographic analysis was performed on Waters 2695 Alliance HPLC system (Waters Co., Milford, MA, USA) with Waters 2998 PDA detector. Chromatographic separation was carried on a SHIMADZU VP-ODS column (4.6 mm × 150 mm, 5 μm; Shimadzu, Kyoto, Japan) at a column temperature of 30°C using methanol (A) and 0.1% phosphoric acid (B) as mobile phase with the gradient elution procedure show in Table 1. The flow rate was set at 1.0 ml/min and the detection wavelengths were 254 nm (for compounds **M1** to **M5**, **M7**, **M8**, and **M17**), 278 nm (for compounds **M12** and **M13**), and 335 nm (for compounds **M6**, **M9** to **M11**, **M14** to **M16**, and **M18**), which were chosen based on the maximum absorption of all the tested compounds. The injection volume was 10 μL, and the analytes were well separated in chromatographic conditions above.

Standard solution preparation

Individual stock solutions were prepared by dissolving the standards in methanol to obtain 3-hydroxy-4,5(R)-dimethyl-2(5H)-furanone (**M1**) 19.920 mg mL^{-1}, isobiflorin (**M2**) 8.330 mg mL^{-1}, vanillic acid (**M3**) 5.802 mg mL^{-1}, biflorin (**M4**) 3.911 mg mL^{-1}, genipin 1-O-β-D-gentiobioside (**M5**) 4.405 mg mL^{-1}, 1-sinapoyl-β-D-glucopyranoside (**M6**) 1.115 mg mL^{-1}, geniposide (**M7**) 23.804 mg mL^{-1}, epijasmnoside A (**M8**) 12.060 mg mL^{-1}, ferulic acid (**M9**) 2.515 mg mL^{-1}, luteolin 8-C-β-glucopyranoside (**M10**) 1.510 mg mL^{-1}, isoorientin (**M11**) 2.203 mg mL^{-1}, nairutin (**M12**) 1.032 mg mL^{-1}, hesperidin (**M13**) 4.801 mg mL^{-1},

6'-O-sinapoylgeniposide (**M14**) 5.312 mg mL^{-1}, 3,5-dihydroxy-3',4',7,8-tetramethoxyl flavones (**M15**) 5.021 mg mL^{-1}, 3',4',3,5,6,8-hexamethoxyl flavone (**M16**) 15.005 mg mL^{-1}, kaempferide (**M17**) 6.408 mg mL^{-1}, and tangeretin (**M18**) 17.155 mg mL^{-1}. A mixed solution containing all the 18 standards was prepared as accurately as 108 μL **M1**, 6.8 μL **M2**, 2.4 μL **M3**, 8.0 μL **M4**, 165 μL **M5**, 96 μL **M6**, 106 μL **M7**, 3.4 μL **M8**, 8.2 μL **M9**, 9.5 μL **M10**, 7.9 μL **M11**, 35 μL **M12**, 40 μL **M13**, 80 μL **M14**, 8.2 μL **M15**, 11 μL **M16**, 12 μL **M17**, and 4.6 μL **M18** and were placed in a 2-mL flask with stopper, diluted with methanol to make sure the volume reached 2 mL. All prepared solutions were respectively stored in a refrigerator at 4°C when not in use.

Treatment for samples

For the analysis, 40 mL of ZYQL were evaporated in vacuum at 50°C to dryness. The dry residue was processed as follows in order to obtain better analytical results: The residue was dissolved with water (10 mL) and applied to an HPD$_{100}$ column eluted with water (150 mL); the water eluent was discarded and then eluted with 95% ethanol (150 mL). The 95% ethanol eluent was condensed and dissolved with methanol and then placed in a 2-mL flask with stopper, with a methanol-metered volume. Prior to HPLC analysis, the sample solution was passed through a 0.22-μm millipore filter.

The 12 crude dried parent plants were pulverized and sifted through 40 mesh sieve, respectively. One gram of the powder from the parent plant was placed in a 50-mL flask with stopper, then weighed again correctly, and extracted by ultrasonic method with 20 mL methanol for 30 min. Then standing, it was cooled down to room temperature (22°C) and the weight was mended to the incipient weight with methanol. Prior to HPLC analysis, the sample solution was passed through a 0.22-μm millipore filter.

Validation of the method
Calibration curves

Linearity was established by the injection of 1, 2, 4, 8, 12, 16, and 20 μL of the mixed reference standard solution prepared, respectively. Calibration graphs were plotted subsequently based on linear regression analysis of the integrated peak (Y) versus content (X, μg).

Limits of detection and quantitation

In order to evaluate the limits of detection (LODs) and the limits of quantification (LOQs) of the compounds, mixed standard stock solution was further diluted serially to provide a series of appropriate concentrations, and an aliquot of the diluted solutions was injected into HPLC for analysis. The LOD and LOQ for each analyte was calculated with corresponding standard solution on the basis of a signal-to-noise ratio (S/N) of 3 and 10, respectively.

Table 1 Time program of the gradient elution

Time (min)	Flow (mL/min)	Methanol (%)	0.1% Phosphatic acid (%)
0	1	5	95
70	1	55	45
75	1	60	40
110	1	80	20
120	1	98	2
125	1	98	2

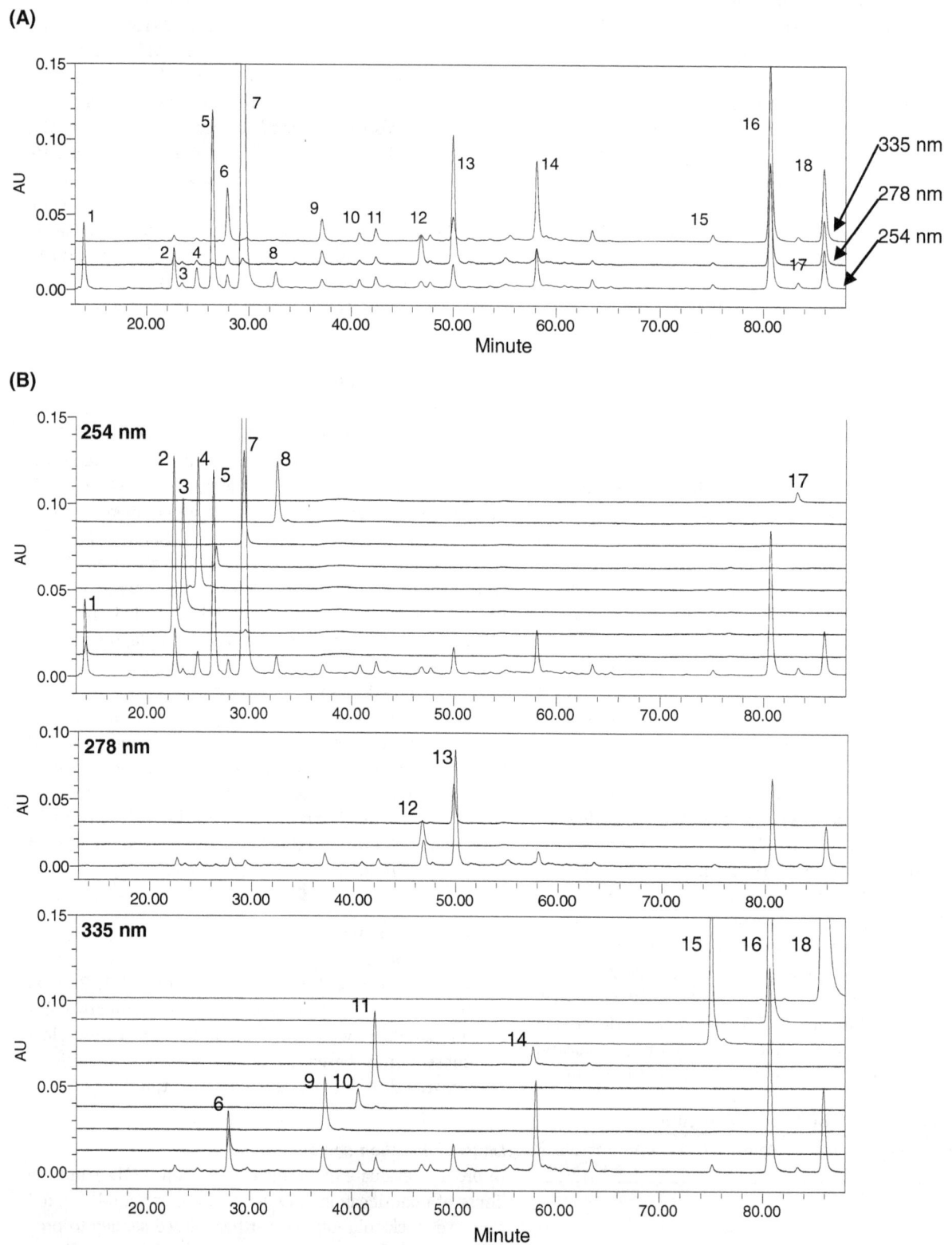

Figure 2 Stack views. (A) Different detector-wavelength HPLC chromatograms of mixed reference standards (from up to down: 335, 278, 254 nm). Column: SHIMADZU VP-ODS column (4.6 mm × 150 mm, 5 μm), temperature of 30℃. **(B)** HPLC chromatograms of **M1** to **M18** and mixed reference standards (from up to down: 254 nm **M1**, **M2**, **M3**, **M4**, **M5**, **M7**, **M8**, **M17**, mixed reference standards; 278 nm **M12**, **M13**, mixed reference standards; 335 nm **M6**, **M9**, **M10**, **M11**, **M14**, **M15**, **M16**, **M18**, mixed reference standards).

Table 2 Optimization of the treatment method of Zhuyeqing Liquor (μg/mL)

Compound[a]	Treatment method				
	Method 1	Method 2	Method 3	Method 4	Method 5
M1	ND	5.2494	3.1331	9.5943	16.4270
M2	0.0888	0.5371	0.5078	0.5818	0.5942
M3	0.0922	0.0906	0.0388	0.0939	0.1063
M4	0.0263	0.4980	0.5098	0.5069	0.5153
M5	ND	44.0545	101.6907	100.6372	101.7888
M6	0.1479	0.1880	0.1902	0.1927	0.1936
M7	45.2421	563.2436	570.3556	566.0022	574.2514
M8	20.8481	24.6279	25.6049	24.5513	25.7319
M9	0.1931	0.1782	0.1906	0.1918	0.1968
M10	0.0324	0.0526	0.0600	0.0705	0.0736
M11	0.2009	0.4601	0.4871	0.4915	0.4954
M12	0.7071	1.0544	1.0683	1.0699	1.0877
M13	1.1371	2.5392	2.8461	2.8029	2.9909
M14	ND	3.4939	4.0839	4.0563	4.3756
M15	ND	0.0641	0.0678	0.0688	0.0689
M16	0.5189	0.6113	0.5842	0.3282	0.6623
M17	2.9599	2.9776	2.1520	1.6446	2.9957
M18	0.4448	0.5307	0.4278	0.4673	0.5413
Sum[b]	72.6396	650.4512	713.9987	713.3521	733.0967

Sample in optimization of the treatment method was 45° Zhuyeqing Liquor (20130207). Method 1, acetoacetate extract; method 2, *n*-butanol extract; method 3, 70% ethanol treatment; method 4, SPE column eluted with methanol; method 5, HPD$_{100}$ column eluted with ethanol. [a]'ND' in the 'Compound' column expressed under LOQ. [b]Total content of the 18 investigated compounds.

Precision and stability

The precision of the chromatographic system was validated by injecting 10 μL of the mixed reference solution six times during 1 day. Stability study was performed with sample solution in 48 h (the time points are 0, 5, 10, 15, 25, 35, and 48 h, respectively). Variations were expressed by relative standard deviations (RSD) of peak area.

Repeatability and recovery

The repeatability test was analyzed by injecting six independently prepared samples (45° ZYQL (20130207), the concentration, and prepared method as the 'Treatment for samples'). The RSD value of concentration was adopted to evaluate repeatability. The recovery tests were studied by adding the proper amount of mixed-reference standard

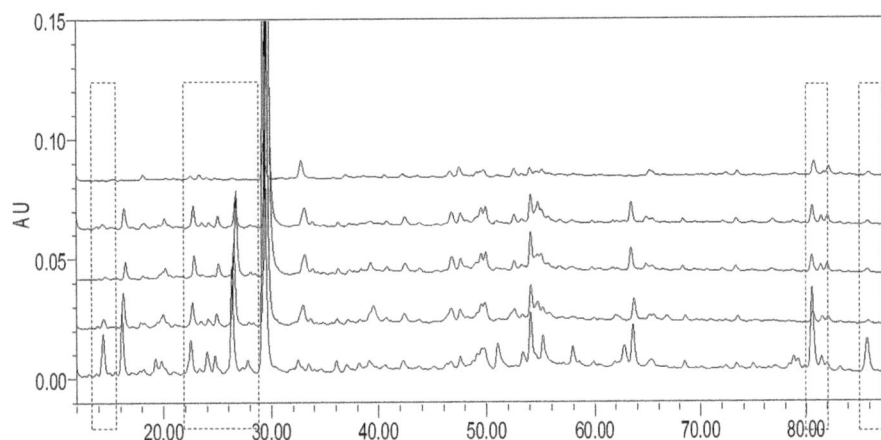

Figure 3 Stack views of 45° Zhuyeqing Liquor preparation method HPLC chromatograms (254 nm, from up to down: method 1, method 2, method 3, method 4 and method 5).

Table 3 Regression equations, correlation coefficients, and linear range for 18 analytes in Zhuyeqing Liquor

Analyte	Time (t_R)	Linear regression				
		Regression equation ($n = 3$)	Correlation coefficients r	Linear range (µg)	LOD (10^{-4} µg/µL)	LOQ (10^{-4} µg/µL)
M1	14.004	$Y = 5.48e + 003X - 1.45e + 003$	0.9998	$1.08 \sim 21.63$	64.90	216.34
M2	22.819	$Y = 2.86e + 006X - 8.84e + 002$	0.9999	$2.80 \times 10^{-2} \sim 5.60 \times 10^{-1}$	1.68	5.60
M3	23.573	$Y = 4.73e + 006X - 5.52e + 003$	0.9991	$1.70 \times 10^{-3} \sim 3.40 \times 10^{-2}$	0.20	0.67
M4	25.069	$Y = 1.62e + 006X - 3.23e + 003$	0.9997	$1.55 \times 10^{-2} \sim 3.10 \times 10^{-1}$	3.10	10.34
M5	26.671	$Y = 5.41e + 005X - 3.44e + 004$	0.9994	$3.63 \times 10^{-1} \sim 7.25$	2.42	8.08
M6	28.087	$Y = 1.31e + 006X - 4.70e + 003$	0.9998	$5.25 \times 10^{-2} \sim 1.05$	8.40	28.0
M7	29.646	$Y = 7.08e + 005X + 4.19e + 003$	0.9998	$1.26 \sim 25.21$	25.46	84.87
M8	32.875	$Y = 9.81e + 005 \ X - 5.87 \ e + 003$	0.9998	$2.04 \times 10^{-2} \sim 4.08 \times 10^{-1}$	4.08	13.60
M9	37.264	$Y = 3.17e + 006X - 1.51e + 004$	0.9992	$1.03 \times 10^{-2} \sim 2.06 \times 10^{-1}$	2.06	6.87
M10	40.883	$Y = 1.65e + 006X - 1.26e + 002$	0.9993	$7.10 \times 10^{-3} \sim 1.42 \times 10^{-1}$	2.13	7.11
M11	42.396	$Y = 2.38e + 006X - 6.23e + 003$	0.9991	$8.70 \times 10^{-3} \sim 1.74 \times 10^{-1}$	1.74	5.83
M12	46.878	$Y = 2.67e + 006X - 3.22e + 002$	0.9998	$1.75 \times 10^{-2} \sim 3.50 \times 10^{-1}$	5.25	17.51
M13	50.008	$Y = 1.69e + 006 \ X + 3.71e + 003$	0.9998	$9.56 \times 10^{-2} \sim 1.91$	5.74	19.12
M14	58.096	$Y = 4.55e + 005X - 2.11e + 003$	0.9998	$2.13 \times 10^{-1} \sim 4.25$	12.75	42.50
M15	74.987	$Y = 2.15e + 006X - 7.62e + 002$	0.9991	$4.10 \times 10^{-3} \sim 8.20 \times 10^{-2}$	3.28	10.95
M16	80.609	$Y = 3.59e + 006X - 2.08e + 004$	0.9999	$8.23 \times 10^{-2} \sim 1.65$	0.55	1.84
M17	83.248	$Y = 1.97e + 005X - 3.36e + 003$	0.9991	$3.85 \times 10^{-2} \sim 7.70 \times 10^{-1}$	15.40	51.32
M18	85.890	$Y = 2.69e + 006X - 1.38e + 004$	0.9996	$3.93 \times 10^{-2} \sim 7.86 \times 10^{-1}$	0.79	2.64

Y is the peak area and X is the content of standard solutions; LOD refers to the limits of detection, S/N = 3; LOQ refers to the limits of quantity, S/N = 10.

Table 4 Precision, stability, recovery, and repeatability data of 18 analytes in Zhuyeqing Liquor

Analyte	Precision ($n = 6$)		Stability RSD (%)	Recovery ($n = 6$)					Repeatability ($n = 6$)	
	Concentrations (mg/mL)	RSD (%)		Original (µg)	Spiked (µg)	Detected (µg)	Recovery (%)	RSD (%)	Average concentration (µg/mL)	RSD (%)
M1	1.08	0.92	1.70	164.16	162.26	326.57	100.17	3.10	16.0862 ± 0.2722	1.50
M2	2.80×10^{-2}	0.82	1.52	7.11	4.20	11.41	102.44	1.84	0.5580 ± 0.0045	1.39
M3	1.70×10^{-3}	1.30	1.66	0.27	0.26	0.52	97.40	2.52	0.1072 ± 0.0020	1.96
M4	1.55×10^{-2}	0.78	1.08	5.88	2.33	8.23	100.88	2.27	0.5154 ± 0.0032	0.64
M5	3.63×10^{-1}	0.87	1.62	1217.80	54.39	1273.19	101.83	3.30	101.1963 ± 0.7206	0.72
M6	5.25×10^{-2}	0.86	1.58	6.00	7.88	14.15	103.44	1.68	0.1940 ± 0.0019	1.00
M7	1.26	0.49	1.18	1270.60	1260.30	2543.54	101.00	0.78	568.4991 ± 2.7722	0.98
M8	2.04×10^{-2}	1.04	1.76	211.50	102.00	316.34	102.79	1.46	25.1942 ± 0.2548	1.45
M9	1.03×10^{-2}	0.30	1.49	2.82	1.55	4.37	100.76	2.63	0.1947 ± 0.0018	1.00
M10	7.10×10^{-3}	1.10	1.67	3.22	1.07	4.28	99.54	3.26	0.0728 ± 0.0011	1.61
M11	8.70×10^{-3}	1.15	1.62	4.65	1.31	5.96	100.66	3.09	0.5101 ± 0.0057	1.28
M12	1.75×10^{-2}	1.09	1.39	2.43	2.63	5.14	103.13	1.48	1.0800 ± 0.0155	1.44
M13	9.56×10^{-2}	1.02	1.36	37.69	14.34	52.20	101.14	1.99	2.8911 ± 0.0351	1.21
M14	2.13×10^{-1}	0.74	1.59	943.73	31.88	975.00	98.10	2.25	4.2790 ± 0.0314	1.30
M15	4.10×10^{-3}	1.47	1.75	1.99	1.23	3.22	100.30	2.66	0.0645 ± 0.0009	1.53
M16	8.23×10^{-2}	0.91	1.58	21.52	12.35	33.77	99.20	2.44	0.6444 ± 0.0052	0.81
M17	3.85×10^{-2}	0.93	1.75	133.12	115.50	249.54	100.79	2.63	2.9457 ± 0.0498	1.36
M18	3.93×10^{-2}	1.11	1.49	10.02	11.79	21.89	100.65	1.58	0.5368 ± 0.0060	1.13

RSD refers to relative standard deviation. Samples in stability, recovery, and repeatability methods were taken from 45°Zhuyeqing Liquor (20130207).

solution to the sample (45° ZYQL (20130207)), and then processed by the method described in the 'Treatment for samples' section to yield the final concentration. The experiment was repeated six times.

Results and discussion

Optimization of chromatographic conditions

To improve resolution and sensitivity of analysis but reduce analytical time, the following chromatographic conditions were optimized (Gao et al. 2013), including different mobile phase compositions (methanol, acetonitrile, and aqueous phosphatic acid of different concentrations), column temperature, and wavelength: To inhibit ionization of the acidic ingredients in the ZYQL sample, phosphatic acid was added in mobile phase. Two mobile phase systems, methanol-phosphatic acid aqueous solution and acetonitrile-phosphatic acid aqueous solution, were examined, and then column temperatures at 25°C, 30°C, 40°C, and 50°C were compared. A sensitive wavelength was determined by PDA with reference compounds. Present researches indicated that better separation and results were obtained using a mobile

phase of water and methanol rather than water and acetonitrile. Therefore, in this work, the optimum resolution was achieved using methanol (A) and 0.1% phosphatic acid (B) as mobile phase, with a column temperature of 30°C at different detection wavelengths, which were described in 'Instrumentation and chromatographic conditions' section, with gradient elution (Table 1). All 18 standard analytes could be eluted with baseline separation in 90 min. Representative chromatograms for the mixed reference standard and 18 standard compounds were shown in Figure 2A,B.

Optimization of sample preparation

In order to eliminate the water-soluble constituents and obtain the liposoluble constituents, the optimization of sample preparation was performed using 45° ZYQL (20130207). Forty milliliters of ZYQL was evaporated in vacuum at 50°C to dryness. And the following five methods were choosen to select the best method for sample preparation. First, the dry residue was suspended with water (10 mL) and extracted with acetoacetate (10 mL). The acetoacetate extract was condensed and then methanol was used to meter the volume

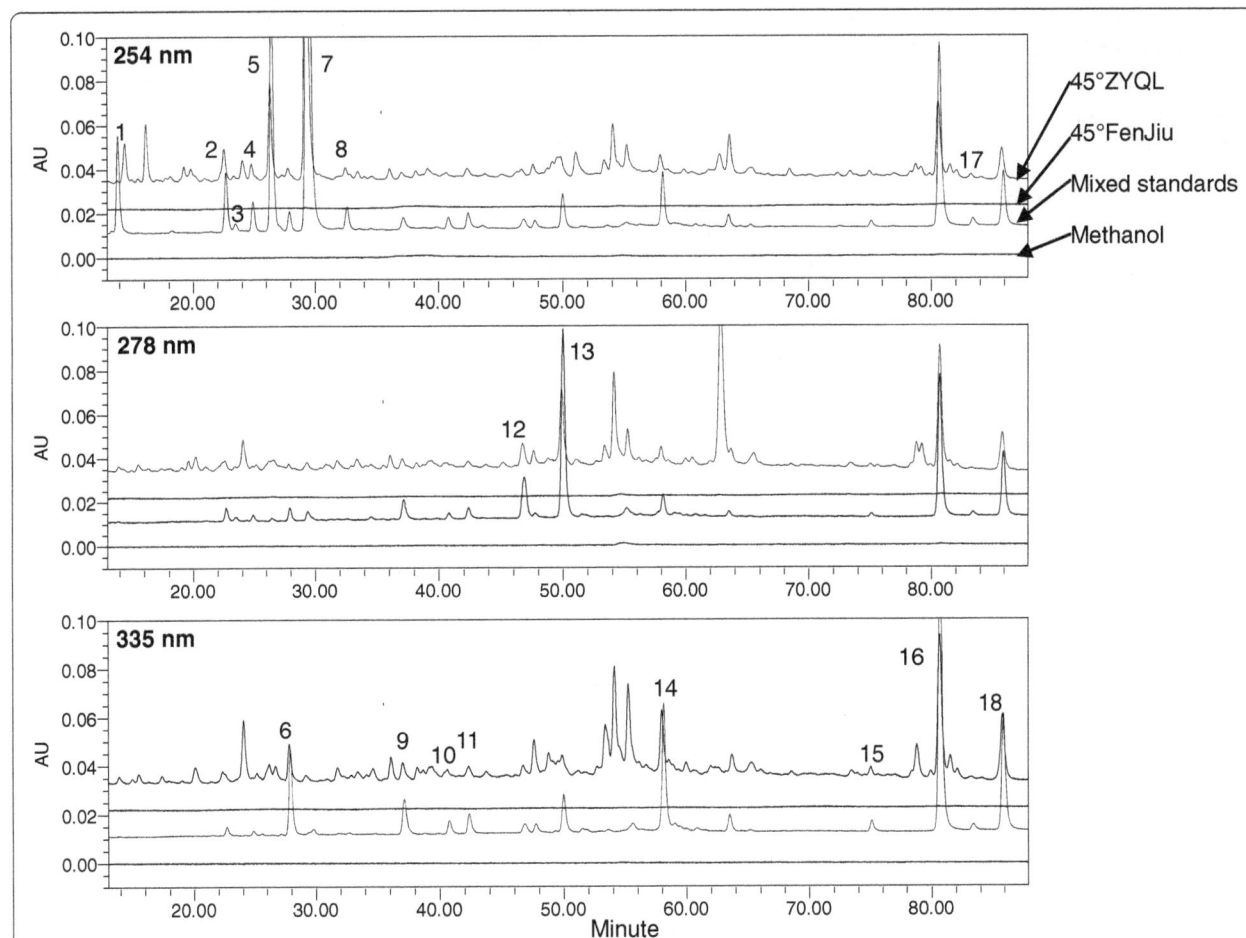

Figure 4 Stack views of different detector-wavelength HPLC chromatograms. (From up to down: 45° Zhuyeqing Liquor, 45° FenJiu, mixed reference standards, blank solvent: methanol, respectively).

(2 mL). Second, the dry residue was suspended with water (10 mL) and extracted with *n*-butanol (10 mL) The *n*-butanol extract was condensed and then methanol was used to meter the volume (2 mL). Third, the dry residues was dissolved with 70% ethanol (20 mL) to precipitate the polysaccharide and then condensed the supernate, use methanol to metered volume (2 mL). Fourth, the dry residues was dissolved with water (10 mL) as fraction A, then the remanent residues was dissolved with methanol (10 mL) as fraction B. Fraction A was applied to an SPE column eluted with water (150 mL); the water eluent was discarded; fraction B was applied to the same SPE column eluted with methanol (150 mL); the methanol eluent was condensed and methanol was used to meter the volume (2 mL). Fifth, the dry residue was dissolved with water (10 mL) and applied to an HPD_{100} column eluted with water (150 mL). The water eluent was discarded and then eluted with 95% ethanol (150 mL). The 95% ethanol eluent was condensed and then methanol was used to meter the volume (2 mL). Comparing the analytical results of the target constituents, though the former three methods proved to be more simple than the other, they could not obtain all the tested constituents and some content too lower to accurately reflect the real content. So, these three methods were deserted. The fourth one although could obtain all the tested constituents but at a lower content. Therefore, the optimized condition was selected, the fifth one (Table 2, Figure 3).

Validation of the method

The method was validated in terms of linearity, LOD and LOQ, precision, repeatability, stability, and recovery test. All calibration curves exhibited good linearity ($r \geq 0.9991$) in a relatively wide linear range as shown in Table 3. For the quantified compounds, the LOD and LOQ were $0.20 \times 10^{-4} \sim 64.90 \times 10^{-4}$ µg/µL and $0.67 \times 10^{-4} \sim 216.34 \times 10^{-4}$ µg/µL, respectively (Table 3), which were calculated with corresponding standard solution on the basic of a signal-to-noise ratio (S/N) of 3 and 10, respectively. Table 4 showed the results of precision, stability, recovery and repeatability of the 18 analytes. It was indicated that the RSD of the precision variations were less than 1.47% for all 18 analytes. The RSD of repeatability was less than 1.96% for all the analysis, which proved that this assay had good reproducibility. Stability test results, with RSD less than 1.76%, indicated that the sample solution was stable at room temperature for at least 48 h. The mean recovery rates, which ranged from 97.40% to 103.44% with RSD values less than 3.30% for the analytes concerned, showed that the developed analytical method had good accuracy. All these values fall within acceptable limits, which indicates this HPLC method is reliable with significant repeatability, recovery rate, and precision. The results proved that HPLC is appropriate for analyzing and assessing the quality of ZYQL.

Table 5 Contents of 18 analytes in different batches and different alcoholicity of Zhuyeqing Liquor (µg/mL)

Compound[a]	45° FenJiu	38°	42°	45°				
	20130207	20130207	20130207	20130207	20120601	20110507	20100417	20090302
M1	ND	15.1395	15.1395	16.7150	16.6674	16.6558	16.6543	16.6239
M2	ND	0.3689	0.3945	0.5851	0.5854	0.5846	0.5839	0.5840
M3	ND	0.0792	0.0830	0.1063	0.1064	0.1058	0.1032	0.1060
M4	ND	0.3028	0.3742	0.5180	0.5139	0.5106	0.5081	0.5111
M5	ND	65.7206	71.8348	101.7175	101.3777	101.0265	101.1293	100.9297
M6	ND	ND	ND	0.1930	0.1904	0.1901	0.1893	0.1934
M7	ND	273.1958	309.9846	574.4770	574.1508	574.1103	573.6367	573.2887
M8	ND	12.6644	19.9230	25.8595	25.4397	25.2009	25.4234	25.1453
M9	ND	0.1395	0.1659	0.1956	0.1934	0.1909	0.1933	0.1938
M10	ND	0.0536	0.0612	0.0715	0.0714	0.0715	0.0715	0.0721
M11	ND	0.2612	0.2963	0.4993	0.4972	0.4983	0.4919	0.5000
M12	ND	0.5690	0.6861	1.0884	1.0835	1.0822	1.0806	1.0830
M13	ND	2.0079	2.0709	2.9740	2.9731	2.9587	2.8982	2.9883
M14	ND	2.5058	2.7162	4.3718	4.3672	4.3615	4.3586	4.3699
M15	ND	0.0549	0.0585	0.0689	0.0686	0.0684	0.0685	0.0686
M16	ND	0.4758	0.5602	0.6434	0.6435	0.6422	0.6427	0.6383
M17	ND	2.6999	2.8338	3.0047	2.9943	2.9927	2.9836	2.9735
M18	ND	0.3534	0.4045	0.5495	0.5432	0.5427	0.5389	0.5433
Sum[b]	ND	376.5922	427.5872	733.6385	732.4671	731.7937	731.5560	730.8129

[a]'ND' in the 'Compound' column expressed under LOQ. [b]Total content of the 18 investigated compounds.

Table 6 Contents of 18 analytes in 12 parent plants (mg/g)

Compound[a]	Zhuye	Zhizi	Paicao	Danggui	Shannai	Chenpi	Juhua	Sharen	Tanxiang	Gongdingxiang	Guangmuxiang	Linglingxiang
M1	ND	2.1057	ND	ND	ND	ND	ND	ND	ND	ND	ND	19.8802
M2	ND	ND	ND	ND	ND	ND	ND	ND	ND	4.7841	ND	0.0790
M3	ND	ND	ND	ND	ND	ND	ND	ND	0.0038	ND	ND	ND
M4	ND	ND	ND	ND	ND	ND	ND	ND	ND	5.4547	ND	0.0498
M5	ND	19.9996	ND	ND	ND	ND	ND	ND	ND	ND	ND	ND
M6	ND	1.6513	ND	ND	ND	0.2580	ND	ND	ND	ND	ND	ND
M7	ND	43.3886	ND	2.5174	ND	ND	ND	ND	ND	ND	ND	ND
M8	ND	5.0078	ND	ND	ND	1.0468	ND	ND	ND	ND	ND	ND
M9	ND	ND	ND	0.6170	ND	0.0805	ND	ND	ND	ND	ND	ND
M10	0.3268	ND	ND	ND	ND	ND	ND	ND	ND	ND	ND	ND
M11	0.4161	ND	ND	ND	ND	ND	ND	ND	ND	ND	ND	ND
M12	ND	0.0111	ND	ND	ND	3.6198	0.2104	ND	ND	ND	0.0217	ND
M13	ND	ND	ND	ND	ND	4.6719	0.3139	ND	ND	ND	ND	ND
M14	ND	8.6950	0.0302	ND	ND	ND	ND	0.0292	0.0025	ND	ND	ND
M15	ND	ND	ND	ND	ND	0.1017	0.0472	ND	ND	ND	ND	ND
M16	0.0638	0.0750	ND	0.2479	17.7933	0.6813	0.4755	ND	ND	ND	ND	0.3214
M17	ND	1.9330	ND	ND	ND	0.8970	1.4942	ND	ND	ND	ND	ND
M18	ND	0.0648	ND	ND	ND	0.5263	0.1625	ND	ND	ND	ND	ND
Sum[b]	0.8067	82.9319	0.0302	3.3823	17.7933	11.8833	2.7037	0.0292	0.0063	10.2388	0.0217	20.3304

[a]'ND' in the 'Compound' column expressed under LOQ. [b]Total content of the 18 investigated compounds.

Sample analysis

The HPLC analytical method described above was subsequently used to simultaneously quantify 18 compounds in seven commercial products and 12 parent plants supplied by Shanxi XinghuaCun Fen Wine Group Co., Ltd. (Shanxi, China). Generally, the 18 compounds were authenticated by comparison of their retention times and MS spectra with those of reference standards. The representative HPLC chromatograms of mixed standard solution and sample solutions are shown in Figure 4. The analytical results are summarized in Tables 5 and 6. According to the chromatographic results shown in Table 5, there was no any constituents to be detected in 45° FenJiu (solvent of ZYQL). Moreover, the concentration of compounds **M1** to **M18** in 45° ZYQL were higher than those in 42° and 38°, which showed that with the increase of alcoholicity, the content of bioactive constituents increased as well. In addition, there was no content difference between the successive 5 years of 45° ZYQL. This indicated that the quality of 45° ZYQL was stable for at least 5 years.

Table 6 showed the content of compounds in 12 parent plants, which exhibited that the major bioactive constituents were mainly from *Gardenia jasminoides* Ellis (Zhizi), *Kaempferia galanga* L. (Shannai), *Citrus reticulata* Blanco (Chenpi), and *Lysimachia foenum-graecum* Hance (Linglingxiang). And this result was greatly useful and helpful for the quality control and further formula optimization of the technical study of Zhuyeqing Liquor.

Conclusions

An HPLC-PDA method has been developed for the simultaneous determination of 18 major compounds extracted from ZYQL for the first time. The validation data indicated that this method is reliable and can be applied to determine the contents of the 18 compounds in different ZYQL products. This valuable information concerning the concentration of these bioactive constituents in ZYQL could be of great importance for the quality assessment and should therefore be useful for the guidance of development of the new health care products. Furthermore, this HPLC-PDA assay supplies a rapidness and effectiveness method for the simultaneous determination of multiple constituents in ZYQL.

Competing interests

The authors declare that they have no competing interests.

Authors' contributions

HYG carried out the whole experiment, SYW participated in the sample preparation, HYG and JHW performed the statistical analysis and drafted the manuscript. All authors read and approved the final manuscript.

Acknowledgements

Grateful acknowledgement is made to the Shanxi XinghuaCun Fen Jiu Group Co., Ltd. (Shanxi Province, China) and National Key Technology R&D Program (2012BAI30B02) for financial support of this work. The authors acknowledge Waters Co. Ltd. for Cooperation Laboratory.

Author details

[1]School of Traditional Chinese Materia Medica 49#, Shenyang Pharmaceutical University, Wenhua Road 103, Shenyang 110016, People's Republic of China. [2]School of Pharmacy, Shihezi University, Shihezi 832002, People's Republic of China. [3]Shanxi Xinghuacun Fen Jiu Group Co., Ltd, Shanxi 450000, People's Republic of China.

References

Ballester AR, Lafuente MT, De Vos RCH, Bovy AG, González-Candelas L (2013) Citrus phenylpropanoids and defence against pathogens. Part I: metabolic profiling in elicited fruits. Food Chem 136:178–185

Chen QC, Youn UJ, Min BS, Bae KH (2008) Pyronane monoterpenoids from the fruit of Gardenia jasminoides. J Nat Prod 71:995–999

Dinda B, Debnath S, Banik R (2011) Naturally occurring iridoids and secoiridoids. An updated review, part 4. Chem Pharm Bull 59:803–833

Esteban MD, González Collado I, Macías FA, Massanet GM, Rodríguez Luis F (1986) Flavonoids from Artemisia lanata. Phytochemistry 25:1502–1504

Gao HY, Huang J, Wang HY, Du XW, Cheng SM, Han Y, Wang LF, Li GY, Wang JH (2013) Protective effect of Zhuyeqing liquor, a Chinese traditional health liquor, on acute alcohol-induced liver injury in mice. J Inflamm 10:30

Han Y (2007) Investigation on the immunoregulation functions of Zhuyeqing Liquor. Liq Making Sci Technol 2:117–119

Hòrie T, Ohtsuru Y, Shibata K, Yamashita K, Tsukayama M, Kawamura Y (1998) ^{13}C NMR spectral assignment of the a-ring of polyoxygenated flavones. Phytochemistry 47:865–874

Huang Y, Chang RJ, Jin HZ, Zhang WD (2012) Phenolic constituents from Tsoongiodendron odorum chun. Tianran Chanwu Yanjiu Yu Kaifa 24:176–178

Ke Y, Jiang Y, Luo SQ (1999) Chemical constituents from Clinopodium chinense (Benth.) O. Ktze. Chin Traditional Herbal Drugs 30:8–10

Kumarasamy Y, Byres M, Cox PJ, Delazar A, Jaspars M, Nahar L, Shoeb M, Sarker SD (2004) Isolation, structure elutiondition, and biological activity of flavone 6-C-glycosides from Alliaria petiolata. Chem Nat Compd 40:122–128

Lin LB, Fu XW, AI CH, Shen J, Wei K, Li W (2006) Studies on chemical constituents in leaves of Mallotus furetianus. China J Chin Materia Medica 31:477–479

Liu XM, Jiang Y, Sun YQ, Xu XW, Tu PF (2011) Chemical constituent study of Herba Cistanches. Chin J Pharm 46:1053–1058

Ma ZT, Yang XW, Zhong GY (2009) A new flavonoid glucoside from Huanglian jiedutang decoction. China J Chin Materia Medica 34:1097–1100

Miyake Y, Mochizuki M, Okada M, Hiramitsu M, Morimitsu Y, Osawa T (2007) Isolation of antioxidative phenolic glucosides from lemon juice and their suppressive effect on the expression of blood adhesion molecules. Biosci Biotech Biochem 71:1911–1919

Okamura N, Hine N, Tateyama Y, Nakazawa M, Fujioka T, Mihashi K, Yagi A (1998) Five chromones from Aloe vera leaves. Phytochemistry 49:219–223

Rayyan S, Fossen T, Solheim Nateland H, Andersen ØM (2005) Isolation and identification of flavonoids, including flavone rotamers, from the herbal drug 'crataegi folium cum flore' (hawthorn). Phytochem Anal 16:334–341

Wang QJ, Wang YS, He L, Lou ZP, Zang S (2010) Study on chemical constituents from Ipomoea Pescaprae (L.) Sweet. Chin J Marine Drugs 29:41–44

Yang HP (2007) Zhuyeqing liquor in different historical period. Liquor Making Sci Technol 2:110–112

Yoo SW, Kim JS, Kang SS, Son KH, Chang HW, Kim HP, Bae KH, Lee CO (2002) Constituents of the fruits and leaves of Euodia daniellii. Arch Pharm Res 25:824–830

Zhang YW, Chen YW (1997) Isobiflorin, a chromone C-glucoside from cloves (Eugenia Caryophyllata). Phytochemistry 45:401–403

23

Novel catalytic fluorescence method for speciative determination of chromium in environmental samples

Sunil Adurty[*] and Jagadeeswara Rao Sabbu

Abstract

Background: Thiourea derivatives act as promising chemosensors for sensing transition metal ions. 1-(2-hydroxyphenyl) thiourea (HPTU) is one such chromophore that has potential for metal ion sensing. The current investigation reports the sensing of chromium species using transition metal-oxo-based reaction of 1,2-hydroxyphenylthiourea.

Methods: The catalytic effect of chromium (III) and chromium (VI) on the oxidation of HPTU was studied. The reaction was followed spectrofluorimetrically by measuring the fluorescence intensities of the reaction product at $\lambda_{ex} = 416$ and $\lambda_{em} = 520$ nm, respectively.

Results: Under the optimum analytical conditions, HPTU acts as a chromogenic sensor for the detection of chromium species in nano-gram levels with a determination range of 0.3 to 250 ng/mL.

Conclusions: The methods are fairly sensitive, and the role of activators and sensitizers in enhancing the catalysis was studied. Interference due to various cations and anions in the experiment was investigated. The proposed method was applied to environmental samples for the analysis of chromium content.

Keywords: 1-(2-hydroxyphenyl) thiourea; Chromium; Speciative determination; Catalytic fluorescence

Background

The toxicological and biological characteristics of many transition metals like chromium are related to their chemical forms. A great interest in chromium speciation originates from applications of this metal in various industrial activities such as tanning of leather, electroplating, pigment production and wood preservation. Owing to these industrial processes, large amounts of chromium compounds discharge into the environment, which can affect biology and ecology of the environment. Therefore, speciation analysis of chromium is of great importance to assess pollution levels. Chromium mainly exist in two oxidation states, i.e. Cr(III) and Cr(VI). Cr(III) appears to be one of the essential elements for the proper functioning of living organisms, effective in the maintenance of normal glucose, cholesterol and fatty acid metabolism, while water soluble Cr(VI) is toxic to human and living organisms and

was found to be carcinogenic. Due to the different toxicities of Cr(III) and Cr(VI), and due to their association in many sample matrices, it is necessary to develop methods where both species can be determined simultaneously (Kotas and Stasicka 2000).

In the past years, various analytical techniques such as atomic absorption spectrometry (AAS) (Karosi et al. 2006; Ren et al. 2007) spectrophotometry (Wu et al. 2007), stripping voltammetry (SV) (Grabarczyk et al. 2006), inductively coupled plasma-mass spectrometry (ICP-MS) (Sun et al. 2006), inductively coupled plasma-optical emission spectrometry (ICPOES) (Schramel et al. 1992), and high performance liquid chromatography (HPLC) (Padarauskas and Naujalis 1998) have been successfully used to determine chromium in various samples. An extensive coverage of the available methods for chromium determination was put forth by Gomez and Callao, including the various types of sample matrices selected for the determination (Gomez and Callao 2006). Reagents such as bis-[2-hydroxy-1-naphthaldehyde] thiourea (Kiran et al. 2008), quercetin (Hosseini and Belador, 2009), chromotropic

---

* Correspondence: asunil@sssihl.edu.in
Department of Chemistry, Sri Sathya Sai Institute of Higher Learning (Deemed to be university), Prasanthi Nilayam-515134, Puttaparthi, Anantapur District, Andhra Pradesh, India

acid (CA) (Themelis et al. 2006), bis (salicylaldehyde) orthophenylenediamine (BSOPD) (Arancibia et al. 2012 and Soomro et al. 2011), etc. have been used. Methods such as solidified floating organic drop microextraction (SFODME) in combination with graphite furnace atomic absorption spectrometry (GFAAS) (Moghadam et al. 2011), cloud point extraction (CPE) using diethyldithiocarbamate (DDTC) as the chelating agent (Yildiz et al. 2011), solid phase extraction procedure using ICP-MS (Guerrero et al. 2012), fluorescence method using tetraphenylphosphonium bromide ($TPP^+ \cdot Br^-$) (El-Shahawi et al. 2011), EPA methods 3060A and 3052 (Martone et al. 2013), a disposable dual screen-printed electrode method using batch and flow analysis (Sánchez-Moreno et al. 2010), p-aminoacetophenone and phloroglucinol (Parmar et al. 2010), ultrasound-assisted cloud point extraction (UACPE) (Hashemi and Daryanavard 2012), electrospray ionization mass spectrometry using CYDTA (Hotta et al. 2012), HPLC and preconcentration by CPE with 1-(2-thiazolylazo)-2-naphthol (TAN) as the chelating agent (Wang et al. 2010), ytterbium (III) hydroxide (Duran et al. 2009), mixed-micelle cloud point extraction using electrothermal atomic absorption spectrometry (ET-AAS) (Ezoddin et al. 2010), room temperature ionic liquids (RTILs) for hollow fiber liquid phase microextraction (HF-LPME) combined with flame atomic absorption spectrometry (FAAS), etc., have been developed (Zeng et al. 2012).

It is evident from the literature that hexavalent chromium compounds are 10 to 100 times more toxic than trivalent chromium compounds when administered orally. The World Health Organization (WHO) and the European Community Directive (ECD) for drinking water has set the limit of total chromium not exceeding 50 µg L^{-1}, while the maximum concentration criterion for Cr(VI) in freshwater is 16 µg L^{-1}, sea water contains between 0.1 and 0.5 µg L^{-1} and unpolluted river water from 0.3 to 0.6 µg L^{-1}. Since the concentration of chromium, mainly Cr(VI), is very low in many natural waters, a highly sensitive method is required for its speciation (Arancibia et al. 2012). Innumerable techniques and methods were reported for chromium speciation using highly sophisticated equipment. Among the plethora of the methods, photometric and fluorometric methods are comparatively simple and relatively sensitive. Hence, speciative determination of chromium by parameter selective approach using a novel thiourea derivative was proposed. This paper is an extension of the work carried out and formerly reported by the authors (Sunil and Rao 2015). Many analytical methods reported for chromium determination are based on the Beer's law and the metal ligand complexation studies by addition of oxidizing agents like hydrogen peroxide. But, the reaction utilized in the current study is a unique catalytic reaction in itself because the reagent synthesis is very

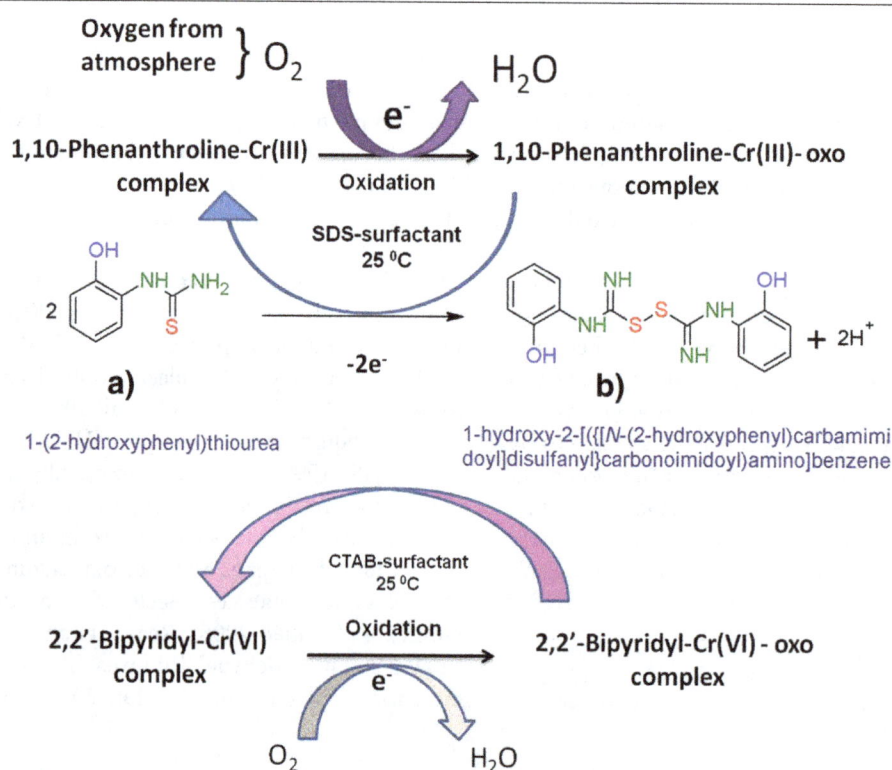

Scheme 1 The proposed mechanism for the process of catalytic effect of chromium species on HPTU.

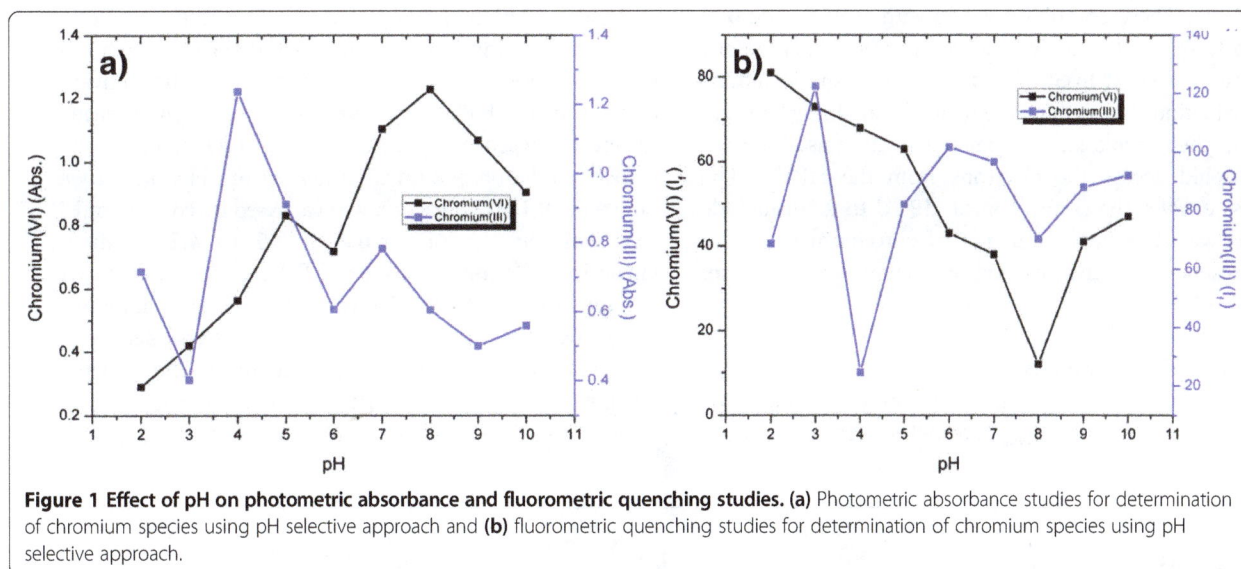

Figure 1 Effect of pH on photometric absorbance and fluorometric quenching studies. (a) Photometric absorbance studies for determination of chromium species using pH selective approach and **(b)** fluorometric quenching studies for determination of chromium species using pH selective approach.

simple and there is no need to add any oxidizing agent. There are no pre-concentration methods involved, the entire study can be done at normal room temperature, no inert atmosphere required, it is bench-stable and the methods are cost-effective.

Results and discussion

Proposed mechanism

The proposed mechanism describes that chromium species complexes with activators like 1,10-phenanthroline, 2,2′-bipyridyl resulting in the formation of organo-metallic

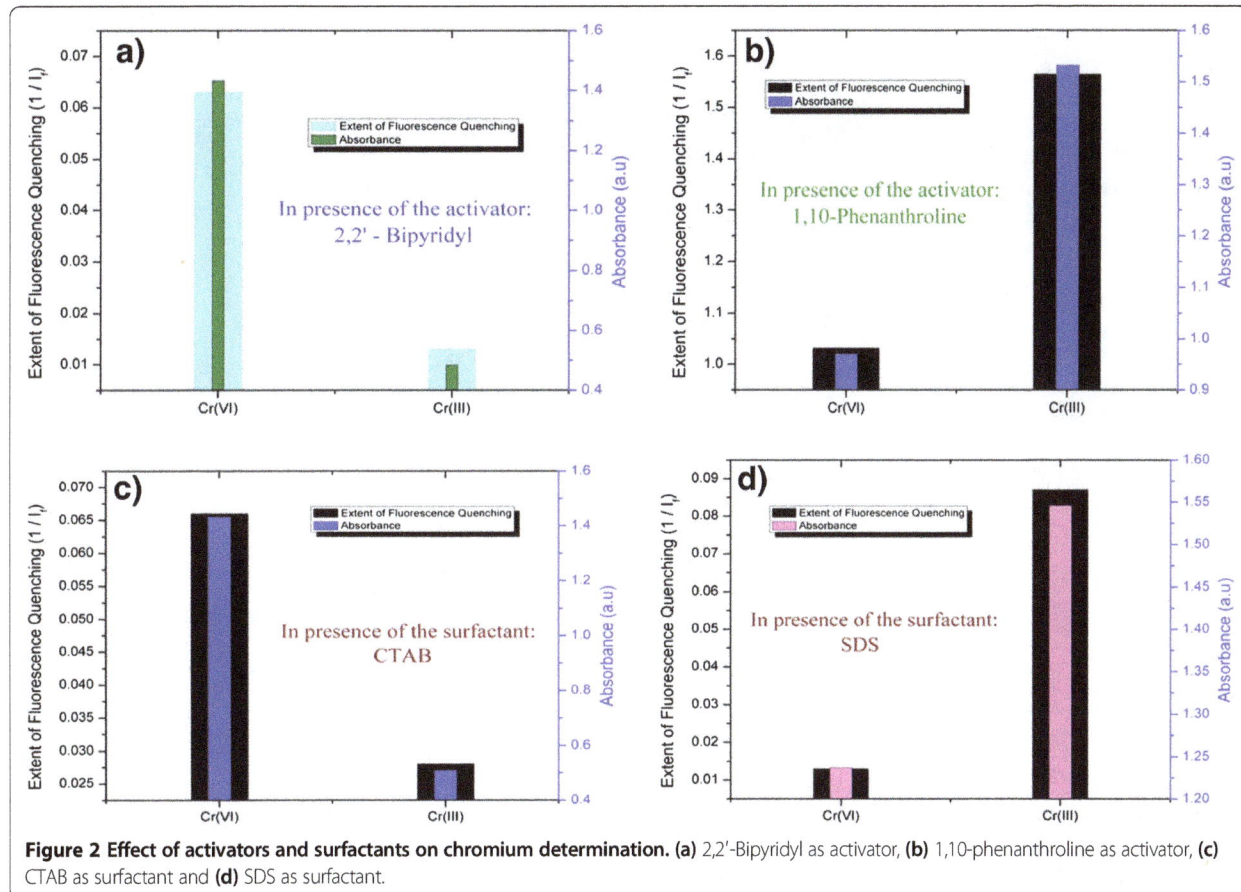

Figure 2 Effect of activators and surfactants on chromium determination. (a) 2,2′-Bipyridyl as activator, **(b)** 1,10-phenanthroline as activator, **(c)** CTAB as surfactant and **(d)** SDS as surfactant.

complexes. These complexes reacts with atmospheric oxygen to form highly reactive oxo-complexes, which are also unstable and short lived. This reactive oxo species attains high oxidation states which are unstable and in order to get back to their stable states from the highly unstable ones, they would accept the electrons from the HPTU. This process enables the conversion of HPTU to its disulphide. In this way, the catalytic action of chromium species in the presence of activators and surfactants was proposed (Scheme 1).

Experimental observations

The speciative determination of chromium(III) and chromium(VI) was done using photometric and fluorometric methods. In both developed methods, three parameters were proposed, which can selectively determine each species. The three parameters are the following: pH, activator and surfactant. HPTU was used as a chromogenic reagent to selectively sense a specific species at a defined pH. With respect to the observations pertaining to pH study, it was noticed that HPTU reaction was catalysed by chromium(III) ions only in the pH range of 3.5 to 4.2. Similarly, chromium(VI) ions catalyse HPTU reaction only in the pH range of 7.8 to 8.4. Thus, by selectively maintaining the respective pH ranges, one can easily and selectively determine the chromium species (Figure 1). The next parameter that was utilized in the determination was the usage of activators like pyridine, 2-aminopyridine, 2,2′-bipyridyl,

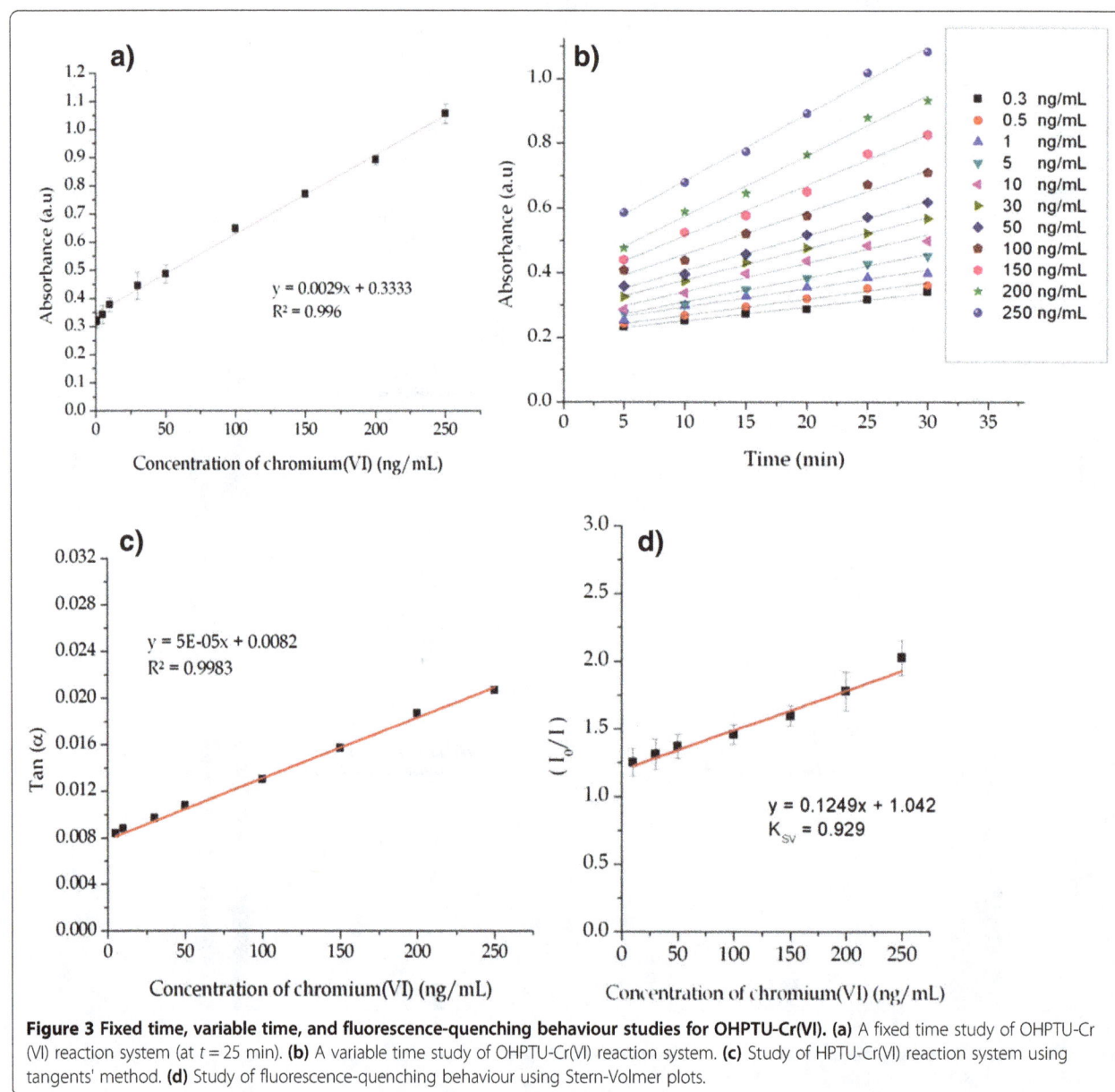

Figure 3 Fixed time, variable time, and fluorescence-quenching behaviour studies for OHPTU-Cr(VI). **(a)** A fixed time study of OHPTU-Cr(VI) reaction system (at $t = 25$ min). **(b)** A variable time study of OHPTU-Cr(VI) reaction system. **(c)** Study of HPTU-Cr(VI) reaction system using tangents' method. **(d)** Study of fluorescence-quenching behaviour using Stern-Volmer plots.

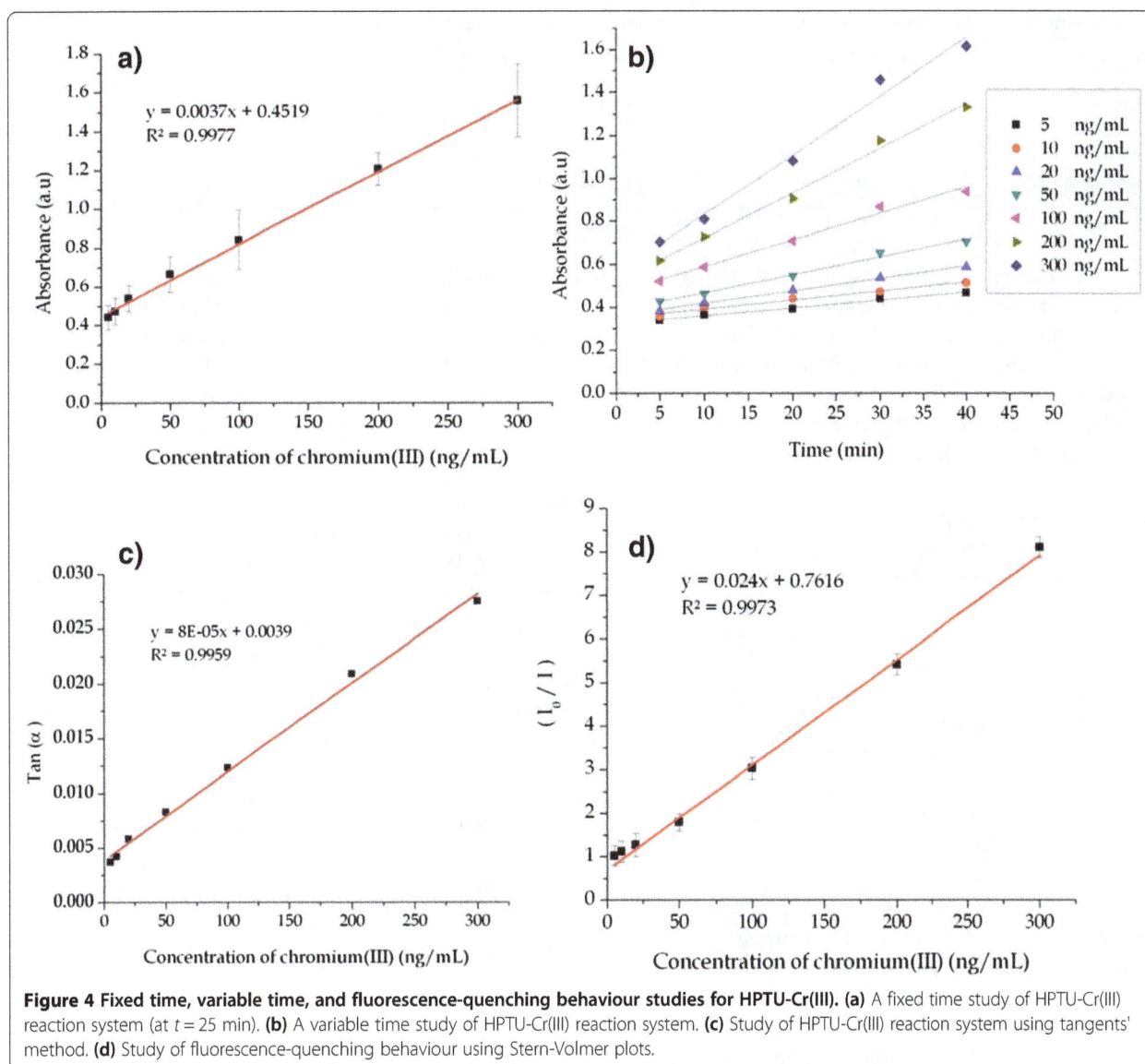

Figure 4 Fixed time, variable time, and fluorescence-quenching behaviour studies for HPTU-Cr(III). (a) A fixed time study of HPTU-Cr(III) reaction system (at $t = 25$ min). **(b)** A variable time study of HPTU-Cr(III) reaction system. **(c)** Study of HPTU-Cr(III) reaction system using tangents' method. **(d)** Study of fluorescence-quenching behaviour using Stern-Volmer plots.

1,10-phenanthroline, etc. In the case of chromium(III), 1,10-phenanthroline acts as a very good activator, by selectively activating chromium(III) in the acidic pH range. Similarly, 2,2′-bipyridyl selectively activates chromium(VI) in the basic pH range. Thus, by the selective activation of HPTU reaction using Cr(III) - 1,10-phenanthroline and Cr(VI) - 2,2′-bipyridyl in acidic and basic pH ranges, the chromium species can be determined (Figure 2a,b).

The third parameter that was followed is the selective use of surfactants in addition to activators. In this study, it was observed that the anionic surfactant, sodium dodecyl sulfate (SDS) encapsules the Cr(III) - 1,10-phenanthroline moiety and acts as a micro reaction centre facilitating the oxidation of HPTU. Similar in the other case, the cationic surfactant, CTAB encapsules the Cr(VI) - 2,2′-bipyridyl moiety in basic pH forming a micro reaction centre and

Table 1 Determination of chromium (III) species using the proposed and standard AAS methods

Sample	Amount of Cr (III) added (ng)	Experimental values for Cr (III) (ng) ($n = 3$)		
		Photometric	Fluorometric	AAS
Tap water	150.0	149.7 ± 0.04	149.4 ± 0.06	149.6 ± 0.06
Rain water	130.0	132.8 ± 0.08	132.5 ± 0.07	132.3 ± 0.05
Drainage water	100.0	108.6 ± 0.04	108.9 ± 0.05	108.9 ± 0.07

Table 2 Determination of chromium (VI) species using proposed and standard AAS methods

Sample	Amount of Cr (VI) added (ng)	Experimental values for Cr (VI) (ng) (n = 3)		
		Photometric	Fluorometric	AAS
Tap water	180.0	179.4 ± 0.05	179.6 ± 0.08	179.5 ± 0.07
Rain water	140.0	166.8 ± 0.08	166.5 ± 0.04	163.7 ± 0.05
Drainage water	120.0	138.4 ± 0.04	138.8 ± 0.05	138.5 ± 0.07

facilitates the catalytic reaction (Figure 2c,d). Using the above parameters, the chromium species were determined selectively in the range of 0.3 to 250 ng/mL. Linear plots were obtained in the absorbance study by following fixed time and tangents' methods. Similarly, fluorescence studies also inferred the quantitative linear relationship through Stern-Volmer plots (Figures 3d and 4d).

Interference by foreign ions

In the speciative determination of chromium, the effect of associated metal ions and anions was studied. It was observed that metal ions like Na(I), Ba(II), Bi(III), Sb(III), V(V), W(VI), Ru(III) and Re(II) showed no interference even up to 1,500-folds. Other metal ions like Mn(II), Co(II), As(V), Pb(II), Hg(II), Cd(II), Zn(II), Pd(II), Ni(II), Zn(II), Ca(II), Mg(II), Zr(IV), Rh(III), Os(IV), Au(III), Mo(IV), Al(III), Ir(III), Pt(II) and Ag(I) do not interfere up to 800-folds at the pH of 8.0. Fe(II) and Cu(II) interfere up to 20-folds, which can be masked by the addition of tartarate and thiosulphate, respectively.

Analytical application of the method

The developed methods were applied to water samples like drainage water, tap water and rain water. It was found that the concentration of chromium(VI) was comparatively more in the case of rain water and drainage water. This can be attributed to the leaching of chromium from paints, pipelines and other sources of pollution. The proposed photometric and fluorometric methods are in agreement with the experimental data obtained from standard atomic absorption spectrometry (AAS) method (Tables 1 and 2).

The obtained results of water sample analysis using absorbance and emission studies have been statistically treated using ANOVA (Tables 3 and 4). A high degree of linear relationship between both methods is very much evident from the Pearson correlation values closer

to 0.999. The F_{Ratio} value calculated from ANOVA: single factor is lesser than 1 and F_{Ratio} is lesser than F_{Crit}, in both the experiments, inferring no significant difference between the three pertained methods. Also, the analysis shows that the type of sample chosen has no significant effect on the developed methods.

Experimental

Chemicals and instrumentation

Analytical grade chemicals with high purity purchased from Sigma-Aldrich Chemicals Ltd., Bengaluru, India, were used in the investigation. Similarly, HPLC-graded solvents purchased from Merck, Mumbai, Maharashtra, India, were used appropriately. The standard prescribed procedure was employed for the preparation of buffer solutions (Vogel 1961). The pH adjustments were appropriately done using Micropro pH meter (Techno Instruments Co., Bangalore, India). The instruments such as PerkinElmer LS-55 fluorimeter (PerkinElmer, Waltham, MA, USA) and Hitachi-2001 spectrophotometer (Hitachi Ltd, Chiyoda-ku, Japan) were utilized for experimental studies. Mettler Toledo AB204-S (Mettler-Toledo, LLC, Columbus, OH, USA) was utilized for weighing purposes.

Reagents

In an 'A' grade 100-mL volumetric flask, 100.0361 mg of chromium (III) chloride was dissolved thoroughly and the resultant solution was standardised with EDTA. A stock solution of 6.3 mM (1.0 mg/mL) of chromium (III) was obtained.

In a 100-mL volumetric flask, 100.0704 mg of potassium dichromate was weighed, and deionised water was added up to the mark. The prepared solution was standardised by titrating with Mohr's salt, using diphenylamine as an indicator along with 2.0 mL of 1:1 H_3PO_4. A stock solution of 3.4 mM (1.0 mg/mL) of chromium (VI) was obtained.

Table 3 Statistical analysis of the obtained data for chromium (III) species

Statistical parameter	Calculated value
F_{Crit}	5.14
F_{Ratio}	2.39E−05
p-value	0.999

Table 4 Statistical analysis of the obtained data for chromium (VI) species

Statistical parameter	Calculated value
F_{Crit}	5.14
F_{Ratio}	0.0023
p-value	0.997

1.0407 g of 1,10-phenanthroline was weighed and dissolved in ethanol in a 100-mL volumetric flask to get 0.58 mM of 1,10-phenanthroline reagent.

By weighing 1.0207 g of 2,2'-bipyridyl in a 100-mL volumetric flask, 0.66 mM of 2,2'-bipyridyl reagent was prepared, and ethanol was added up to the mark.

In distilled water using a 100-mL volumetric flask to get a 1% SDS solution, 1.0 g of SDS was weighed and dissolved. In a 100-mL volumetric flask, 1.0 g of CTAB was weighed and dissolved in distilled water to get 1% CTAB solution.

The above recommended experimental procedures were followed as described (Vogel 1961).

Methods

The reagent, HPTU, was synthesized (Sunil and Rao 2012a) and the analytical parameters, and their influence was studied systematically. A solution of 1.0 mL of acetate buffer of pH = 4.0 and appropriate amount (5 to 300 ng/mL) of chromium (III) solution was pipetted in to a 10.0-mL volumetric flask, followed by 1.0 mL of 1,10-phenanthroline and 1.0 mL of 1% SDS. The temperature was maintained at 25°C. Two milliliters of HPTU (1 mg/mL) solution was then added, and millipore water was added up to the mark. The resulting reaction mixture was transferred into 10-mm quartz cuvettes. The photometric measurements were recorded at $\lambda_{Max} = 416$ nm. The fluorescence emission measurements were recorded at $\lambda_{Em} = 520$ nm upon excitation at $\lambda_{Ex} = 416$ nm, respectively, at 30 min. The blank experiments were repeated by following the same procedure to obtain relative fluorescence intensity I_0 and the value of I_0/I was calculated. The calibration graph was plotted and the method was applied for the determination of chromium in waste water and plant samples. Water samples were boiled, treated with concentrated nitric acid and then filtered to remove organic particulate matter before using for the analysis. The digestion for plant sample was done as mentioned (Sunil and Rao 2012b). The same procedure was followed by taking chromium (VI) solution. Here, 2,2'-bipyridyl was used as an activator and 1% CTAB as a surfactant.

Conclusions

The results suggest that the speciative determination of chromium species was achieved by the selective application of parameters of pH, activators and surfactants using photometric and fluorometric techniques. Chromium(III) was determined in the pH range of 3.5 to 4.2 and in presence of 1,10-phenanthroline as an activator and SDS as a surfactant. Chromium(VI) was determined in the pH range of 7.8 to 8.4 and in presence of 2,2'-bipyridyl as an activator and CTAB as a surfactant. The methods are fairly sensitive with a determination range of 0.3 to

250 ng/mL. The proposed methods were applied to tap water, rain water and drainage water with satisfactory results.

Competing interests
The authors declare no competing interests.

Authors' contributions
Sunil has performed the entire experimental, analytical work and prepared the draft of the manuscript. The supervision of this work was done by Rao. All authors read and approved the final manuscript.

Acknowledgements
Authors are thankful to the Founder Chancellor, Bhagawan Sri Sathya Sai Baba and the authorities of Sri Sathya Sai Institute of Higher Learning for the support and guidance.

References
Arancibia V, Nagles E, Gomez M, Rojas C (2012) Speciation of Cr (VI) and Cr (III) in water samples by adsorptive stripping voltammetry in the presence of pyrogallol red applying a selective accumulation potential. Int J Electrochem Sci 7:11444–11455

Duran A, Tuzen M, Soylak M (2009) Preconcentration of some trace elements via using multiwalled carbon nanotubes as solid phase extraction adsorbent. J Haz Mat 169:466–471

El-Shahawi MS, Al-Saidi HM, Bashammakh AS, Al-Sibaai AA, Abdelfadeel MA (2011) Spectrofluorometric determination and chemical speciation of trace concentrations of chromium (III and VI) species in water using the ion pairing reagent tetraphenyl-phosphonium bromide. Talanta 84:175–179

Ezoddin M, Shemirani F, Khani R (2010) Application of mixed-micelle cloud point extraction for speciation analysis of chromium in water samples by electrothermal atomic absorption spectrometry. Desalination 262:183–187

Gomez V, Callao MP (2006) Chromium determination and speciation since 2000. TrAC Trend Anal Chem 25:1006–1015

Grabarczyk M, Tyszczuk K, Korolczuk M (2006) Catalytic adsorptive stripping voltammetric procedure for determination of total chromium in environmental materials. Electroanalysis 18:1223–1226

Guerrero ML, Alonso EV, Pavon JC, Cordero MS, De Torres AG (2012) On-line preconcentration using chelating and ion-exchange minicolumns for the speciation of chromium (III) and chromium (VI) and their quantitative determination in natural waters by inductively coupled plasma mass spectrometry. J Anal At Spectrom 27:682–688

Hashemi M, Daryanavard SM (2012) Ultrasound-assisted cloud point extraction for speciation and indirect spectrophotometric determination of chromium (III) and (VI) in water samples. Spectrochim Acta Mol Biomol Spectros 92:189–193

Hosseini MS, Belador F (2009) Cr (III)/Cr (VI) speciation determination of chromium in water samples by luminescence quenching of quercetin. J Haz Mat 165:1062–1067

Hotta H, Yata K, Kamarudin KFB, Kurihara S, Tsunoda KI, Fukumoto N, Kinugasa SI (2012) Determination of chromium (III), chromium (VI) and total chromium in chromate and trivalent chromium conversion coatings by electrospray ionization mass spectrometry. Talanta 88:533–536

Karosi R, Andruch V, Posta J, Balogh J (2006) Separation of chromium (VI) using complexation and its determination with GFAAS. Microchem J 82:61–65

Kiran K, Kumar KS, Prasad B, Suvardhan K, Lekkala RB, Janardhanam K (2008) Speciation determination of chromium (III) and (VI) using preconcentration cloud point extraction with flame atomic absorption spectrometry (FAAS). J Haz Mat 150:582–586

Kotas J, Stasicka Z (2000) Chromium occurrence in the environment and methods of its speciation. Environ Pollut 107:263–283

Martone N, Rahman GM, Pamuku M, Kingston HS (2013) Determination of chromium species in dietary supplements using speciated isotope dilution mass spectrometry with mass balance. J Agr Food Chem 61:9966–9976

Moghadam MR, Dadfarnia S, Haji Shabani AM (2011) Speciation and determination of ultra trace amounts of chromium by solidified floating organic drop microextraction (SFODME) and graphite furnace atomic absorption spectrometry. J Haz Mat 186:169–174

Padarauskas A, Naujalis E (1998) On-line preconcentration and determination of chromium (VI) in waters by high-performance liquid chromatography using pre-column complexation with 1,5-diphenylcarbazide. J Chromatogr A 808:193–199

Parmar P, Pillai AK, Gupta VK (2010) An improved colorimetric determination of micro amounts of chromium (VI) and chromium (III) using p-aminoacetophenone and phloroglucinol in different samples. J Anal Chem 65:582–587

Ren Y, Fan Z, Wang J (2007) Speciation analysis of chromium in natural water samples by electrothermal atomic absorbance spectrometry after separation/ preconcentration with nanometer-sized zirconium oxide immobilized on silica gel. Microchim Acta 158:227–231

Sánchez-Moreno RA, Gismera MJ, Sevilla MT, Procopio JR (2010) Direct and rapid determination of ultratrace heavy metals in solid plant materials by ET-AAS ultrasonic-assisted slurry sampling. Phytochem Anal 21:340–347

Schramel P, Xu LQ, Knapp G, Michaelis M (1992) Application of an on-line preconcentration system in simultaneous ICP-AES. Microchim Acta 106:191–201

Soomro R, Ahmed MJ, Memon N (2011) Simple and rapid spectrophotometric determination of trace level chromium using bis (salicylaldehyde) orthophenylenediamine in nonionic micellar media. Turk J Chem 35:155–170

Sun YC, Lin CY, Wu SF, Chung YT (2006) Evaluation of on-line desalter-inductively coupled plasma-mass spectrometry system for determination of Cr (III), Cr (VI), and total chromium concentrations in natural water and urine samples. Spectrochim Acta B 61:230–234

Sunil A, Rao SJ (2012a) Eco-friendly approach for a facile synthesis of o-hydroxyphenylthiourea and its property as an analytical reagent in sensing mercury (II). Res J Chem Sci 2:30–40

Sunil A, Rao SJ (2012b) Surfactant based fluorometric sensing of copper at picogram level using o-hydroxyphenylthiourea with pyridine as activator. Int J Chem Anal Sci 3:1318–1321

Sunil A, Rao SJ (2015) Photometric and fluorimetric determination of chromium (VI) using metal-oxo mediated reaction of 1-(2-hydroxyphenyl)thiourea in micellar medium. J Anal Chem 70:159–165

Themelis DG, Kika FS, Economou A (2006) Flow injection direct spectrophotometric assay for the speciation of trace chromium (III) and chromium (VI) using chromotropic acid as chromogenic reagent. Talanta 69:615–620

Vogel AI (1961) A text book of quantitative inorganic analysis. Longman Group Limited, London

Wang LL, Wang JQ, Zheng ZX, Xiao P (2010) Cloud point extraction combined with high-performance liquid chromatography for speciation of chromium (III) and chromium (VI) in environmental sediment samples. J Haz Mat 177:114–118

Wu Y, Jiang Y, Han D, Wang F, Zhu J (2007) Speciation of chromium in water using crosslinked chitosan-bound FeC nanoparticles as solid-phase extractant, and determination by flame atomic absorption spectrometry. Microchim Acta 159:333–339

Yildiz Z, Arslan G, Tor A (2011) Preconcentrative separation of chromium (III) species from chromium (VI) by cloud point extraction and determination by flame atomic absorption spectrometry. Microchim Acta 174:399–405

Zeng C, Lin Y, Zhou N, Zheng J, Zhang W (2012) Room temperature ionic liquids enhanced the speciation of Cr (VI) and Cr (III) by hollow fiber liquid phase microextraction combined with flame atomic absorption spectrometry. J Haz Mat 237:365–370

24

Trace determination of cadmium in water using anodic stripping voltammetry at a carbon paste electrode modified with coconut shell powder

Deepak Singh Rajawat[1], Nitin Kumar[2] and Soami Piara Satsangee[3*]

Abstract

Background: Increasing awareness on the environmental impact of heavy metals has increased a considerable interest in the determination of metals in natural water bodies. The present paper describes the development and electrochemical application of carbon paste electrode modified with fibrous part of coconut shell for the determination of cadmium in water samples.

Methods: Determination was carried out using anodic stripping voltammetry. It is a two-step process. First, the metal ions get accumulated at the electrode surface at open-circuit potential, followed by a potential scan for voltammetric determination of cadmium.

Results: Different parameters affecting the determination of Cd (II) were optimized and are as follows: HCl as stripping solvent, acetate buffer of pH 5 as accumulating solvent, and 15-min accumulation time. Triton X-100, cetyltrimethylammonium bromide, and sodium dodecyl sulfate were used as representative for neutral, cationic, and anionic surfactants, respectively, to see the effect of surface active macromolecules. Interference caused by other metal ions on the determination of cadmium was also studied.

Conclusions: The method shows the development of a sensor for the sensitive determination of cadmium with limit of detection at 105 µg L^{-1}. This technique does not use mercury and, therefore, has a positive environmental benefit.

Keywords: Plant-modified carbon paste electrode; *Cocos nucifera*; Cadmium; Stripping voltammetry

Background

Cadmium is classified as one of the priority pollutants which entered water streams through various industrial operations (Pan et al. 2012). It is ranked seventh by the Environmental Protection Agency in 'Top hazardous substances priority list'. Cadmium can easily be dissolved and transported by water (Li et al. 2009a, b). However, due to anthropogenic activities, its content can be elevated at the site of the action. High concentrations of cadmium ions can injure human health and pollute the environment. It is carcinogenic to human by damaging human immune and central nervous systems and causes diseases such as renal dysfunction and liver damage. Hence, the identification of cadmium-polluted sites is needed by society (Eshaghi et al. 2011).

Different analytical methods for the determination of Cd (II) ions have been reviewed several times (Sneddon and Vincent 2008; Ferreira et al. 2007; Pyrzynska 2007; Davis et al. 2006). Among them is stripping voltammetric determination of cadmium using mercury-based electrodes which is one of the very sensitive analytical methods available. But, due to different issues related to its harmful effects and disposal, it is strongly recommended to replace mercury with another electrode material. Recently, modified carbon paste electrodes can be a better substitute of mercury-based electrodes due to its simplicity of preparation, the versatility of chemical modification, rapid renewal of the electrode surface, and sensitivity equivalent to that of mercury-based electrodes (Roa et al. 2003; Sar et al. 2008; Heitzmann et al. 2005; Lu et al. 2011; Li et al. 2009a, b; Bagheri et al. 2012). Thus, modified carbon paste electrodes (MCPEs) and related sensors using different types of modifiers (chemicals, enzymes, and extracts) have been developed (Chow and Gooding 2006; Heitzmann et al. 2005; Ensafi et al. 2010; Portaccio et al. 2010).

* Correspondence: electrochemanal@gmail.com
[3]Remote Instrumentation Lab, USIC, Dayalbagh Educational Institute, Agra 282010, India
Full list of author information is available at the end of the article

In comparison to the conventionally used MCPEs, plant-modified carbon paste electrodes represent a green approach in the environmental perspectives. The use of plant agricultural wastes as a modifier in carbon paste electrodes is due to the high metal hyper-accumulating properties in certain plants (Rajawat et al. 2013b; Mojica et al. 2006, 2007). They possess an electrochemically or chemically active moiety. These moieties could be any of the following: redox or ligand sites and ion-exchange sites, which possess certain functionalities or donor groups (Rajawat and Satsangee 2011).

Coconut shell has been widely used as an agricultural waste material for the sorption of Cd (II) from aqueous solution (Pino et al. 2006; Okafor et al. 2012). In continuation of our previous research work on modified carbon paste electrodes (Rajawat et al. 2012, 2013a) and keeping the above views in mind, the powder of coconut shell (*Cocos nucifera*) was used to modify the carbon paste electrode with the main goal of using it as a modifier material for the development of a sensor for the determination of cadmium.

Methods
Chemicals and reagents
For DPASV study, first, accumulation was done under open-circuit potential by placing the electrode in a metal solution with stirring for a certain time, rinsed with deionized distilled water followed by medium exchange for stripping analysis.

All chemicals were of analytical reagent grade. A 1,000-ppm stock solution of Cd (II) was prepared by dissolving an appropriate amount of cadmium nitrate (Merck & Co., Inc., Whitehouse Station, NJ, USA). The working solution was prepared daily by the dilutions from the stock solution. Graphite powder (<20 μm) and mineral oil Nujol (light, density 0.838) were obtained from Aldrich (Wyoming, IL, USA). Triply distilled water (ELGA, Millipore Co., Billerica, MA, USA) was used throughout the experiment.

Procedure
All quantitative measurements were carried out in anodic stripping voltammetry using differential pulse (DP) to achieve the sensitivity required for trace analysis. Each DPASV run was made up of two steps: accumulation under open circuit where the modified electrode is immersed in metal solution for a certain time. The electrode was then rinsed with deionized distilled water, followed by medium ex-change for stripping analysis. All measurements were carried out at room temperature (24 ± 1°C). Finally, the calibration curves were plotted and the influence of various substances as potential interference compounds on the determination of Cd(II) ions was studied under the optimum conditions.

Preparation of coconut shell powder-modified carbon paste electrode
Coconut was purchased from the local market of Agra. The coconut shell between the outer layer and inner layer

(i.e., mesocarp) was separated. It was properly washed with water and dried in an oven at 50°C. The dried material was grounded and passed through the sieve. Fraction of the particles with size less than 150 μm was selected for electrode preparation. Unmodified carbon paste electrode was prepared by mixing the graphite powder with the mineral oil (80:20 w/w ratio) using mortar and pestle. MCPEs of different proportions (5%, 10%, 15%, 20%, 25%, and 30% w/w) were prepared by substituting the corresponding amount of graphite powder with coconut shell powder. The mixture is thoroughly hand-mixed in a mortar and pestle. The paste was pressed in a glass tube with an inner diameter of 3 mm and a depth of 4 cm to form a target electrode surface. A copper wire was inserted from the backside for electrical contact. A smooth and fresh electrode surface was obtained by pushing the electrode material from the backside, removing a small amount of paste from the electrode tip, and polishing the electrode surface on a photo paper.

Results and discussion
Characterization of the coconut shell powder-modified carbon paste electrode
Morphological characterization
Scanning electron microscope images of (a) unmodified and (b) coconut shell powder-modified carbon paste electrodes are shown in Figure 1. The comparison of the scanning electron microscope (SEM) images shows a difference in the morphology which was observed after the modification with coconut shell powder. The unmodified CPE surface is homogenous and fine pores are uniformly distributed, whereas after the modification, the pore size was increased and the distribution of the pores on the surface become uneven.

Increase in the roughness value for coconut shell powder-modified CPE (CS-MCPE) was justified with the atomic force microscope (AFM) results. 3-D atomic force microscope images of unmodified CPE (a) and CS-MCPE (b) are shown in Figure 2. Increase in the average surface roughness of the prepared electrode was observed after modification with coconut shell powder from 84.55 to 183.96 nm, respectively, for unmodified and modified CPE. It justifies the results obtained from SEM.

Electrochemical characterization
All the electrochemical experiments were performed using a μAutolab Type III potentiostat (Eco Chemie, Utrecht, Netherlands) controlled by a PC using the NOVA 1.8 software. A three-electrode system containing the modified carbon paste electrode as working, an Ag/AgCl (3.0 mol L^{-1} KCl) as reference, and a platinum wire as auxiliary electrodes was used.

Electrochemical characterization of the prepared electrodes was carried out using cyclic voltammetry. Cyclic voltammetry study for the CS-MCPE in K$_3$[Fe(CN)$_6$] solution with 0.1 M KCl as supporting electrolyte was

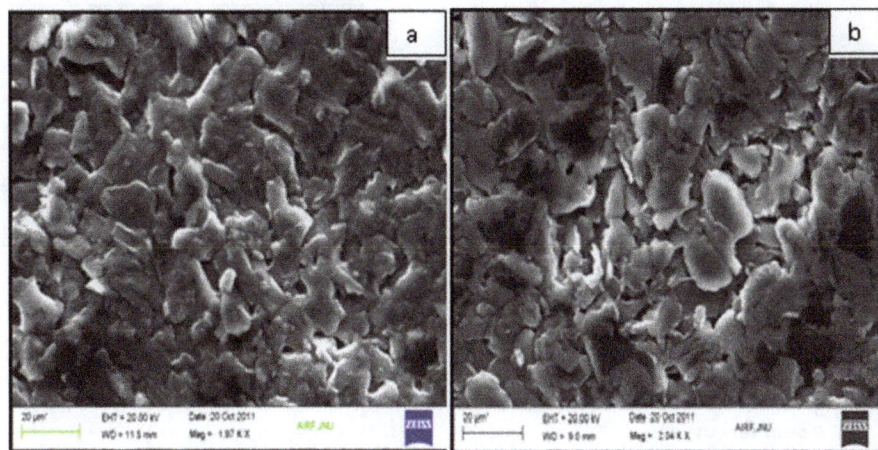

Figure 1 SEM images of the carbon paste electrode. (a) Unmodified carbon paste electrode and **(b)** coconut shell powder-modified carbon paste electrode.

carried out to determine the effect of modification on charge transfer activity, and the results are given in Figure 3. Well-defined cyclic voltammogram (CV) was observed for the unmodified CPE (Figure 3, curve a). With modified CPE, the redox peak currents decreased, and the peak potential difference increased remarkably (Figure 3, curve b). Because the coconut shell powder acts as an inert electron and a mass transfer blocking layer, it hinders the diffusion of ferricyanide toward the electrode surface. Based on the CV results, we can conclude that the coconut shell powder was successfully embedded on the surface of the modified electrode.

Electrochemical impedance study
The FRA module with the same instrumental setup was used for the electrochemical impedance study (EIS). For EIS, the AC amplitude of 10 mV at a frequency of 1 to 100,000 Hz was applied. Figure 4 shows the Nyquist plots of the electrochemical impedance study for the unmodified (Figure 4 curve a) and modified (Figure 4 curve b) CPE in

10 mM $K_3[Fe(CN)_6]$ solution. The results of the EIS studies were analyzed by fitting the appropriate equivalent circuit. The circuit which most fitted to the unmodified CPE is given in Figure 4b. Here, R_s is the solution resistance, Q is a constant phase element, and R_p is charge transfer resistance. The same circuit was used to analyze the EIS results of the CS-MCPE. The charge transfer resistance value observed for CS-MCPE is high in comparison to the unmodified CPE. The electrochemical impedance spectroscopy results are in accordance with the cyclic voltammetry results.

Electrochemical studies for metal determination
The ability of the CS-MCPE for the determination of Cd (II) was evaluated by the accumulation of Cd (II) at open-circuit potential and its electrochemical stripping in electrolyte solution using cyclic voltammetry. The CV curves recorded in 0.1 M hydrochloric acid did not show any peak without accumulation (Figure 5 curve a), whereas after accumulation for 10 min, well-defined oxidation peak appears (Figure 5 curve b), corresponding to the

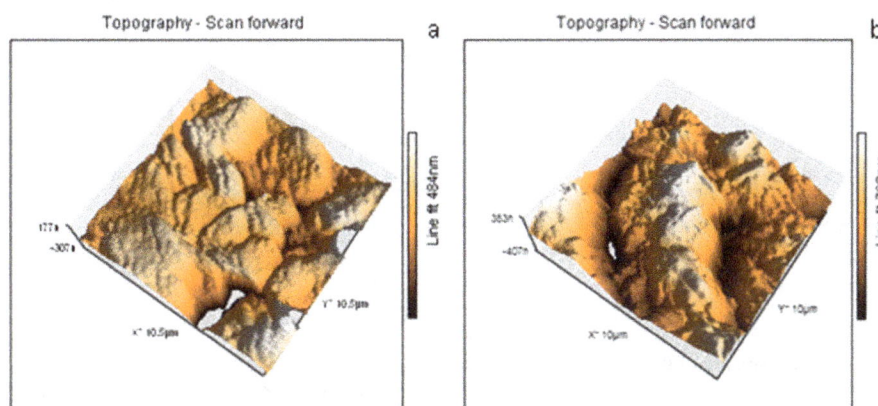

Figure 2 AFM images of the carbon paste electrode. (a) Unmodified carbon paste electrode and **(b)** coconut shell powder-modified carbon paste electrode.

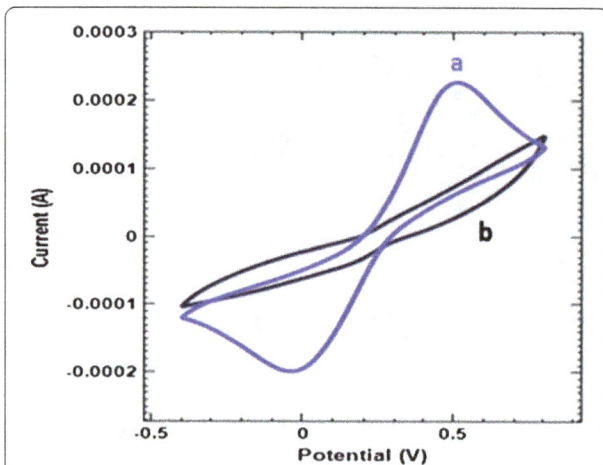

Figure 3 Cyclic voltammograms of the carbon paste electrode.
(a) Unmodified CPE and **(b)** coconut shell powder-modified carbon paste electrode in potassium ferricyanide solution with 0.1 M KCl at the scan rate of 100 mV/S.

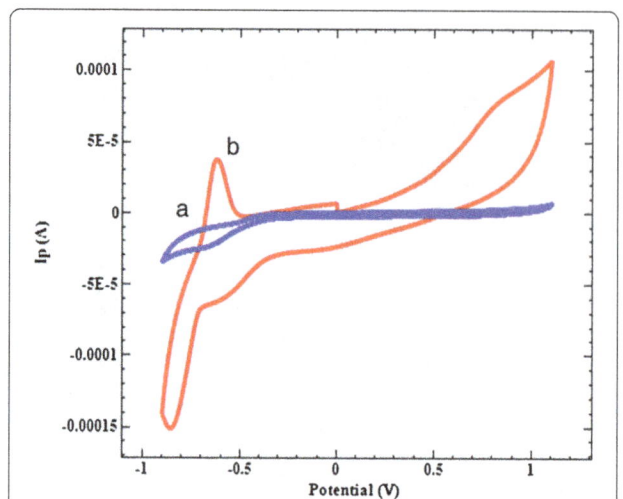

Figure 5 Cyclic voltammograms of coconut shell powder-modified carbon paste electrode in 0.1 M HCl. (a) Without accumulation and **(b)** after 10-min accumulation of Cd (II) at a scan rate of 50 mV/S.

oxidation of the accumulated cadmium which proves that cadmium metal gets accumulated at the electrode surface.

Based on the cyclic voltammetric results, differential pulse anodic stripping voltammetry (DPASV) was used. The DPASV result show a well-defined and symmetric anodic stripping peak with peak potential at −713 mV after the accumulation (Figure 6 curve b), whereas a flat curve was attained without accumulation (Figure 6 curve a). So, the peak was because of the oxidation of Cd (II) in the stripping step. Therefore, differential pulse anodic stripping

voltammetry was employed in the present work for the determination of cadmium. All measurements were carried out at room temperature (24°C ± 1°C). During the experiment, the electrodes were stored in 0.1 M HCl solution.

FTIR studies

The accumulation of metals at the electrode surface is thought to be mainly by electrostatic attraction between different functional groups on the coconut shell powder embedded at the electrode surface and cationic metal ions. Ionization of these groups in aqueous solution enables them to participate in cation binding. The binding of the metal ions at the electrode surface was studied using FTIR. An FTIR spectrum of the coconut shell powder before and after its treatment with metal solution

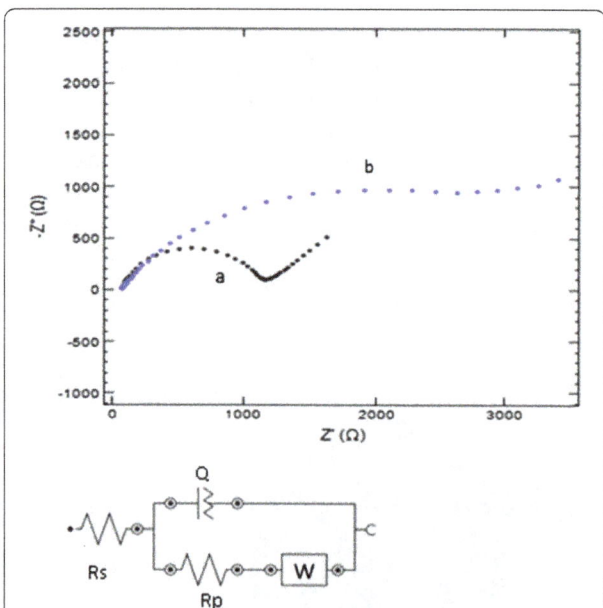

Figure 4 Electrochemical impedance spectra of the carbon paste electrode. (a) Unmodified (curve a) and coconut shell powder-modified (curve b) CPE in potassium ferricyanide solution. **(b)** The circuit which most fitted to the unmodified CPE.

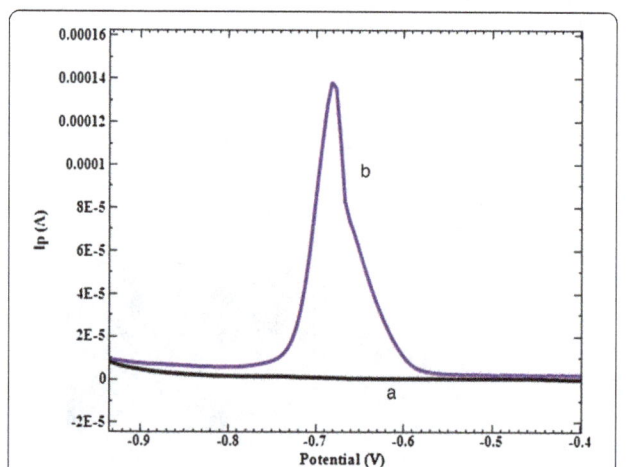

Figure 6 Differential pulse anodic stripping voltammograms for determination of Cd (II). (a) Before accumulation and **(b)** after 10-min accumulation; stripping solution is 0.1 M HCl.

is shown in Figure 7. The FTIR spectroscope analysis of the finely powdered and dried coconut shell powder (native) indicates broad absorption bands at 3,425.5, 1,625.9, 1,448.5, and 1,261.4 cm^{-1} (Figure 7 spectrum a) representing –OH stretching, C = O stretching, OH bending, and C-O stretching vibrations. After treatment with Cd (II), the adsorption bands shifted to 3,431.3, 1,631.7, 1,446.6, and 1,257.7 cm^{-1}, respectively (Figure 7 spectrum b). Slight shift in the adsorption bands indicated that the –OH, C-O, and C = O groups are involved in the Cd (II) binding.

Mechanism of accumulation

Based on our experimental findings and pertinent information available on the relevant topic, a mechanism for metal binding at the modified electrode surface is proposed. Coconut shell powder contains oxygen-containing functional groups in lignins and cellulose. These groups may constitute a physiologically active group to interact with the Cd (II) ions. The mechanism of the accumulation at the modified electrode is as follows:

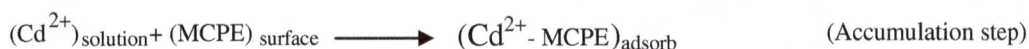

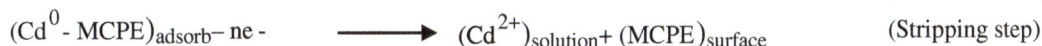

Optimization parameters

Different parameters affecting the voltammetric determination of Cd (II) such as amount of modifier, accumulation media, accumulation time, and stripping media were investigated.

Amount of modifier

The effect of the amount of coconut shell powder in the modified CPE on the peak current was investigated and shown in Figure 8. The results indicated that the peak current increases with the increase in the amount of coconut shell powder initially because the increase in the amount of the coconut shell powder results in the increase of binding sites, which facilitates the accumulation of cadmium at the electrode surface. However, at 20% of coconut shell powder relative to the mass of carbon powder, the highest peak currents were obtained. After 20% modifier, the continuous increase in the amount of the modifier causes a decrease of the peak current, because excessive coconut shell powder results in the decrease of conductivity of the modified electrode.

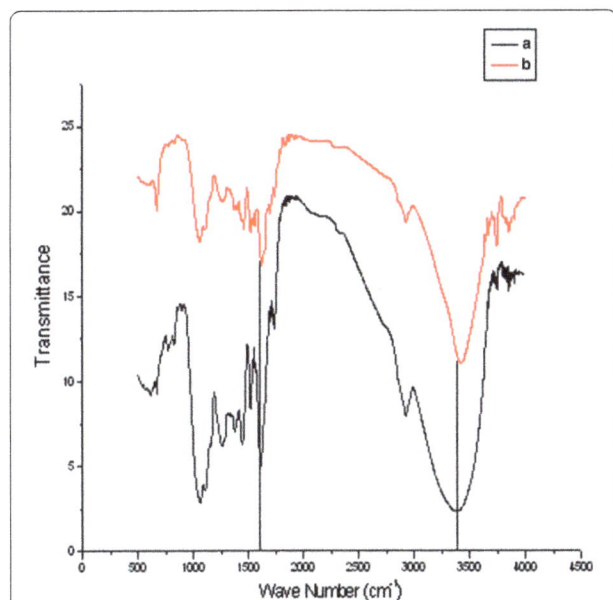

$$(Cd^{2+})_{solution} + (MCPE)_{surface} \longrightarrow (Cd^{2+} - MCPE)_{adsorb} \qquad \text{(Accumulation step)}$$

$$(Cd^{2+} - MCPE)_{adsorb} + ne- \longrightarrow (Cd^{0} , MCPE)_{adsorb} \qquad \text{(Reduction step)}$$

$$(Cd^{0} - MCPE)_{adsorb} - ne- \longrightarrow (Cd^{2+})_{solution} + (MCPE)_{surface} \qquad \text{(Stripping step)}$$

Figure 7 FTIR spectra of coconut shell powder before (a) and after treatment with Cd (II) (b).

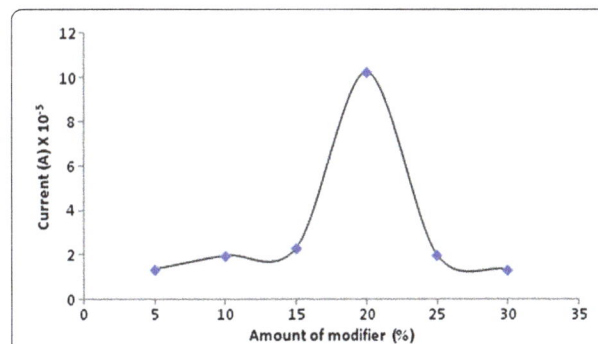

Figure 8 Effect of the amount of modifier on DPASV current response.

So, the best ratio of the modifier in the carbon paste composition is 20% (*w/w*) for the determination of Cd (II).

Accumulation media

The accumulation of cadmium was examined in supporting electrolytes such as acid solution, base solution, and different buffers like acetate buffer, phosphate buffer, and Britton-Robinson buffer. Voltammetric peaks were observed in most of these electrolytes; however, in acetate buffer solution, the anodic peak current was higher, and better defined peak shape was observed for Cd (II). A baseline for the determination of metals in acetate buffer is comparatively low, suggesting acetate buffer as the best accumulating medium.

The effect of the pH of accumulating media on the voltammetric response of the CS-MCPE was studied in a pH range between 3.0 to 6.0 in a solution containing 1 ppm Cd (II) with 1 mM sodium acetate buffer. As shown in Figure 9, anodic peaks current was increased as the pH is changed from 3.0 to 6.0, reaching a maximum at pH 5 and then decreases. The results show that maximum binding of these metal ions on the electrode surface occurs at pH 5. Since the ionization of functional groups depends on the accumulating solvents and its pH, at pH > pKa, most of these functional groups are mainly in ionized form and can exchange H$^+$ with metal ions in a solution. The concentration of acetate buffer was also varied from 0.1 to 100 mM. The maximum current value was obtained for 1 mM acetate buffer.

Accumulation time

The influence of accumulation time on the stripping peak currents of 0.1, 1, and 10 ppm of Cd (II) in 1 mM sodium acetate buffer solution was investigated. An increase in the current response was observed with increasing preconcentration time initially, which indicates that cadmium was rapidly adsorbed on the modified electrode surface while further prolonged accumulation did not improve the peak height. In comparison to the three concentrations of

metal ions selected for this experiment, the electrode surface gets saturated early for higher concentration compared to the low concentration. For further experiment, 1 ppm Cd (II) solution was used with accumulation time of 10 min.

Stripping media

The influence of the nature of stripping medium on the current response was investigated using hydrochloric acid, nitric acid, sulfuric acid, perchloric acid, phosphoric acid, sodium hydroxide, ammonium hydroxide, sodium chloride, and potassium chloride; the results are shown in Figure 10. The maximum current response was observed for hydrochloric acid solution as a stripping media, since in acidic solutions, its protons have the ability to displace the Cd (II) ions; in addition, the chloride ions of HCl promotes the stripping of metals. It is evident from the previous studies that the Cl$^-$ ion is the best migrating ligand for Cd (II) (Rajawat et al. 2013a). The effect of the stripping solvent concentration was also studied by varying the concentration within the range 0.01 to 1 M. The optimum current response was observed at 0.1 M HCl.

Determination of metals in the presence of surfactants

The effect of the presence of the surfactants on the determination of Cd (II) was carried out using Triton X-100, SDS, and CTAB as representatives of non-ionic, anionic, and cationic surfactants, respectively. Typical voltammograms in the presence of increasing amounts of each surfactant were recorded, and the effect of surfactant on current value was presented in Figure 11. An increase in the current response was observed for the anionic surfactants, whereas for the cationic

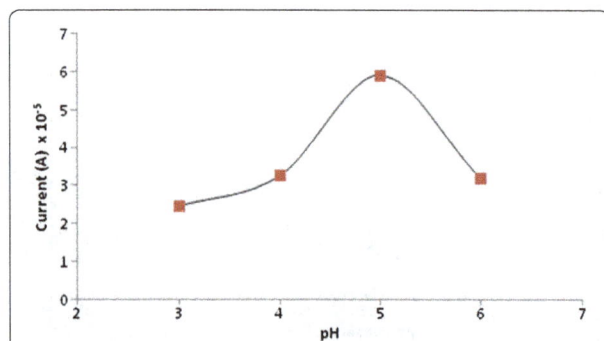

Figure 9 Effect of pH of accumulating solvent on the DPASV response of the modified carbon paste electrode.

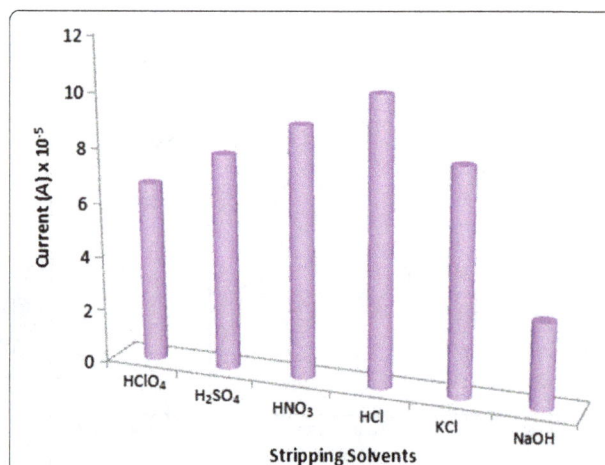

Figure 10 Effect of different stripping solvents on DPASV current response.

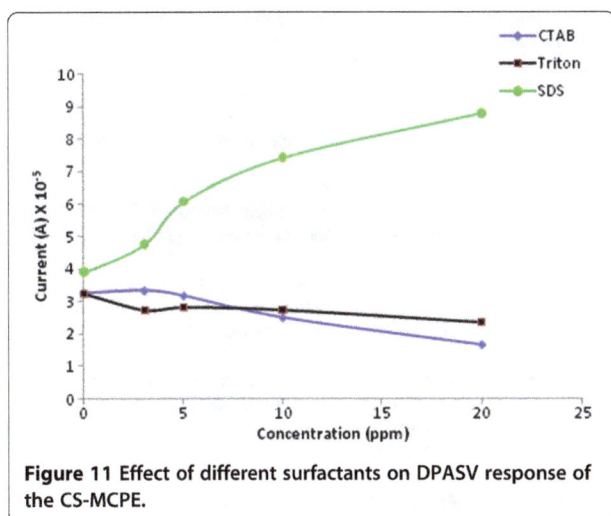

Figure 11 Effect of different surfactants on DPASV response of the CS-MCPE.

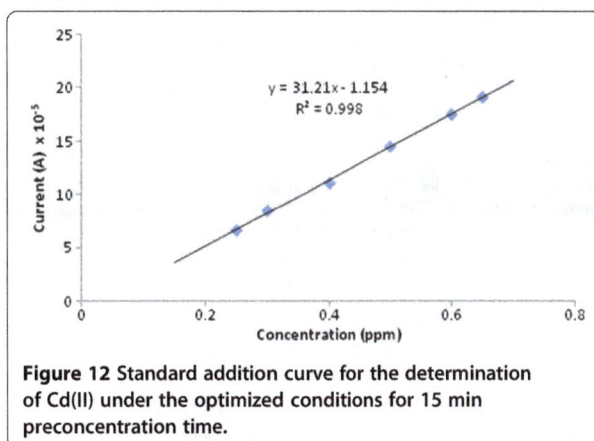

Figure 12 Standard addition curve for the determination of Cd(II) under the optimized conditions for 15 min preconcentration time.

surfactants, a decrease in the current response was observed. For neutral surfactant, not so much change was observed. No variation in the peak potential value was observed for Cd (II) determination on increasing amount of surfactant.

Interference study

The effect of the presence of other metal ions present in the solution on the anodic peak currents of Cd (II) was evaluated. A 10% change in the current response is assumed as a constant current response. No effect on the determination of Cd (II) was observed up to 250-fold of Pb (II), 5-fold of Cu (II), 25-fold Ni (II), and 30-fold Cr (VI). The results of interference study can be justified by hard-soft acid base theory. The main binding sites in these modifiers are COOH and OH groups, which are hard bases, whereas the metals Pb (II) and Cu (II) are intermediate; Cd (II), Hg (I), and Zn (II) are soft acids. According to hard-soft acid base theory, hard acid tends to form complexes with hard bases, and soft acids tend to form complexes with soft bases. Since copper is strongly bonded on electrode surface, so it is the most interfering metal.

Analytical characteristics

Analytical performance of the CS-MCPE was evaluated with standard addition of Cd (II) under the optimum conditions determined above, and the corresponding results are shown in Figure 12. A linear relationship was observed between the analytical signal and concentration ranging from 200 to 650 μgL^{-1}. In this region, the resulting equation is $y = 31.21x - 1.154$, with a correlation coefficient 0.998.

The detection limit is evaluated to be about 105 ppb ($S/N = 3$) after a 15-min accumulation. The limit of detection of the prepared electrodes was compared with

the previously prepared electrodes; it is comparatively better with some of the previously reported electrodes (Beltagi et al. 2011; Roa et al. 2003). The stability of the prepared electrode was determined using the DPASV for the same modified electrode with an interval of 2 weeks over 6 months, and it was found 6 months, assuming 5% change in the current response as a constant current response.

Conclusion

The present paper demonstrates a simple, ecofriendly, and sensitive electrochemical method for the determination of cadmium based on the coconut shell powder-modified carbon paste electrode. Cyclic voltammetry and electrochemical impedance spectroscopy study results confirm the incorporation of coconut shell powder at the electrode surface. Open-circuit accumulation, followed by anodic stripping voltammetry, was used for the determination of cadmium. An enhancement in the current response was observed in the presence of anionic surfactants. Different factors affecting the sensitivity of the prepared electrode were optimized. The optimized conditions for the determination of Cd (II) using CS-MCPE are acetate buffer of pH = 5 as accumulating solvent, 15 min accumulation time and hydrochloric acid as the stripping solvent. Despite some mutual interference effects, cadmium ions can be reliably determined with low detection limits using the standard addition procedure. The utilization of plant-based electrode in place of mercury-based electrodes is an attempt to perform environment friendly electrochemical determination of cadmium.

Abbreviations

CPE: carbon paste electrode; CS-MCPE: coconut shell powder-modified carbon paste electrode; CTAB: cetyltrimethylammonium bromide; CV: cyclic voltammogram; DPASV: differential pulse anodic stripping voltammetry; FTIR: Fourier transform infrared; MCPE: modified carbon paste electrode; SDS: sodium dodecyl sulfate; SEM: scanning electron microscope.

Competing interests

The authors gratefully acknowledge Prof. V.G. Dass, Director, Dayalbagh Educational Institute, Dayalbagh, Agra, for providing necessary research facilities. The authors also thank Ministry of Human Resource and Development, New Delhi for rendering financial assistance.

Author details

[1]Department of Chemistry, IIS University, Jaipur 302020, India. [2]Department of Chemistry, MLS University, Udaipur 313001, India. [3]Remote Instrumentation Lab, USIC, Dayalbagh Educational Institute, Agra 282010, India.

References

Bagheri H, Afkhami A, Shirzadmehr A, Khoshsafar H, Ghaedi H (2012) Novel potentiometric sensor for the determination of Cd^{2+} based on a new nano-composite. Int J Environ Anal Chem, doi:10.1080/03067 319.2011.6497 41

Beltagi AM, Ghoneim EM, Ghoneim MM (2011) Simultaneous determination of cadmium (II), lead (II), copper (II) and mercury (II) by square wave anodic stripping voltammetry at a montmorillonite calcium modified carbon paste electrode. Int J Environ Anal Chem 91(1):17–32

Chow E, Gooding JJ (2006) Peptide modified electrodes as electrochemical metal Ion sensors. Electroanal 18:1437–1448

Davis AC, Wu P, Zhang XF, Hou XD, Jones BT (2006) Determination of cadmium in biological samples. Appl Spectrosc Rev 41:35–75

Ensafi AA, Arabzadeh A, Karimi-Maleh H (2010) Simultaneous determination of dopamine and uric acid by electrocatalytic oxidation on a carbon paste electrode using pyrogallol red as a mediator. Anal Lett 43:1976–1988

Eshaghi Z, Khalili M, Khazaeifar A, Rounaghi GH (2011) Simultaneous extraction and determination of lead, cadmium and copper in rice samples by a new pre-concentration technique: hollow fiber solid phase microextraction combined with differential pulse anodic stripping voltammetry. Electrochim Acta 56:3139–3146

Ferreira SLC, Andrade JBD, Korn MDA, Pereira MD, Lemos VA, Dos Santos WNL, Rodrigues FD, Souza AS, Ferreira HS, Da Silva EGP (2007) Review of procedures involving separation and preconcentration for the determination of cadmium using spectrometric techniques. J Hazard Mater 145:358–367

Heitzmann M, Basaez L, Brovelli F, Bucher C, Limosin D, Pereira E, Rivas BL, Royal G, Aman ES, Moutet JC (2005) Voltammetric sensing of trace metals at a poly (pyrrole-malonic acid) film modified carbon electrode. Electroanal 17:1970–1976

Li J, Guo S, Zhai Y, Wang E (2009a) Nafion–graphene nanocomposite film as an enhanced sensing platform for ultrasensitive determination of cadmium. Electrochem Commun 11:1085–1088

Li Y, Liu X, Zeng X, Liu Y, Liu X, Wei W, Luo S (2009b) Simultaneous determination of ultra-trace lead and cadmium at a hydroxyl apatite-modified carbon ionic liquid electrode by square-wave stripping voltammetry. Sensor Actuat B-Chemical 139:604–610

Lu M, Toghill KE, Compton RG (2011) Simultaneous detection of trace cadmium (II) and lead (II) using an unmodified edge plane pyrolytic graphite electrode. Electroanal 23:1089–1094

Mojica ERE, Micor JRL, Gomez SP, Deocaris CC (2006) Lead detection using a pineapple bioelectrode. Philipp Agric Sci 89:134–140

Mojica ERE, Vidal JM, Pelegrina AB, Micor JRL (2007) Voltammetric determination of lead (ii) ions at carbon paste electrode modified with banana tissue. J Appl Sci 7:1286–1292

Okafor PC, Okon PU, Daniel EF, Ebenso EE (2012) Adsorption capacity of coconut (Cocos nucifera L.) shell for lead, copper, cadmium and arsenic from aqueous solutions. Int J Electrochem Sci 7:12354–12369

Pan D, Zhang L, Zhuang J, Lu W, Zhu R, Qin W (2012) New application of tin–bismuth alloy for electrochemical determination of cadmium. Mater Lett 68:472–474

Pino GH, Mesquita LMS, Torem ML, Pinto GAS (2006) Biosorption of cadmium by green coconut shell powder. Min Eng 19:380–387

Portaccio M, Tuoro DD, Arduini F, Lepore M, Mita DG, Diano N, Mita L, Moscone D (2010) A thionine-modified carbon paste amperometric biosensor for catechol and bisphenol A determination. Biosens Bioelectron 25:2003–2008

Pyrzynska K (2007) Online sample pretreatment systems for determination of cadmium by the ETAAS method. Crit Rev Anal Chem 37:39–49

Rajawat DS, Satsangee SP (2011) Voltammetric determination of Pb (II) ions by carbon paste electrode modified with lemon grass powder. Res J Chem Environ 15:55–60

Rajawat DS, Srivastava S, Satsangee SP (2012) Electro Chemical Determination of Pb (II) Ions by Carbon Paste Electrode Modified with Coconut Powder. In: Chemistry of Phytopotentials: Health, Energy and Environmental Perspectives, Springer, Berlin Heidelberg, pp 293–297

Rajawat DS, Kardam A, Srivastava S, Satsangee SP (2013a) Nano cellulosic fibers modified carbon paste electrode for ultra trace determination of Cd (II) and Pb (II) in aqueous solution. Environ Sci Pollut Res Environ Sci Pollut Res 20:3068–3076

Rajawat DS, Kardam A, Srivastava S, Satsangee SP (2013b) Adsorptive stripping voltammetric technique for monitoring of mercury ions in aqueous solution using nano cellulosic fibers modified carbon paste electrode. Nat Aca Sci Lett 36:181–189

Roa G, Ramírez-Silva MT, Romero-Romo MA, Galicia L (2003) Determination of lead and cadmium using a polycyclodextrin-modified carbon paste electrode with anodic stripping voltammetry. Anal Bioanal Chem 377:763–769

Sar E, Berber H, Asc B, Cankurtaran H (2008) Determination of some heavy metal ions with a carbon paste electrode modified by poly (glycidylmethacrylate-methylmethacrylate-divinylbenzene) microspheres functionalized by 2-aminothiazole. Electroanal 20:1533–1541

Sneddon J, Vincent MD (2008) ICP-OES and ICP-MS for the determination of metals: application to oysters. Anal Lett 41:1291–1303

25

Environmentally preferable solvents promoted resolution of multi-component mixtures of amino acids: an approach to perform green chromatography

Ali Mohammad[1*], Asma Siddiq[1] and Gaber E El-Desoky[2]

Abstract

Background: Amino acids are the building blocks of proteins and found in structural tissues of the human body. Their deficiency may cause a number of diseases.

Methods: Amino acids were analysed by the proposed thin layer chromatographic system using silica gel as stationary phase and Ethanol/70% aq. ethylene glycol/acetone (5:3:1) as mobile phase.

Results: Separations of some closely related amino acids from their quaternary, ternary and binary mixtures were achieved. The limits of detection of the separated amino acids have been calculated. The effects of presence of metal cations and anions on the separation have been examined. The developed method has been successfully applied for identifying the amino acids (lysine and methionine) in pharmaceutical formulations.

Conclusion: Thin layer chromatographic technique using silica as stationary phase in combination with mixtures of environmentally preferable solvents as mobile phase has been used for the analysis of amino acids of pharmaceutical importance.

Keywords: Thin layer chromatography; Amino acids; Environmentally preferable solvents; Separation; Pharmaceutical formulation

Background

Amino acids are the building blocks of proteins and found in structural tissues of the human body (Bromley et al. 2008; Castellino et al., 1987). A number of fascinating roles associated with amino acids include pH regulation (Wang et al. 2008) neurotransmitter functioning (Roberts 1974) pain control, cholesterol mechanism, inflammation control, detoxification, regulators of gene expression (Wu 2009) etc. Their deficiency may cause certain diseases such as, gastrointestinal insufficiencies, inadequacy of proteins, inflammatory responses, detoxification impairments, cardiovascular disease, neurological dysfunction, and inborn errors of metabolism etc. Because of these reasons the analysis of amino acids to ascertain their presence in matrices is of great importance.

Solvents being important components of chemical reactions and processes have an impact on cost, safety and health issues (Reichardt 2007). The toxic organic solvents used in any chemical process if released in the atmosphere, adversely affect the quality of environment. Therefore, the term green chemistry came into existence in 1990 to address environmental issues in an economically profitable manner. Green chemistry is commonly presented as a set of twelve principles proposed by *Anastas* and *Warner* (1998). The reason of using green solvents is to reduce the environmental impact resulting from the use of solvents in chemical production.

* Correspondence: alimohammad08@gmail.com
[1]Department of Applied Chemistry, Faculty of Engineering & Technology, Aligarh Muslim University, Aligarh- 202002, India
Full list of author information is available at the end of the article

Table 1 Mobility of amino acids in terms of R_F values using different mobile phases

Amino acids	Mobile phases (M)											
	M_1	M_2	M_3	M_4	M_5	M_6	M_7	M_8	M_9	M_{10}	M_{11}	M_{12}
Leucine	0.98	0.14	0.01	0.91	0.71	0.92	0.77	0.82	0.82	0.75	0.87	0.87
Isoleucine	0.97	0.21	0.01	0.92	0.72	0.92	0.76	0.81	0.81	0.77	0.85	0.81
Phenyl alanine	0.99	0.26	0.01	0.90	0.74	0.90	0.78	0.81	0.88	0.78	0.88	0.89
Tyrosine	0.97	0.22	0.02	0.92	0.75	0.91	0.80	0.83	0.86	0.77	0.90	0.88
Alanine	0.93	0.14	0.01	0.90	0.54	0.91	0.66	0.67	0.70	0.70	0.77	0.68
Lysine	0.97	0.06	0.01	0.27	0.07	0.31	0.60	0.10	0.08	0.11	0.08	0.07
Proline	0.92	0.10	0.01	0.75	0.47	0.76	0.36	0.45	0.46	0.53	0.58	0.58
Serine	0.94	0.14	0.01	0.88	0.61	0.91	0.59	0.65	0.70	0.71	0.75	0.68
Glutamic acid	0.95	0.17	0.01	0.70	0.68	0.77	0.54	0.47	0.41	0.58	0.64	0.49
Methionine	0.92	0.10	0.01	0.84	0.68	0.89	0.69	0.72	0.75	0.66	0.85	0.81
Arginine	0.92	0.07	0.01	0.24	0.05	0.28	0.09	0.10	0.06	0.05	0.07	0.05
Histidine	0.92	0.11	0.01	0.79	0.40	0.85	0.55	0.64	0.48	0.16	0.66	0.54
Tryptophan	0.99	0.14	0.01	0.83	0.77	0.89	0.63	0.83	0.88	0.78	0.88	0.83

From the ancient times, ethanol has been produced by fermentation of sugars (Yan & Shuzo 2006). Zymase, the enzyme from yeast changes the simple sugar molecules into ethanol and carbon dioxide by following reaction:

$$C_6H_{12}O_6 \rightarrow 2CH_3CH_2OH + 2CO_2$$

Being biosolvent, ethanol has been safely used in perfumes, make-up products, wine and liquor (Buchanan et al. 1988; Scherger et al. 1988). Thus, it can be a better alternative as a solvent for Thin Layer Chromatography (TLC). Acetone is naturally found in trees, plants and as a by-product of the breakdown of body fat showing low toxicity (Macdonald and Fall 1993). Ethylene glycol can be produced using renewable biomass (cellulose and polyols) which could be a burgeoning area for future research due to its environmental friendliness and long-term economical advantages (Zhao et al. 2013). It is a simple diol containing two hydroxyl groups in adjacent positions along a hydrocarbon chain, non-volatile and completely miscible with most of the polar solvents and slightly soluble with some non polar solvents showing low toxicity. Being an excellent solvent, its several attractive applications in various chemical fields such as plasticizers, elastomers, corrosion inhibitors, detergents for equipment cleaning and inks for ball point pens, etc. Considering low toxicity, we have used these solvents for achieving separations of amino acids.

Due to various attractive features of thin layer chromatography such as open and disposable nature of TLC pates, wider choice of selection of stationary and mobile phases, reduced need of modern laboratory facilities, minimal sample cleanup, use of small amount of sample for analysis, this technique has been used for the analysis of amino acids.

During present study, we have designed the best combination of environmentally preferable solvents as mobile phase for silica thin layer chromatography of amino acids. Chromatographic parameters have been calculated for achieved separations of amino acids and the practical applicability of developed TLC method has been illustrated by identifying the presence of amino acids in pharmaceutical formulations.

Methods
All experiments were performed at $25 \pm 2°C$.

Apparatus
A TLC applicator (Toshniwal, India) was used for coating SIG on 20 cm × 3.5 cm glass plates. The TLC was performed in 24 cm × 6 cm glass jars. Micropipette (Tripette, Germany) (0.1-1.0 µL) was used for spotting of analytes. A glass sprayer (Borosil, India) was used to spray

Table 2 Effect of interferences on ΔR_F values of separated amino acids

Foreign substances		Lysine-proline	Proline-alanine	Alanine-tyrosine
		(0.43)	(0.17)	(0.17)
Cations	Zn^{2+}	0.43	0.21	0.11
	Cu^{2+}	-	-	-
	Mg^{2+}	0.44	0.15	0.09
	Th^{4+}	-	-	-
Anions	Br^-	0.41	0.18	0.14
	CH_3COO^-	0.41	0.17	0.18
	CO_3^-	0.38	0.18	0.17
	NO_3^{2-}	0.42	0.17	0.16

*Each value is an average of four replicates.

reagent on the plates to locate the positions of the spots of analytes.

Chemicals and reagents

All chemicals were of analytical reagent grade. Silica gel G, methanol, ethanol, propanol, butanol, pentanol, acetone and ethylene glycol were purchased from Merck, India. Ninhydrin, amino acids (leucine, isoleucine, phenylalanine, tyrosine, alanine, lysine, proline, serine, glutamic acid, methionine, arginine, histidine, tryptophan) were purchased from Central Drug House, New Delhi, India. Water used in these experiments was double distilled water.

Test solutions

Test solutions of amino acids (1% w/v) were prepared by dissolving 0.1 g of amino acids in 10 mL of double distilled water. Ninhydrin solution (0.3% w/v) in acetone was used to detect all amino acids.

Stationary phase and mobile phase

Silica gel G was used as stationary phase and following solvent systems were used as mobile phase in all experiments.

Symbol composition

M_1 Ethylene glycol
M_2 Ethanol
M_3 Acetone
M_4 Ethanol/ethylene glycol (5:5)
M_5 Ethanol/ethylene glycol (7:3)
M_6 Ethanol/ethylene glycol (3:7)
M_7 Ethanol/ethylene glycol/Acetone (5:3:2)
M_8 Ethanol/30% aq. ethylene glycol/acetone (5:3:2)
M_9 Ethanol/50% aq. ethylene glycol/ acetone (5:3:2)
M_{10} Ethanol/70% aq. ethylene glycol/acetone (5:3:2)
M_{11} Ethanol/70% aq. ethylene glycol/acetone (5:3:1)
M_{12} Ethanol/70% aq. ethylene glycol/ethyl acetate (5:3:3)

Thin layer chromatographic separations

The details of preparation of TLC plates and chromatographic procedure can be seen elsewhere [14]. For separation, equal volumes of amino acids to be separated were mixed and an aliquot (0.1 µL) of the resultant mixture was loaded onto the activated TLC plate. The plates were developed with selected mobile phase M_{11}, the spots were detected and the R_F values of the separated amino acids were calculated.

The limits of detection of separated amino acids were determined by spotting 0.10 µL of amino acids solutions on the TLC plates which were developed with mobile phase M_{11} and the spots were visualized using ninhydrin. This process was repeated with successive reduction in

Table 3 Effect of interferences on separation factor (α) values of separated amino acids

Foreign substances		Lysine-proline	Proline-alanine	Alanine-tyrosine
		(10.22)	(2.09)	(3.0)
Cations	Zn^{2+}	10.9	2.55	1.89
	Cu^{2+}	-	-	-
	Mg^{2+}	11.48	1.87	1.62
	Th^{4+}	-	-	-
Anions	Br^-	8.79	2.19	2.21
	CH_3COO^-	8.79	2.09	3.14
	CO_3^-	6.96	2.18	2.75
	NO_3^{2-}	9.19	2.09	2.62

*Each value is an average of four replicates.

the loading volume. The amount of amino acid just detectable was taken as the detection limit.

Interference

It is estimated that half of all proteins contain a metal. These metalloprotiens play different functions in cell such as storage, transport of proteins, enzymes etc. Hence, chromatographic behaviour of amino acids was supposed to change in the presence of metal cations, anions and vitamins. Therefore it is important to observe the effect of these species as interferences on the separation of amino acids. To examine this, an aliquot of 0.10 µL of foreign substance (cations, anions or vitamins) was spotted on silica layer followed by the spotting of the mixture (0.10 µL) of amino acids. After drying the spot, TLC was performed with mobile phase M_{11} and R_F values of separated amino acids were calculated.

Detection of lysine in original and spiked drug sample

Syrup (0.10 ml) of All-B (containing lysine) and drug Beplex-forte (containing methionine) were dissolved in 10 ml of demineralised water. Aliquot (0.10 mL) of

Table 4 Effect of interferences on resolution (Rs) values of separated amino acids

Foreign substances		Lysine-proline	Proline-alanine	Alanine-tyrosine
		(3.18)	(3.40)	(1.88)
Cations	Zn^{2+}	2.0	1.20	1.22
	Cu^{2+}	-	-	-
	Mg^{2+}	2.31	1.0	1.0
	Th^{4+}	-	-	-
Anions	Br^-	2.0	1.24	1.78
	CH_3COO^-	1.95	1.13	1.44
	CO_3^-	1.72	1.24	1.78
	NO_3^{2-}	2.0	1.13	1.77

*Each value is an average of four replicates.

the resultant sample was loaded onto TLC plates and developed with mobile phase (M_{11}), followed by drying and spraying of ninhydrin solution. Similarly, drug sample solutions (1.0 mL) of All-B syrup was mixed with solutions (1.0 mL) each of proline, alanine, and tyrosine and beplex-forte with solutions (1.0 mL) each of arginine and glutamic acid to prepare synthetic samples and an aliquot (0.10 mL) of the resultant mixture was loaded onto TLC plates and developed with M_{11}, chromatography was performed on these spiked drugs samples for the identification and separation of amino acids in quaternary mixture.

Composition of syrup All-B (Merril Pharma Pvt. Ltd. Ahemdabad, India)

Each 10 ml contains: Thiamine hydrochloride (2 mg), riboflavin sodium phosphate (2 mg), pyridoxine hydrochloride (0.75 mg), nicotinamide (15 mg), D- panthenol (3 mg), L- lysine hydrochloride (70 mg), sorbitol solution 70% qs and sunset yellow FCF.

Composition of drug beplex forte (Anglo-French Drugs & Industries Ltd. Bangalore, India)

Each 10 ml contains: Choline chloride (9.1 mg), DL- methionine (9.1 mg), D-panthenol (9.1 mg), inositol (4.5 mg), niacinamide (90.9 mg), vitamin B_1 (90.9 mg), vitamin B_{12} (90.9mcg/1 ml), vitamin B_2 (4.5 mg), vitamin B_6 (4.5 mg).

Results and discussion

Results of this study have been summarized in Tables 1, 2, 3 and 4 and shown in Figures 1, 2, 3, 4 and 5. Several combinations of environmentally preferable solvents have been used for observing the migration behaviour of amino acids and some closely related amino acids have been identified and separated from their multi component mixtures.

When pure ethylene glycol (M_1) is used as mobile phase for examining migration behaviour of amino acids on silica gel, all amino acids reached the maximum allowed distance, showing R_F in the range of 0.92-0.99. In case of ethanol (M_2) as mobile phase, tailing in the spots of all amino acids was observed while none of the amino acids leaves the point of application. It may be probably due to competitive interactions of amino acids with ethanol and silica gel. With acetone (M_3) as eluent, all amino acids remained near the point of application as compact spots.

It appears that the mobility of amino acids strongly depends on the presence of hydroxyl groups in the molecule of mobile phase as evident by following mobility (R_F value) trend:

R_F in ethylene glycol containing two OH groups in the molecule > R_F in ethanol containing one OH group > R_F in acetone with absence of OH group.

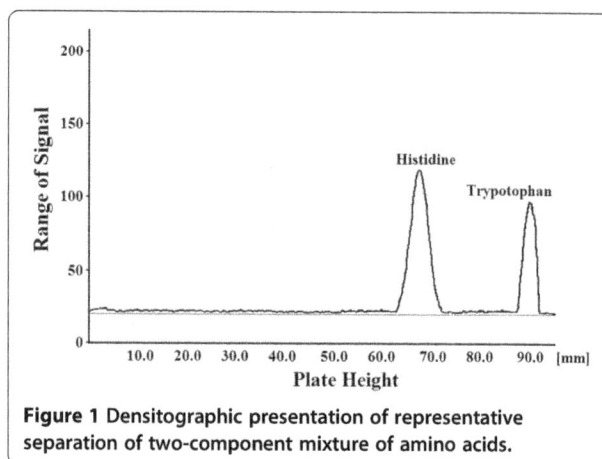

Figure 1 Densitographic presentation of representative separation of two-component mixture of amino acids.

This trend is in consonance with the dielectric constant of the mobile phase. The dielectric constant values given in parenthesis is in the order ethylene glycol (37) > ethanol (24.2) > acetone (19.8) which shows the highest solubility of amino acids in ethylene glycol as reflected also by the high R_F values of all amino acids.

From the above discussion, it is clear that single component mobile phase is not capable to resolve amino acids from their mixtures but from the differential migration of amino acids in individual mobile phase, it is expected that mixed solvent systems consisting of different proportions of components may prove useful for resolving amino acids from their multi-component mixtures. With this hope, different combinations of mixed solvent systems were examined to realize most appropriate system for reliable separation of amino acids.

Different combinations of selected solvents were tested for the analysis of amino acids which were characterized as follows:

Ethanol and ethylene glycol mixed systems in different combinations (M_4-M_6) were used and obtained results are listed in Table 1. From the results it was observed

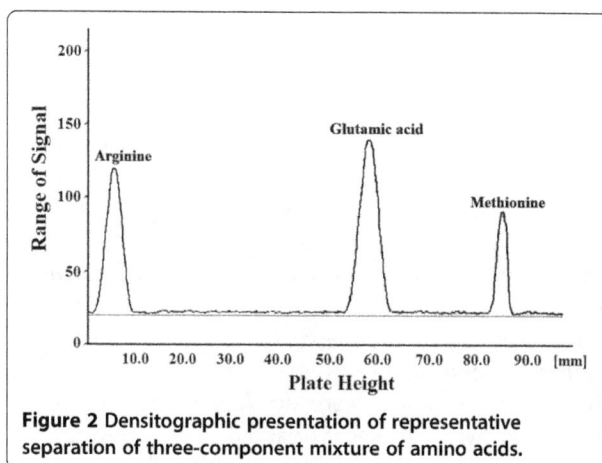

Figure 2 Densitographic presentation of representative separation of three-component mixture of amino acids.

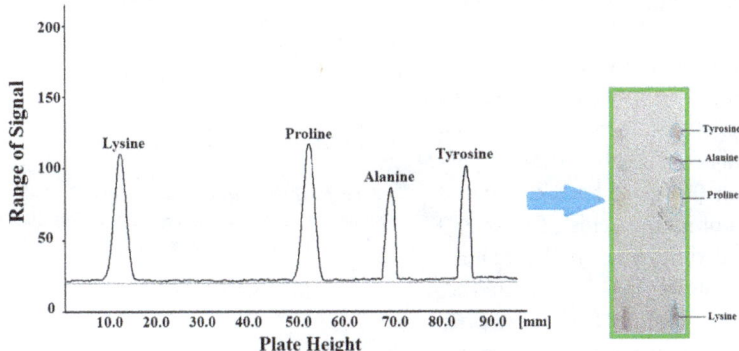

Figure 3 Densitographic presentation of representative separation of four-component mixture of amino acids (Digital photograph of the separation achieved alongside).

that ethanol-ethylene glycol (7:3, M_6) gives the most satisfactory results in terms of differential mobility pattern of amino acids. With this mobile phase, amino acids with positively charged side chain remained at the point of application except histidine which is due to the presence of imidazolium ring in its structure which is responsible for little mobility. By using this mobile phase, little mobility was observed for polar amino acids while non polar amino acids shows high mobility showing R_F in the range 0.54-0.78. But the spots obtained with this mobile phase shows diffused boundaries. To overcome this problem small amount of acetone was added in ratio 5:3:2, to make the volume 10 ml. By adding acetone, compact spots were obtained showing improved separation possibilities.

Due to the viscous nature of ethylene glycol, time taken by the mobile phase M_7 was too much (4 h). In order to decrease the development time, aqueous solution of ethylene glycol in different concentrations (M_8-M_{10}) were used in place of pure ethylene glycol, along with ethanol and acetone. With M_8 and M_9 as mobile phase, spots of amino acids were large and

tailing in case of serine was observed. Among them, M_{10} was most effective because of achieving compact spots with no tailing in case of serine. Improved separations of amino acids from their ternary and binary mixtures were possible with this mobile phase.

But our aim is to resolve amino acids from their quaternary mixture along with ternary and binary mixtures, therefore volume of acetone was varied from 1 ml to 3 ml (M_{10}-M_{12}) (Table 1). From the table it was clear that only mobile phase M_{11} was able to separate four component mixture of amino acids along with ternary and binary one which was not possible in case of M_{10} and M_{12}. Because of getting differential mobility pattern of amino acids and maximum possibilities of separations, M_{11} was selected as the best mobile phase for resolving amino acids from their quaternary, ternary and binary mixtures (Figures 1, 2 and 3). The amino acids present in each mixture are easily separable.

The lowest possible detectable amounts by using M_{11} as mobile phase were found to be 0.10 µg/spot for proline, alanine and 0.05 µg/spot for lysine and glutamic

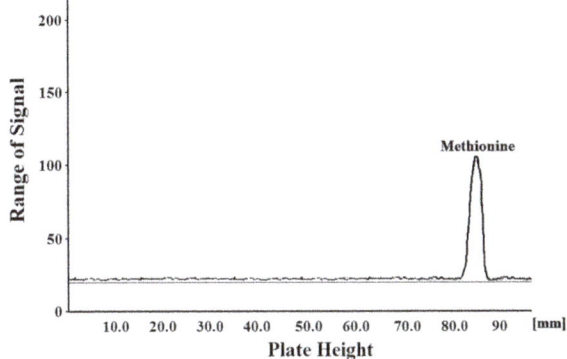

Figure 4 Densitographic presentation of detection of methionine in Beplex forte drug sample.

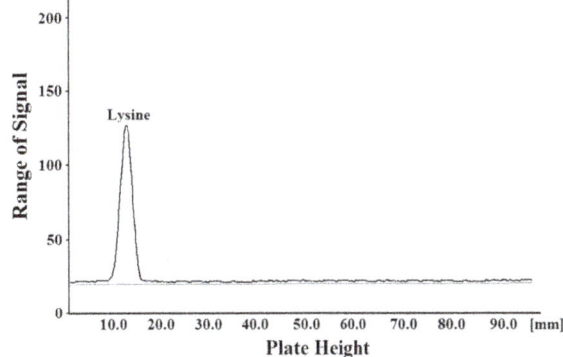

Figure 5 Densitographic presentation of detection of lysine in All-B syrup sample.

acid, 0.03 for tyrosine, histidine, methionine, tryptophan and serine.

Effect of foreign substances such as metal cations, inorganic anions on separation of amino acids have been examined and chromatographic parameters, Difference between R_F values of two adjacent spots (ΔR_F), Separation factor (α) and Resolution (Rs) have been calculated for representative separations of amino acids (Tables 2, 3 and 4). There was a marginal difference in the magnitude of parameters but separation was not hampered. The result was not the same in case of Cu^{2+} and Th^{4+} because these metal ions form complexes with amino acids to form metalloprotiens. This is due to the reason that these metal ions form complexes with the amino acids (Sovago et al. 2003; Nourmand & Meissami 1983).

Application

The proposed method was applicable for the identification of lysine in syrup All-B and methionine in drug Beplex forte (Figures 4 and 5). It was also applied for the separation of lysine in syrup All-B and spiked sample (lysine, proline, alanine and tyrosine) and methionine from Beplex forte and spiked sample (arginine, glutamic acid and methionine).

Conclusion

TLC system consisting of silica gel as stationary phase and combination of environmentally preferable solvents such as ethanol-70% aqs ethylene glycol-acetone in ratio (5:3:1, M_{11}) as an eluent was found to be most effective for the resolution of closely related amino acids from their multi-component mixtures. The proposed method could be applied for the identification of amino acids in drug samples of pharmaceutical importance.

Competing interests
The authors declare that they have no competing interests.

Authors' contributions
Experimental work was done by AS. All authors equally contributed in experimental design, framing, writing, proofing and approval of the manuscript.

Author details
[1]Department of Applied Chemistry, Faculty of Engineering & Technology, Aligarh Muslim University, Aligarh- 202002, India. [2]Department of Chemistry, College of Science, Building 5, King Saud University, Riyadh- 11451, Saudia Arabia.

References
Anastas PT, Warner JC (1998) Green Chemistry: Theory and Practice. Oxford University Press, New York

Bromley EHC, Channon K, Moutevelis E, Woolfson DN (2008) Peptide and protein building blocks for synthetic biology: from programming biomolecules to self-organized biomolecular systems. ACS Chem Biol 3(1):38–50

Buchanan BR, Honigs DE, Lee CJ, Roth W (1988) Detection of ethanol in wines using optical-fiber measurements and near-infrared analysis. Appl Specrosc 42(6):1106–1111

Castellino P, Luzi L, Simonson DC, Haymond M, DeFronzo RA (1987) Effect of insulin and plasma amino acid concentrations on leucine metabolism in man. Role of substrate availability on estimates of whole body protein synthesis. J Clin Invest 80(6):1784–1793

Macdonald CR, Fall R (1993) Acetone emission from conifer buds. Phytochem The Int J Plant Biochem 34(4):991–994

Nourmand M, Meissami N (1983) Complex formation between uranium(VI) and thorium(IV) ions with some α-amino-acids. J Chem Soc Dalton Trans 8:1529–1533

Reichardt C (2007) Solvents and solvent effects: an introduction. Org Process Res Dev 11(1):105–113

Roberts PJ (1974) The release of amino acids with proposed neurotransmitter function from the cuneate and gracile nuclei of the rat in vivo. Brain research 67(3):419–428

Scherger DL, Wruk KM, Kulig KW, Rumack BH (1988) Ethyl alcohol (ethanol)-containing cologne, perfume, and after-shave ingestions in children. Am J Dis Child 142(6):630–632

Sóvágó I, Ősz K, Várnagy K (2003) Copper(II) complexes of amino acids and peptides containing chelating bis(imidazolyl) residues. Bioinorg Chem Appl 1(2):123–139

Wang J, Zhao LY, Uyama T, Tsuboi K, Tonai T, Ueda N (2008) Amino acid residues crucial in pH regulation and proteolytic activation of N-acylethanolamine-hydrolyzing acid amidase. Biochim Biophys Acta 1781(11–12):710–717

Wu G (2009) Amino acids: metabolism, functions, and nutrition. Amino acids 37(1):1–17

Yan L, Shuzo T (2006) Ethanol fermentation from biomass resources: current state and prospects. Appl Microbiol Biot 69(6):627–642

Zhao G, Zheng M, Zhang J, Wang A, Zhang T (2013) Catalytic conversion of concentrated glucose to ethylene glycol with semicontinuous reaction system. Ind Eng Chem Res. doi:10.1021/ie400989a

26

Antimicrobial activities of silver nanoparticles synthesized from *Lycopersicon esculentum* extract

Swarnali Maiti[1], Deepak Krishnan[2], Gadadhar Barman[1], Sudip Kumar Ghosh[2] and Jayasree Konar Laha[1*]

Abstract

Background: It has been known for quite some time now that silver nanoparticles (AgNP) can inhibit microbial growth and even kill microbes. Our investigation reports the antimicrobial activity of AgNP against a model bacterium, *Escherichia coli*.

Methods: The aqueous extract of *Lycopersicon esculentum* (red tomato) was used for the rapid synthesis of AgNP, which is very simple and eco-friendly in nature. The UV-visible spectroscopy technique was employed to establish the formation of AgNP.

Results: The transmission electron microscopic images showed that the particles were of mostly spherical shape. For the bacteriological tests, the microorganism *E. coli* was inoculated on Luria broth (LB) agar plate in the presence of varied amounts of AgNP. The antibacterial activity was obvious from the zone of inhibition. At concentration 20 µg/ml and above, the AgNP showed a clear zone of inhibition and the minimum inhibitory concentration of AgNP to *E. coli* was 50 µg/ml. Growth rates and bacterial concentrations were determined by measuring optical density at 600 nm at different time points.

Conclusions: From the slope of the bacterial growth curve, it has been concluded that the nanoparticles are bacteriostatic at low concentration and bactericidal at high concentration. So these nanoparticles are believed to act as preventive for bacterial contamination.

Keywords: Silver nanoparticle; Green synthesis; *Lycopersicon esculentum*; Antibacterial activity; *Escherichia coli*

Background

Disease-causing microbes that have become resistant to drug therapy are an increasing public health problem. Many researchers are now engaged in developing new effective antimicrobial reagents with the emergence and increase of microbial organisms resistant to multiple antibiotics, which will increase the cost of health care. Therefore, there is an urgent need to develop new bactericides. Silver has been used for years in the medical field for antimicrobial applications such as burn treatment (Parikh et al. 2005; Ulkur et al 2005), elimination of microorganisms on textile fabrics (Jeong et al. 2005; Lee et al. 2007; Yuranova et al. 2003), disinfection in water treatment (Russell and Hugo 1994; Chou et al. 2005), prevention of bacteria colonization on catheters (Samuel and Guggenbichler 2004; Alt et al. 2004; Rupp et al. 2004), etc. It has also been found to prevent HIV from binding to host cells (Sun et al. 2005), but the effects of silver nanoparticles (AgNP) on microorganisms have not been developed fully. Nanosilver, being less reactive than silver ions, is expected to be more suitable for medical applications. Reducing the particle size of metals is also an efficient and reliable tool for improving their biocompatibility, which facilitates their applications in different fields such as bioscience and medicine. The mechanism of the bacterial effect of AgNP as proposed is due to the attachment of AgNP to the surface of the cell membrane, thus disrupting permeability and respiration functions of the cell (Kevitec et al. 2008). It is also proposed that AgNP not only interact with the surface of a membrane but can also penetrate inside the bacteria (Morones et al. 2005). The antibacterial activity of AgNP is significantly enhanced when it is modified with sodium dodecyl sulfate (SDS) (Kevitec et al. 2005; Carpenter 1972).

* Correspondence: j.laha@yahoo.co.in
[1]Midnapore College, Midnapore, West Bengal, India
Full list of author information is available at the end of the article

In this study, we have investigated the antimicrobial effects of silver nanoparticles prepared by a biosynthesis method. The chemical reduction method is widely used to synthesize AgNP because AgNP could be synthesized under a mild as well as on a large scale (Cao et al. 2010). However, the use of environmentally benign materials like plant leaf extract, bacteria, and fungi for the synthesis of silver nanoparticles is more acceptable as they offer several benefits over chemical methods like conditions of high temperature, pressure, and toxic chemicals which are not required in the synthesis protocol (Singh et al. 2010). Therefore, preparation of AgNP by a green synthesis approach has compatibility for pharmaceutical and biomedical applications.

In the present work, the synthesis of silver nanoparticles has been carried out using the aqueous extract of *Lycopersicon esculentum* (red tomato). The water extract of tomato juice mostly contains proteins and water-soluble organic acids (Gould 1983) which are believed to act as stabilizing and reducing agents, respectively. With these nanoparticles, a preliminary test for antibacterial activity was carried out by cup diffusion method and the effects of AgNP on bacterial growth has been studied by employing minimum inhibitory concentration (MIC) method. Results obtained by us prove that AgNP prepared by the green method is suitable for the formulation of new types of bactericidal materials.

Results and discussion
Results
Characterization and optimization of AgNP preparation
The absorbance spectra of the AgNP were analyzed by using a 'SHIMADZU' UV 1800 spectrophotometer (Shimadzu Corporation, Kyoto, Japan). AgNP exhibited a reddish yellow color in water due to the excitation of the localized surface plasmon vibrations of the metal nanoparticles. Generally, the surface plasmon resonance (SPR) bands are influenced by the size, shape, morphology, composition, and dielectric environment of the synthesized nanoparticles (Kelly et al. 2003; Stepanov 1997). Previous studies showed that spherical AgNP contribute to the absorption bands at around 400 nm in the UV-visible spectra (Maiti et al. 2013; Barman et al. 2014). The SPR band due to AgNP was observed in our case at around 410 nm (Figure 1) when 3×10^{-3} M silver nitrate solution was used. This strongly suggested that AgNP were nearly spherical in shape and it was confirmed by the transmission electron microscopy (TEM) results.

In this study, AgNP have been synthesized both in the presence and in the absence of a stabilizer and both anionic and neutral surfactants were used one at a time. Though soluble proteins and amino acids present in *L. esculentum* extract were expected to act as stabilizer for

nanoparticles (Barman et al. 2013), a smooth and narrow absorption band of AgNP at 410 nm was observed only in the presence of SDS of 3×10^{-3} M (Figure 1A). So we preferred to synthesize AgNP using SDS as the stabilizing agent.

An absorption band was observed at 410 nm for 1:1 extract composition. The plasmon band shifted to higher values with the increase of the concentrations of tomato in aqueous extracts and reached to 415 nm for 3:2 composition (Figure 1B). At concentrations higher than 3:2 composition, the plasmon band shifted to higher values and the extinction coefficient of the band decreased appreciably. However, tomato extract of 1:1 composition was used throughout the work.

A bathochromic shift of the SPR bands from 388 to 445 nm was observed while the concentration of AgNO₃ varied from 3×10^{-3} to 5×10^{-2} M keeping the extract composition constant at 1:1 using SDS of 3×10^{-3} M (Figure 1C). When the particle size increased, the absorption peak shifted towards the red wavelength, which indicated the formation of larger sized nanoparticles (Peng et al. 2010).

The shape and size distribution of the synthesized AgNP were characterized by TEM study. The TEM images were taken using JEOL JEM-2100 high-resolution transmission electron microscope (HR-TEM; JEOL Ltd., Akishima-shi, Japan). Samples for the TEM studies were prepared by placing a drop of the aqueous suspension of particles on carbon-coated copper grids followed by solvent evaporation under vacuum. The TEM images of AgNP produced from 1:1 composition of tomato extract showed that the particles were mostly spherical and their sizes varied from 10 to 40 nm. Selected area electron diffraction (SAED) pattern illustrated the crystalline nature of AgNP (Figure 2).

Antibacterial activity of AgNP against the microorganism
Preliminary test for antibacterial activity The antimicrobial activity of AgNP was evaluated against *Escherichia coli* by cup diffusion method. Approximately 10^6 colony-forming units (CFU) of the microorganism *E. coli* were inoculated on Luria broth (LB) agar plate, and then different concentrations of AgNP (1, 2, 5, 10, 20, 50, 100, and 200 µg/ml) were added to the well present in the LB agar plate. A reaction mixture containing no AgNP was put in the well in the LB plate and cultured under the same condition as the control test. All the LB plates were incubated at 37°C overnight. After incubation, the plates were observed for the presence of a zone of inhibition. The antibacterial activity of AgNP was proved from the zone of inhibition (Figure 3). At concentration 20 µg/ml and above, the AgNP showed a clear zone of inhibition. No zone of inhibition was found in the vehicle control well (spot in the middle of the plate) which suggested that the antimicrobial activity was specifically due to AgNP.

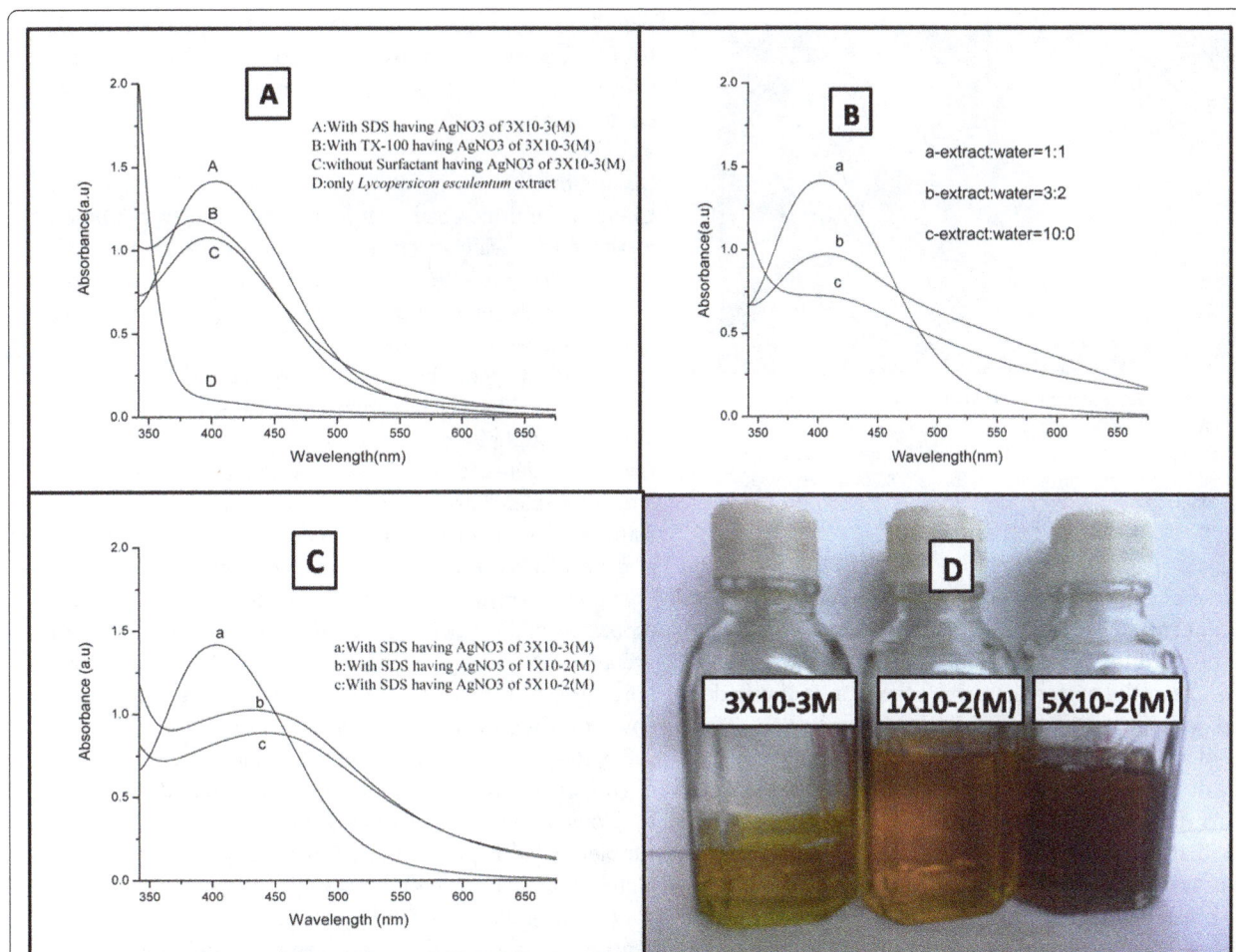

Figure 1 UV-Vis spectra and digital photographic images of AgNP. (A) UV-Vis spectra of AgNP: spectrum 1A-A with surfactant SDS, spectrum 1A-B with surfactant TX-100, spectrum 1A-C without any surfactant, and spectrum 1A-D is for pure *Lycopersicon esculentum* extract. **(B)** UV-Vis spectra of AgNP at different compositions of *Lycopersicon esculentum* extract. **(C)** UV-Vis spectra of AgNP with varying concentrations of silver nitrate (a) at 3×10^{-3} M, (b) at 1×10^{-2} M, and (c) at 5×10^{-2} M using 1:1 extract composition and 3×10^{-3} M SDS solution in each case. **(D)** Digital photographic images of AgNP produced from different concentrations of silver nitrate.

Evaluation of antibacterial effectiveness using minimum inhibitory concentration method The antimicrobial activity of AgNP was evaluated using the MIC method. The antimicrobial effectiveness was determined against the bacterial concentration of 10^6 CFU/ml with different

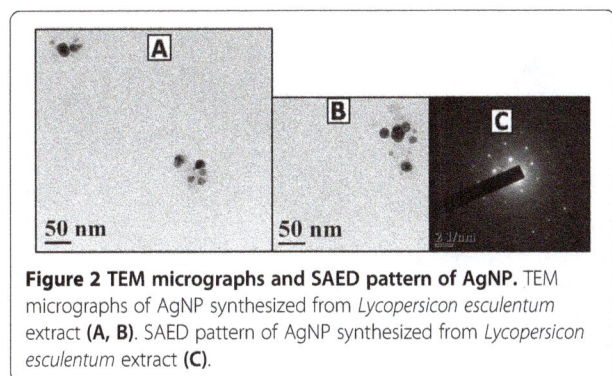

Figure 2 TEM micrographs and SAED pattern of AgNP. TEM micrographs of AgNP synthesized from *Lycopersicon esculentum* extract **(A, B)**. SAED pattern of AgNP synthesized from *Lycopersicon esculentum* extract **(C)**.

concentrations of AgNP (0.2, 0.5, 1, 2, 5, 10, 20, 50, and 100 µg/ml). The cultures were incubated at 37°C at 250 rpm. Bacterial concentrations were determined by measuring optical density (OD) at 600 nm (0.1 OD600 corresponding to 10^8 cells per milliliter). With the increase of concentration of nanoparticles, the final bacterial concentration decreased. When the concentration of AgNP was 50 µg/ml, growth of *E. coli* was completely inhibited, which indicated that the MIC of AgNP to *E. coli* was 50 µg/ml (Figure 4).

Effect of AgNP on bacterial growth To determine the growth curve in the presence of silver nanoparticles, *E. coli* bacteria were grown in liquid LB medium till they reached the log phase. Then they were diluted in fresh LB liquid medium to optical density (OD600) 0.05, 0.1, and 0.2. AgNP solution was added into the cell culture medium at different concentrations, and the culture was

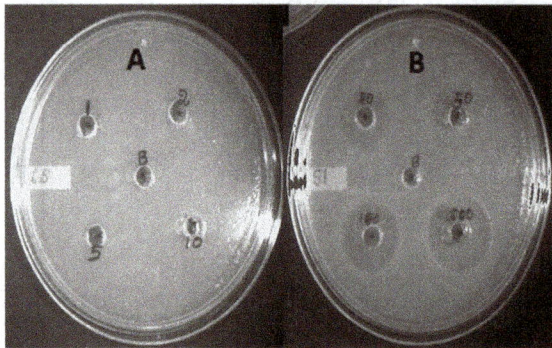

Figure 3 Antibacterial activity of AgNP. Antibacterial activity of AgNP having different concentrations: **(A)** 1, 2, 5, and 10 μg/ml and **(B)** 20, 50, 100, and 200 μg/ml, with 10^6 CFU of *E. coli* inoculated on Luria broth agar plate. The 'B' spot in the middle of the agar plate is for the blank test, having no AgNP.

incubated at 37°C and 250 rpm. Growth rates and bacterial concentrations were determined by measuring OD at 600 nm at different time points (Figure 5A,B,C).

The slope of the bacterial growth curve continuously decreased with increasing nanoparticle concentration. This means that at low concentration of nanoparticles, the growth of bacteria was delayed and at higher concentration, growth was completely inhibited. So it can be concluded that the nanoparticles are bacteriostatic at low concentration and bactericidal at high concentration. It was also clear from the graphs that the bacterial growth was dependent on the initial number of cells present in the medium. It was observed that at lower initial OD, the

AgNP concentration necessary to completely inhibit bacterial growth was also low. So silver nanoparticles produced by us will be suitable for preventing bacterial contamination.

Discussion

Chemical antimicrobial agents are increasingly becoming resistant to a wide spectrum of antibiotics. An alternative way to overcome the drug resistance of various microorganisms is therefore urgently needed. Ag ions and silver salts have been used for decades (Silver and Phung 1996) as antimicrobial agents in various fields due to their growth-inhibitory abilities against microorganisms. However, there are some limitations in using Ag ions or Ag salts as antimicrobial agents. Probable reasons include the interfering effects of salts. This type of limitation can be removed by using silver in nano form. Due to the increase of the surface area in nano state, the contact area between Ag(0) and that of the microorganism increases. To use AgNP against microbes in various fields, it is important and necessary to prepare AgNP in a green environment. In this study, we report a green method for the preparation of AgNP which is environmentally benign and cost-effective.

For the assessment of the antimicrobial effects of AgNP, *E. coli* was used in our study. The effect was investigated by growing *E. coli* on agar plates and in liquid LB medium, supplemented with AgNP. The bacterial growth was completely inhibited in the presence of AgNP on the LB agar plate. The inhibition solely depended upon the AgNP concentration. It showed a clear zone of inhibition at and above the concentration 20 μg/ml.

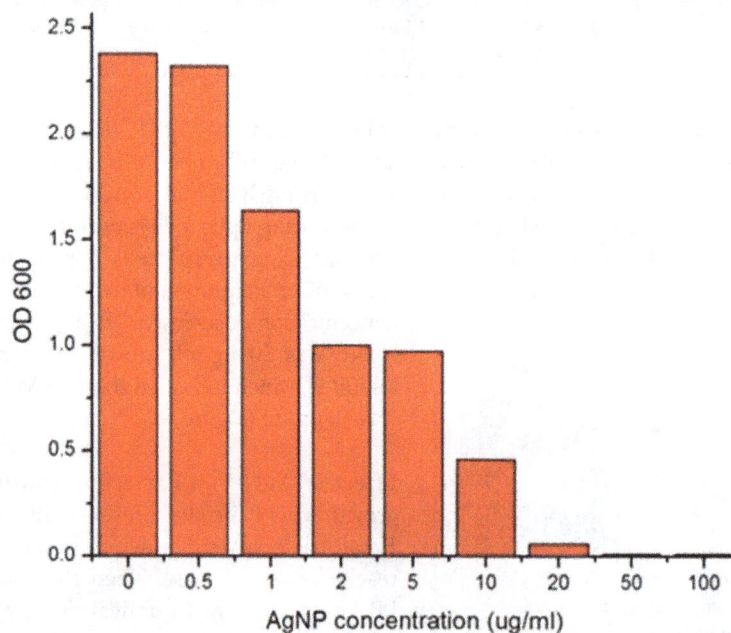

Figure 4 Optical density vs concentration of AgNP. MIC assay 50 μg/ml.

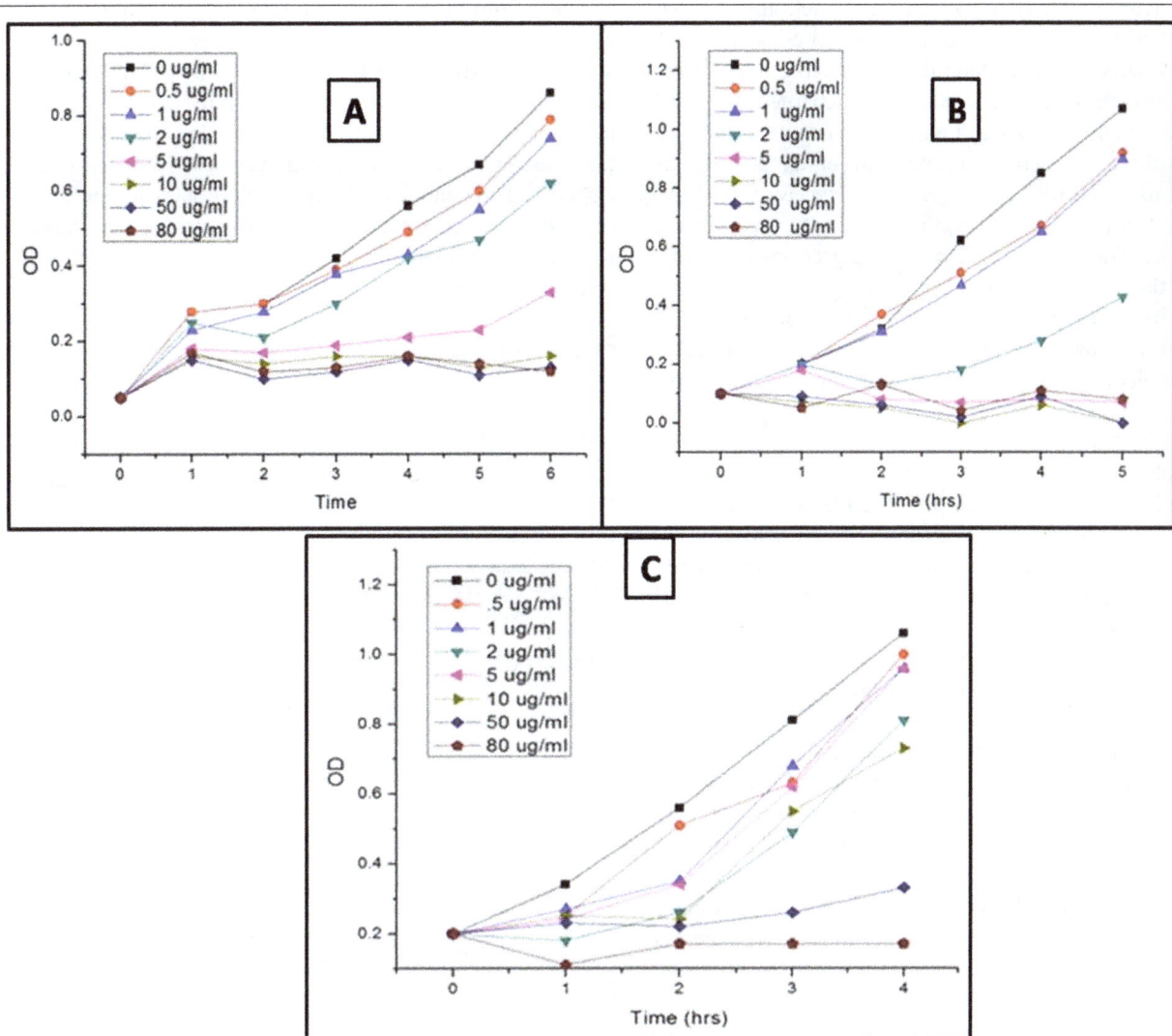

Figure 5 Growth curves with initial OD 0.05 (A), 0.10 (B), and 0.20 (C).

To study the antimicrobial effectiveness of AgNP, we treated a bacterial concentration at high CFU (10^6/ml) with varying concentrations of AgNP from 0.2 to 100 μg/ml. When the concentration of AgNP was increased, the bacterial concentration was found to decrease. At concentration 50 μg/ml of AgNP, the growth of *E. coli* was completely inhibited, which indicated that the minimum inhibitory concentration was 50 μg/ml (Figure 4). Since high CFU are seldom found in real-life systems, it may be concluded that these AgNP have a biocidal effect and effectiveness in delaying bacterial growth, findings which may lead to valuable inventions in the future in various fields like in antimicrobial systems as well as medical devices.

The slope of the bacterial growth curves (Figure 5A,B,C) continuously decreased with increasing nanoparticle concentration. This indicated that at low concentration of

nanoparticles, bacterial growth was delayed and growth was completely inhibited at higher concentrations. So it appears that these particles are bacteriostatic at low concentration and bactericidal at high concentration. It is also clear from the graphs that the bacterial growth is dependent on the initial number of cells present in the medium. It was observed that at lower initial OD, the AgNP concentration necessary to completely inhibit bacterial growth was also low. So it is confirmed that these nanoparticles may be used to prevent bacterial contamination.

The mechanism of the inhibitory effects of Ag ions on microorganisms is partially known. It is reported that the positive charge on the silver ion is the reason for antimicrobial activity as it can attract the negatively charged cell membrane of microorganisms through the electrostatic interaction (Dibrov et al. 2002; Hamouda et al. 2000). Due to their unique size and greater surface area, silver nanoparticles can easily reach the nuclear content of bacteria (Chen et al. 2010; Chudasama et al. 2009). A survey of the literature showed that the electrostatic attraction between negatively charged bacterial cells and positively charged nanoparticles was crucial for the antibacterial activity (Stoimenov et al. 2000). The AgNP used in this study, however, received negative charge from SDS, an anionic surfactant, used during synthesis. The bacterium *E. coli* being gram-negative, the interaction with the negatively charged nanoparticles might have occurred through 'pit' formation in the cell wall of the bacteria (Sondi and Salopek-Sondi 2004) which helped the permeability and resulted in cell death.

Experimental
Materials
Silver nitrate and SDS, both of AR grade, were purchased from Sigma-Aldrich Chemical Ltd. (St Louis, MO, USA) Sodium hydroxide was purchased from Merck (Darmstadt, Germany). Double-distilled de-ionized water was used in all experiments.

Green synthesis of silver nanoparticles by *L. esculentum* extract
Silver nanoparticles were made according to the recipe described below. For this purpose, red tomato (*L. esculentum*) was collected from the local market and washed with double-distilled de-ionized water. The tomato skin was removed and the whole mass was squeezed to get tomato juice. Then it was diluted two times and filtered using a Whatman filter paper to get the aqueous extract of the red tomato.

Method
AgNP were produced by reduction of silver nitrate solution by using red tomato extract. Ten milliliters of aqueous red tomato extract was mixed with 10 ml of 3×10^{-3} M SDS solution and cooled in ice-cold water for few minutes. The solution was made alkaline (pH 9) with 0.15 N sodium hydroxide solution. After that, 8 ml of 3×10^{-3} M aqueous silver nitrate was added into it dropwise with continuous stirring. The mixture was then heated for 20 min at 80°C. The color of the solution gradually changed from colorless to reddish yellow. The reddish yellow color indicated the formation of AgNP.

Conclusions
We have described a simple and green method for the synthesis of AgNP by using the aqueous extract of red tomato. The formation of AgNP was confirmed by UV-visible spectroscopy. The TEM images showed that the particles were mostly spherical. These biosynthesized AgNP were then used to demonstrate antimicrobial activity against a model bacterium, *E. coli*. The antibacterial activity of AgNP was

apparent from the zone of inhibition. At concentrations 20 µg/ml and above, the AgNP showed a clear zone of inhibition and the MIC of AgNP to *E. coli* was 50 µg/ml. Growth rates and bacterial concentrations were determined by measuring OD at 600 nm at different time points. From the slope of the bacterial growth curve, it has been concluded that the nanoparticles are bacteriostatic at low concentration and bactericidal at high concentration. So these nanoparticles are believed to act as preventive for bacterial contamination.

Competing interests
The authors declare that they have no competing interests.

Authors' contributions
SM and DK carried out the experiments. SM, GB, and DK drafted the manuscript. JKL and SKG guided the research and modified the manuscript. All authors read and approved the final manuscript.

Acknowledgements
We are thankful to Central Research Facility at IIT Kharagpur, India, for the HR-TEM measurements.

Author details
[1]Midnapore College, Midnapore, West Bengal, India. [2]Department of Biotechnology, Indian Institute of Technology, Kharagpur, India.

References
Alt V, Bechert T, Streinrucke P, Wagener M, Seidel P, Dingeldein E, Domanne E, Schnettler R (2004) An in vitro assessment of the antibacterial properties and cytotoxicity of nanoparticulate silver bone cement. Biomaterials 25:4383–4391

Barman G, Maiti S, Konar LJ (2013) Bio-fabrication of gold nanoparticles using aqueous extract of red tomato and its use as a colorimetric sensor. Nanoscale Res Lett 8:181–189

Barman G, Samanta A, Maiti S, Konar LJ (2014) Detection of Cu^{+2} ion by the synthesis of bio-mass-silver nanoparticle nanocomposite. Int J Scientific and Engg Res 5(6):1086–1097

Cao XL, Cheng C, Ma YL, Zhao CS (2010) Preparation of silver nanoparticles with antimicrobial activities and the researches of their biocompatibilities. J Mater Sci Mater M 21:2861–2868

Carpenter PL (1972) Microbiology. W.B. Saunders Company, Philadelphia, p 245

Chou WL, Yu DG, Yang MC (2005) The preparation and characterization of silver-loading cellulose acetate hollow fibre membrane for water treatment. Polym Adv Technol 16:600–607

Chen SF, Li JP, Quin K, Xu WP (2010) Large scale photochemical synthesis of M@TiO$_2$ nanocomposites (M = Ag, Pd, Au, Pt) and their optical properties, Co oxidation performance and antibacterial effect. Nano Res 3:244–255

Chudasama B, Vala AV, Andharya N, Upadhyay RV, Mehta RV (2009) Enhanced antibacterial activity of bifunctional Fe$_3$O$_4$ core-shell nanostructures. Nano Res 2:955–965

Dibrov P, Dzoiba J, Gosink KK, Häse CC (2002) Chemiosmotic mechanism of antimicrobial activity of Ag(+) in vibrio cholera. Antimicrob Agents Chemother 46:2668–2770

Gould WA (1983) Tomato production, processing and quality evaluation, 2nd edition. AVI Publishing Company, Inc, Westport, CT

Hamouda T, Myc A, Donovan B, Shih A, Reuter JD, Baker RJ (2000) A novel surfactant nanoemulsion with a unique non-irritant topical antimicrobial against bacteria, enveloped viruses and fungi. Microbial Res 156:1

Jeong SH, Yeo SY, Yi SC (2005) The effect of filler particle size on the antibacterial properties of compounded polymer/silver fibres. J Mater Sci 40:5407–5411

Kelly KL, Coronado E, Zhao LL, Schatz GC (2003) The optical properties of metal nanoparticles: the influence of size, shape and dielectric environment. J Phys Chem B 107:668–677

Kevitec L, Panacek A, Soukupova J, Kolar M, Vecerova R, Prucek R (2008) Effect of surfactants and polymers on stability and antibacterial activity of silver nanoparticles (NPs). J Phys Chem 112(15):5825–5834

Lee HY, Park HK, Lee YM, Kim K, Park SB (2007) A practical procedure for producing silver nanocoated fabric and its antibacterial evaluation for biomedical applications. Chem Commun 2959–2961

Maiti S, Barman G, Konar LJ (2013) Synthesis of silver nanoparticles having different morphologies and its application in estimation of chlorpyrifos. Adv Sci Focus 1:145–150

Morones JR, Elechiguerra JL, Camacho A, Holt K, Kouri JT, Ramirez JT (2005) The bactericidal effect of silver nanoparticles. Nanotechnology 16:2346–2361

Parikh DV, Fink T, Rajasekharan K, Sachinvala ND, Sawhney APS, Calamari TA, Parikh AD (2005) Antimicrobial silver/sodium carboxymethyl cotton dressings for burn wounds. Text Res J 75:134–138

Peng S, Mc Mahon JM, Schatz GC, Gray SK, Sun Y (2010) Reversing the size-dependence of surface plasmon resonances. PNAS 107:14530–14534

Rupp ME, Fitzgerald T, Marion N, Helget V, Puumala S, Anderson JR, Fey PD (2004) Effect of silver-coated urinary catheters: efficacy, cost-effectiveness antimicrobial resistance. Am J Infect Control 32:445–450

Russell AD, Hugo WB (1994) Antimicrobial activity and action of silver. Prog Med Chem 31:351–370

Samuel U, Guggenbichler JP (2004) Prevention of catheter-related infections: the potential of a new nanosilver impregnated catheter. Int J Antimicrob Agents 23:75–78

Silver S, Phung LT (1996) Bacterial heavy metal resistance: new surprises. Annu Rev Microbiol 50:753–789

Singh A, Jain D, Upadhyay MK, Khandlewal N, Verma HN (2010) Green synthesis of silver nanoparticles using Argemone mexicana leaf extract and evaluation of their antimicrobial activities. Dig J Nanomater Bios 5:483–489

Sondi I, Salopek-Sondi B (2004) Silver nanoparticles as antimicrobial agent: a case study on E. coli as a model for Gram-negative bacteria. J Colloid Interface Sci 275:177–182

Stepanov AL (1997) Optical properties of metal nanoparticles synthesized in a polymer by ion implantation: a review. Tech Phys 49:143

Stoimenov PK, Klinger RL, Marchin GL, Klabunde KJ (2000) Metal oxide nanoparticles as bactericidal agents. Langmuir 18:6679–6686

Sun RW, Chen R, Chung NP, Ho CM, Lin CL, Che CM (2005) Silver nanoparticles fabricated in Hepes buffer exhibit cytoprotective activities toward HIV-1 infected cells. Chem Commun 40:5059–5061

Ulkur E, Oncul O, Karagoz H, Yeniz E, Celikoz B (2005) Comparison of silver-coated dressing (Acticoat™), chlorhexidine acetate 0.5% (Bactigrass®), and fusidic acid 2% (Fucidin®) for topical antibacterial effect in methicillin-resistant Staphylococci-contaminated, full-skin thickness rat burn wounds. Burns 31:874–877

Yuranova T, Rincon AG, Bozzi A, Parra S, Pulgarin C, Albers P, Kiwi J (2003) Antibacterial textiles prepared by RF-plasma and vacuum-UV mediated deposition of silver. Photochem Photobiol A 161:27–34

Permissions

All chapters in this book were first published in JAST, by Springer; hereby published with permission under the Creative Commons Attribution License or equivalent. Every chapter published in this book has been scrutinized by our experts. Their significance has been extensively debated. The topics covered herein carry significant findings which will fuel the growth of the discipline. They may even be implemented as practical applications or may be referred to as a beginning point for another development.

The contributors of this book come from diverse backgrounds, making this book a truly international effort. This book will bring forth new frontiers with its revolutionizing research information and detailed analysis of the nascent developments around the world.

We would like to thank all the contributing authors for lending their expertise to make the book truly unique. They have played a crucial role in the development of this book. Without their invaluable contributions this book wouldn't have been possible. They have made vital efforts to compile up to date information on the varied aspects of this subject to make this book a valuable addition to the collection of many professionals and students.

This book was conceptualized with the vision of imparting up-to-date information and advanced data in this field. To ensure the same, a matchless editorial board was set up. Every individual on the board went through rigorous rounds of assessment to prove their worth. After which they invested a large part of their time researching and compiling the most relevant data for our readers.

The editorial board has been involved in producing this book since its inception. They have spent rigorous hours researching and exploring the diverse topics which have resulted in the successful publishing of this book. They have passed on their knowledge of decades through this book. To expedite this challenging task, the publisher supported the team at every step. A small team of assistant editors was also appointed to further simplify the editing procedure and attain best results for the readers.

Apart from the editorial board, the designing team has also invested a significant amount of their time in understanding the subject and creating the most relevant covers. They scrutinized every image to scout for the most suitable representation of the subject and create an appropriate cover for the book.

The publishing team has been an ardent support to the editorial, designing and production team. Their endless efforts to recruit the best for this project, has resulted in the accomplishment of this book. They are a veteran in the field of academics and their pool of knowledge is as vast as their experience in printing. Their expertise and guidance has proved useful at every step. Their uncompromising quality standards have made this book an exceptional effort. Their encouragement from time to time has been an inspiration for everyone.

The publisher and the editorial board hope that this book will prove to be a valuable piece of knowledge for researchers, students, practitioners and scholars across the globe.

List of Contributors

Hye-Jin Cho
Division of Electron Microscopic Research, Korea Basic Science Institute, 113 Gwahangno, Daejeon 305-333, Korea

Jae-Kyung Hyun
Division of Electron Microscopic Research, Korea Basic Science Institute, 113 Gwahangno, Daejeon 305-333, Korea

Jin-Gyu Kim
Division of Electron Microscopic Research, Korea Basic Science Institute, 113 Gwahangno, Daejeon 305-333, Korea

Hyeong Seop Jeong
Division of Electron Microscopic Research, Korea Basic Science Institute, 113 Gwahangno, Daejeon 305-333, Korea

Hyo Nam Park
Division of Electron Microscopic Research, Korea Basic Science Institute, 113 Gwahangno, Daejeon 305-333, Korea

Dong-Ju You
Division of Electron Microscopic Research, Korea Basic Science Institute, 113 Gwahangno, Daejeon 305-333, Korea

Hyun Suk Jung
Division of Electron Microscopic Research, Korea Basic Science Institute, 113 Gwahangno, Daejeon 305-333, Korea

Preeti S Kulkarni
Department of Chemistry, Postgraduate and Research Centre, MES Abasaheb Garware College, Pune-411005, India

Satish D Dhar
Modern College, Shivajinagar, Pune 411005, India

Sunil D Kulkarni
Department of Chemistry, Sir Parashurambhau, Pune-411030, India

Gadadhar Barman
Department of Chemistry, Midnapore College, Midnapore 721101, W.B, India

Swarnali Maiti
Department of Chemistry, Midnapore College, Midnapore 721101, W.B, India

Jayasree Konar Laha
Department of Chemistry, Midnapore College, Midnapore 721101, W.B, India

Faisal Asif
Department of Pharmacy, State University of Bangladesh, Dhaka 1205, Bangladesh

Arshida Zaman Boby
Department of Pharmacy, State University of Bangladesh, Dhaka 1205, Bangladesh

Nur Alam
Department of Pharmacy, State University of Bangladesh, Dhaka 1205, Bangladesh

Muhammad Taraquzzaman
Department of Pharmacy, State University of Bangladesh, Dhaka 1205, Bangladesh

Sharmin Reza Chowdhury
Department of Pharmacy, State University of Bangladesh, Dhaka 1205, Bangladesh

Mohammad Abdur Rashid
Phytochemical Research Laboratory, Department of Pharmaceutical Chemistry, Faculty of Pharmacy, University of Dhaka, Dhaka 1000, Bangladesh

Youri Lee
Center for Inflammation, Immunity & Infection, Institute for Biomedical Sciences, Georgia State University, Atlanta, GA 30303, USA

Yu-Jin Kim
Center for Inflammation, Immunity & Infection, Institute for Biomedical Sciences, Georgia State University, Atlanta, GA 30303, USA

Yu-Jin Jung
Center for Inflammation, Immunity & Infection, Institute for Biomedical Sciences, Georgia State University, Atlanta, GA 30303, USA

Ki-Hye Kim
Center for Inflammation, Immunity & Infection, Institute for Biomedical Sciences, Georgia State University, Atlanta, GA 30303, USA

Young-Man Kwon
Center for Inflammation, Immunity & Infection, Institute for Biomedical Sciences, Georgia State University, Atlanta, GA 30303, USA

Seung Il Kim
Division of Life Science, Korea Basic Science Institute, Daejeon 305-333, South Korea

Sang-Moo Kang
Center for Inflammation, Immunity & Infection, Institute for Biomedical Sciences, Georgia State University, Atlanta, GA 30303, USA

Cristina MI Theil
Escola de Química e Alimentos, Federal University of Rio Grande – FURG, Av. Italia, km 8, 96201-900, Rio Grande, RS, Brazil

Luis FH Niencheski
Instituto de Oceanografia, Federal University of Rio Grande – FURG, Av. Italia, km 8, 96201-900, Rio Grande, RS, Brazil

Gilberto Fillmann
Instituto de Oceanografia, Federal University of Rio Grande – FURG, Av. Italia, km 8, 96201-900, Rio Grande, RS, Brazil

Marcio R Milani
Escola de Química e Alimentos, Federal University of Rio Grande – FURG, Av. Italia, km 8, 96201-900, Rio Grande, RS, Brazil

Isam A Mohamed Ahmed
Department of Food Science and Technology, Faculty of Agriculture, University of Khartoum, Shambat 13314, Sudan
Arid Land Research Center, Tottori University, 1390 Hamasaka, Tottori 680-0001, Japan

Ailijiang Maimaiti
Arid Land Research Center, Tottori University, 1390 Hamasaka, Tottori 680-0001, Japan

Nobuhiro Mori
School of Agricultural, Biological, and Environmental Sciences, Faculty of Agriculture,
Tottori University, Koyama, Tottori 680-8553, Japan

Norikazu Yamanaka
Arid Land Research Center, Tottori University, 1390 Hamasaka, Tottori 680-0001, Japan

Takeshi Taniguchi
Arid Land Research Center, Tottori University, 1390 Hamasaka, Tottori 680-0001, Japan

Iain M McIntyre
Forensic Toxicology Laboratory Manager, County of San Diego Medical Examiner's Office, 5570 Overland Ave., Suite 101, San Diego, CA 92123, USA

Mansour Arab Chamjangali
Department of Chemistry, Shahrood University, Shahrood, P.O. Box 36155–316, Iran

Gadamali Bagherian
Department of Chemistry, Shahrood University, Shahrood, P.O. Box 36155–316, Iran

Nasser Goudarzi
Department of Chemistry, Shahrood University, Shahrood, P.O. Box 36155–316, Iran

Shima Mehrjoo-Irani
Department of Chemistry, Shahrood University, Shahrood, P.O. Box 36155–316, Iran

Pravin U Singare
Department of Chemistry, Bhavan's College, Munshi Nagar, Andheri (West), Mumbai 400 058, India

Yang Hoon Huh
Division of Electron Microscopic Research, Korea Basic Science Institute, 169-148 Gwahangno, Yuseong-gu, 305-806, Daejeon, Republic of Korea
Boston Biomedical Research Institute, 64 Grove St, 02472, Watertown, MA, USA

Hee-Seok Kweon
Division of Electron Microscopic Research, Korea Basic Science Institute, 169-148 Gwahangno, Yuseong-gu, 305-806, Daejeon, Republic of Korea

Toshio Kitazawa
Boston Biomedical Research Institute, 64 Grove St, 02472, Watertown, MA, USA
Department of Microbiology and Physiological Systems, University of Massachusetts Medical School, 55 Lake Avenue North, 01655, Worcester, MA, USA

Sayan Bhattacharya
Department of Environmental Science, University of Calcutta, Kolkata, West Bengal, India

Gunjan Guha
Department of Pharmaceutical Sciences, College of Pharmacy, Oregon State University, Corvallis, OR, USA

Dhrubajyoti Chattopadhyay
B. C. Guha Centre for Genetic Engineering and Biotechnology, University of Calcutta, Kolkata, West Bengal, India

Aniruddha Mukhopadhyay
Department of Environmental Science, University of Calcutta, Kolkata, West Bengal, India

Purnendu K Dasgupta
Department of Chemistry and Biochemistry, University of Texas at Arlington, Arlington, TX, USA

Mrinal K Sengupta
Department of Chemistry and Biochemistry, University of Texas at Arlington, Arlington, TX, USA

Uday C Ghosh
Department of Chemistry, Presidency University, Kolkata, West Bengal, India

Javed Iqbal
Department of Chemistry, Quaid-i-Azam University, Islamabad 45320, Pakistan

Munir H Shah
Department of Chemistry, Quaid-i-Azam University, Islamabad 45320, Pakistan

Nader Shokoufi
Analytical Instrumentation & Spectroscopy Laboratory, Chemistry & Chemical Engineering Research Center of Iran, Tehran, Iran

Rasoul Jafari Atrabi
Analytical Instrumentation & Spectroscopy Laboratory, Chemistry & Chemical Engineering Research Center of Iran, Tehran, Iran

Kazem Kargosha
Analytical Instrumentation & Spectroscopy Laboratory, Chemistry & Chemical Engineering Research Center of Iran, Tehran, Iran

Jeongmin Kim
Division of Earth and Environmental Sciences, Korea Basic Science Institute, 162 Yeongudanji-ro, Ochang-eup, Cheongwon-gun, Chungcheongbuk-do 363-886, Korea

Su-in Jeon
Division of Earth and Environmental Sciences, Korea Basic Science Institute, 162 Yeongudanji-ro, Ochang-eup, Cheongwon-gun, Chungcheongbuk-do 363-886, Korea

Marin Senila
INCDO-INOE 2000, Research Institute for Analytical Instrumentation, 67 Donath, Cluj-Napoca 400293, Romania

Andreja Drolc
Laboratory for Environmental Sciences and Engineering, National Institute of Chemistry, 19 Hajdrihova, Ljubljana SI-1000, Slovenia

Albin Pintar
Laboratory for Environmental Sciences and Engineering, National Institute of Chemistry, 19 Hajdrihova, Ljubljana SI-1000, Slovenia

Lacrimioara Senila
INCDO-INOE 2000, Research Institute for Analytical Instrumentation, 67 Donath, Cluj-Napoca 400293, Romania

Erika Levei
INCDO-INOE 2000, Research Institute for Analytical Instrumentation, 67 Donath, Cluj-Napoca 400293, Romania

Kumara Shanthamma Kavitha
Department of Studies in Microbiology, Herbal Drug Technological, Laboratory, University of Mysore, Manasagangotri, Mysore 570 006, Karnataka, India

Sreedharamurthy Satish
Department of Studies in Microbiology, Herbal Drug Technological, Laboratory, University of Mysore, Manasagangotri, Mysore 570 006, Karnataka, India

Yeonhee Hong
Division of Life Science, Korea Basic Science Institute, Daejeon 305-333, Republic of Korea
Pioneer Research Center for Protein Network Exploration, Korea Basic Science Institute, Daejeon 305-333, Republic of Korea

Edmond Changkyun Park
Division of Life Science, Korea Basic Science Institute, Daejeon 305-333, Republic of Korea
Pioneer Research Center for Protein Network Exploration, Korea Basic Science Institute, Daejeon 305-333, Republic of Korea

Eun-Young Shin
Division of Life Science, Korea Basic Science Institute, Daejeon 305-333, Republic of Korea
Pioneer Research Center for Protein Network Exploration, Korea Basic Science Institute, Daejeon 305-333, Republic of Korea
Department of Functional Genomics, University of Science and Technology (UST), Daejeon 305-333, Republic of Korea

Sang-Oh Kwon
Division of Life Science, Korea Basic Science Institute, Daejeon 305-333, Republic of Korea

Young-Taek Oh
Department of Radiation Oncology, Ajou University School of Medicine, Suwon 443-721, Republic of Korea

Byung-Ock Choi
Department of Radiation Oncology, Seoul St. Mary's Hospital, The Catholic University of Korea, Seoul 137-701, Republic of Korea

Giwon Kim
Department of Radiation Oncology, Ajou University School of Medicine, Suwon 443-721, Republic of Korea
Department of Radiation Oncology, Seoul St. Mary's Hospital, The Catholic University of Korea, Seoul 137-701, Republic of Korea

Gun-Hwa Kim
Division of Life Science, Korea Basic Science Institute, Daejeon 305-333, Republic of Korea
Pioneer Research Center for Protein Network Exploration, Korea Basic Science Institute, Daejeon 305-333, Republic of Korea
Department of Functional Genomics, University of Science and Technology (UST), Daejeon 305-333, Republic of Korea

Mekala Suneetha
Department of Chemistry, Acharya Nagarjuna University, 522 510 Guntur Dt., AP, India

Bethanabhatla Syama Sundar
Department of Chemistry, Acharya Nagarjuna University, 522 510 Guntur Dt., AP, India

Kunta Ravindhranath
Department of Chemistry, K L University, Vaddeswaram, 522 502 Guntur Dt., AP, India

Tae Yeon Kang
Gangneung Center, Korea Basic Science Institute (KBSI), Gangneung 210-702, Republic of Korea

Ki Soo Chang
Center for Analytical Instrumentation Development, Korea Basic Science Institute (KBSI), Daejeon 305-806, Republic of Korea

Jae Young Kim
Center for Analytical Instrumentation Development, Korea Basic Science Institute (KBSI), Daejeon 305-806, Republic of Korea

Seon-Kang Choi
Tongyang Life Science Corp., Seoul 135-995, Republic of Korea

Weon-Sik Chae
Gangneung Center, Korea Basic Science Institute (KBSI), Gangneung 210-702, Republic of Korea

Hong-ying Gao
School of Traditional Chinese Materia Medica 49#, Shenyang Pharmaceutical University, Wenhua Road 103, Shenyang 110016, People's Republic of China

Shu-yun Wang
School of Traditional Chinese Materia Medica 49#, Shenyang Pharmaceutical University, Wenhua Road 103, Shenyang 110016, People's Republic of China

Hang-yu Wang
School of Pharmacy, Shihezi University, Shihezi 832002, People's Republic of China

Guo-yu Li
School of Pharmacy, Shihezi University, Shihezi 832002, People's Republic of China

Li-fei Wang
Shanxi Xinghuacun Fen Jiu Group Co., Ltd, Shanxi 450000, People's Republic of China

Xiao-wei Du
Shanxi Xinghuacun Fen Jiu Group Co., Ltd, Shanxi 450000, People's Republic of China

Ying Han
Shanxi Xinghuacun Fen Jiu Group Co., Ltd, Shanxi 450000, People's Republic of China

Jian Huang
School of Pharmacy, Shihezi University, Shihezi 832002, People's Republic of China

Jin-hui Wang
School of Traditional Chinese Materia Medica 49#, Shenyang Pharmaceutical University, Wenhua Road 103, Shenyang 110016, People's Republic of China
School of Pharmacy, Shihezi University, Shihezi 832002, People's Republic of China

Sunil Adurty
Department of Chemistry, Sri Sathya Sai Institute of Higher Learning (Deemed to be university), Prasanthi Nilayam-515134, Puttaparthi, Anantapur District, Andhra Pradesh, India

Jagadeeswara Rao Sabbu
Department of Chemistry, Sri Sathya Sai Institute of Higher Learning (Deemed to be university), Prasanthi Nilayam-515134, Puttaparthi, Anantapur District, Andhra Pradesh, India

Deepak Singh Rajawat
Department of Chemistry, IIS University, Jaipur 302020, India

Nitin Kumar
Department of Chemistry, MLS University, Udaipur 313001, India

Soami Piara Satsangee
Remote Instrumentation Lab, USIC, Dayalbagh Educational Institute, Agra 282010, India

Ali Mohammad
Department of Applied Chemistry, Faculty of Engineering & Technology, Aligarh Muslim University, Aligarh-202002, India

Asma Siddiq
Department of Applied Chemistry, Faculty of Engineering & Technology, Aligarh Muslim University, Aligarh-202002, India

Gaber E El-Desoky
Department of Chemistry, College of Science, Building 5, King Saud University, Riyadh- 11451, Saudia Arabia

Swarnali Maiti
Midnapore College, Midnapore, West Bengal, India

Deepak Krishnan
Department of Biotechnology, Indian Institute of Technology, Kharagpur, India

Gadadhar Barman
Midnapore College, Midnapore, West Bengal, India

Sudip Kumar Ghosh
Department of Biotechnology, Indian Institute of Technology, Kharagpur, India

Jayasree Konar Laha
Midnapore College, Midnapore, West Bengal, India